Probiotic Bacteria and Enteric Infections

Baode Sun · Jun Wang · Da Shu

Precision Forming Technology of Large Superalloy Castings for Aircraft Engines

上海交通大學出版
SHANGHAI JIAO TONG UNIVERSITY PRESS

Springer

Baode Sun
Shanghai Jiao Tong University
Shanghai, China

Jun Wang
Shanghai Jiao Tong University
Shanghai, China

Da Shu
Shanghai Jiao Tong University
Shanghai, China

ISBN 978-981-33-6219-2 ISBN 978-981-33-6220-8 (eBook)
https://doi.org/10.1007/978-981-33-6220-8

Jointly published with Shanghai Jiao Tong University Press
The print edition is not for sale in China (Mainland). Customers from China (Mainland) please order the print book from: Shanghai Jiao Tong University Press.

This Springer imprint is published by the registered company Springer Nature Singapore Pte Ltd.
The registered company address is: 152 Beach Road, #21-01/04 Gateway East, Singapore 189721, Singapore

Preface

The aircraft engine is known as the "flower of industry" and the "crown jewel" in the high-end manufacturing industry. It is an important indicator of the strength of science and technology, industry, and the national defense of a country. In addition, the space of aero-engine industry is also very broad. According to the prediction of the British Rolls-Royce company, the global market of civil aero-engine will reach US$1.4 trillion in the next 20 years and the market of military aero-engine will reach US$430 billion. Developed countries have always listed the aero-engine as a strategic industry of priority development, and have strictly controlled the transfer of the core technology to foreign countries. At present, only a few countries in the world, such as the United Kingdom, the United States, and Russia, can independently develop aero-engines. The global civil aero-engine market is almost completely monopolized by European and American enterprises.

Being the heart of aircraft, the aero-engine is one of the determining factors of its mobility, range, reliability, economy, and environmental impact. Nowadays, a high thrust-to-weight ratio, supersonic cruise, low life-cycle cost, unconventional maneuverability, low fuel consumption, low pollution, and other excellent performance are the development trends and goals of aircraft in the world. This goal is mainly achieved by improving the performance of the engine. In order to improve the thrust-to-weight ratio and reduce the fuel consumption, the most effective way is to increase the turbine inlet temperature and reduce the structural mass, which requires that the key materials and manufacturing technology of the engine hot end components must be constantly innovated. Therefore, the United States, Western Europe, and other aviation-developed countries have formulated comprehensive strategy, basic research, and technology development plans to actively carry out scientific research, exploring and accumulating abundant material data and manufacturing technology.

China has been carrying out the research and development of large-scale civil aircraft since 2006 and the design and manufacture of large-scale commercial aero-engines has also begun recently. With the promotion of national security strategy and localization substitution, it is expected that the research and development of aero-engine will be supported by major national policies, which will bring huge pulling effect on the industrial chain. However, at present, the foundation of commercial aero-engine design and manufacturing in China is still weak, especially in terms of materials, key parts, manufacturing equipment, etc., which result in lagging behind the international top level.

As the key load-bearing component of civil aircraft power system, the structural superalloy castings are developing rapidly towards the direction of large-scale size, complex structure, thin-walled surface and integrated components. It is of great technical challenges to control the defects, such as porosity, deformation, and oversize. The lack of theoretical and technical basis of precision forming of such large castings has become one of the bottlenecks limiting the development of aero-engines in China.

Since 2008, researchers at the Shanghai Jiao Tong University have aimed at the major national demand for large superalloy castings and have conducted a series of fundamental research and technological development. With the support of the Shanghai Science and Technology Committee, the Shanghai Key Laboratory of Advanced High-temperature Materials and Precision Forming was established in 2013. Through continuous efforts in the last decade, we have made theoretical achievements and key technological breakthroughs in the precision forming of large superalloy castings, and have successfully produced qualified large-scale, complex-structured, and thin-walled castings for domestically-developed commercial aero-engines.

This book is a systematic summary of the research work of our group over the years, which has been published in Chinese by the Shanghai Jiao Tong University Press in 2016. The content not only covers the basic principles, key technologies, and control methods of the whole process of precision casting of superalloys, but also introduces a series of new technologies, new methods, and new progresses in the field, such as synchrotron radiation X-ray radiation of metal solidification process, three-dimensional computerized tomography (3D CT) image reconstruction of porosity, error flow modeling of casting process, etc.

This book consists of contributions of many of my students, colleagues, and collaborators. I thank them for their diligence, efforts, intelligence, and great teamwork. I would also like to thank my postdocs and graduate students who helped me in the text translation and typing. They are Lishibao Ling, Tao Liu, Yongfei Jun, Yujie Wang, Ya Zhang, Jiang Ju, Jiahao Liu, Zhen Zhang, Yun Wu, Xuan Zhang, Ting Zhang, Zhengyi Ding, Jiangping Yu, and Zhicheng Shi.

In our research and in writing this book, we have benefited from encouragement of many experts and scholars, particularly Chinese Academician of Engineering Jianxin Xie and Zhongqin Lin. Finally, this book would have never come about had it not been for the continual editorial support from Ms. Fangzhen Qian of Shanghai Jiao Tong University Press.

It is a pleasure to dedicate the book to two particularly distinguished scholars, Chinese Academician of Sciences Yaohe Zhou and Zhuangqi Hu. They have made great contributions to China's academy and industry of casting and superalloys and will be remembered by us forever.

Shanghai, China
June 2020

Baode Sun

Contents

About the Authors

Baode Sun is a chair professor of Materials Science and Engineering at Shanghai Jiao Tong University (SJTU). He received his Ph.D. degree from Northwestern Polytechnic University, China, in 1993. He became the "Yangtze River Scholar" distinguished professor in 2007, and the Dean of School of Materials Science and Engineering in 2013. His research is focused on the fundamental and applied aspects of liquid metal processing and investment casting of large-size thin-wall structural components. He has published more than 200 technical papers and he holds more than 100 Chinese patents. He won numerous prestigious research awards, such as second prize of the National Technology Invention Award of China, and first prize of the Shanghai Science and Technology Invention Award.

Jun Wang is a professor of Materials Processing at SJTU. He received his B.Sc. (1991), M.Sc. (1994), and Ph.D. (1997) at Southeast University. He was a research fellow in the Venture Business Laboratory in Saga University and senior researcher in Department of Materials Engineering at the University of Tokyo in Japan. His research focuses on investment casting process. He has published more than 200 peer-reviewed articles, and he holds 95 China patents and one US patent. He received a number of awards, including the 2nd Prize of the National Technology Invention Award, the New-century Talents of Ministry of Education, China, and Shanghai Excellent Scientific Leader.

Da Shu is a professor of Materials Processing at SJTU. He received his B.Sc. (1994) at Nanjing University of Aeronautics and Astronautics and his M.Sc. (1997) and Ph.D. (2001) at SJTU. He was an academic visitor at the Department of Materials, Oxford University, in 2009–2010. His research focuses on processing of advanced metallic materials and control of solidified structure. He has published more than 100 peer-reviewed articles, and he holds 30 China patents and one US patent. He received a number of awards, including the 2nd Prize of the National Technology Invention Award, Rising-Star of Shanghai Municipal Science and Technology Commission, and the New-century Talents of Ministry of Education, China.

Contributors

Chapter 1 Baode Sun, Da Shu, Maodong Kang
Chapter 2 Guoxiang Wang
Chapter 3 Donghong Wang
Chapter 4 Fei Li, Fei Wang
Chapter 5 Jiao Zhang, Yongbing Dai, Jianbo Ma, Qing Dong, Yanfeng Han, Taiwen Huang, Guangyu Yang
Chapter 6 Jun Wang, Haiyan Gao, Xinhua Tang, Maodong Kang, Yang Zhou
Chapter 7 Changhui Liu, Sun Jin, Xinmin Lai
Chapter 8 Anping Dong, Dafan Du, Hui Xing, Guoliang Zhu

Chapter 1
Introduction

Baode Sun, Da Shu, and Maodong Kang

1.1 Requirements of Aircraft Engines on Large Superalloy Castings

Aeroengines are recognized as the most sophisticated and complex mechanical systems ever built. Known as the "heart" of an aircraft, the aeroengine represents a comprehensive embodiment of a country's scientific and industrial achievements [1]. The progress in materials, structural design and manufacture are all controlling factors in the development of aviation industry.

Since the first batch of superalloy GH3030 was successfully made in 1956, the research, production and application of superalloys in China have undergone over 60 years of development [2]. Most of the designations of superalloys have been directly related to various types of engines, and the development of aeroengine and aerospace technology has actively promoted the research and advancement of superalloys in China. Along with the development of aeroengines, various materials used in the engine are constantly changing. Although a variety of high-performance new materials have continuously been used, the dominant use of steel and aluminum have gradually given way to the age of "Three Kingdoms" of nickel, titanium and steel, with titanium being the major material for the cold end parts and nickel for the hot end parts of aeroengines. The K-series nickel casting alloys, e.g., K4169, K417G, K405, K419, K419H, and K4537, are a type of materials that can serve for a long time under the condition of large complex stress at temperatures above 650 °C. Some of the great advantages of this type of alloys include high initial melting temperatures, excellent cold and hot fatigue resistance, and good anti-oxidation and corrosion resistance, etc., facilitating their common use in the hot end parts of engines. For example, the turbine rear frame (TRF), which carries all the thrust and vibration loads of the engine, is the most important bearing structural component of an aeroengine, and is listed as one of the key components of the aeroengine in parallel with the blade, turbine disk and turbine shaft. As a result, the quality of TRF greatly affects the performance, long-term life, and reliability of the aeroengine,

B. Sun et al., *Precision Forming Technology of Large Superalloy Castings for Aircraft Engines*, https://doi.org/10.1007/978-981-33-6220-8_1

and currently more than 80% of them are made from Inconel 718 (IN718) alloy (virtually the same composition as K4169 alloy).

In the early stage, limited by the forming technique, engineers used to increase the casting wall thickness, simplify the casting design, and hand over the complex thin-walled structure to the follow-up machining in order to manufacture the key load-bearing parts of the engine. As a result, the large complex structural parts were often welded by the separate conventional castings, forgings and sheet metal parts. The disadvantages of this processing route includes the long development cycle, the waste of raw materials, and poor overall performance. In the 1970s, leading countries in industry, such as the United States, adopted the investment casting technology to produce large-scale castings with a diameter of more than 1 m in batch. The remarkable advantages of this technology are good integrity, high reliability, low manufacturing cost, and significant weight reduction effect. With the continuous improvement of performance and reliability of aeroengines, intensified requirements on structural design must be met for the new generation of high thrust-weight ratio aeroengines, such as the usage of a large number of lightweight, integral parts of hollow, thin-wall complex structures [3]. It puts forward higher requirements for the structural rigidity, lightweight, reliability, and operation temperature of large-scale structural parts such as turbine cases. As a result, the precision casting technology has developed into the world's mainstream technology route of manufacturing large-scale structural aeroengine parts, and has become one of the important technical bases for the development of advanced aeroengines towards lightweight, precision, and long service life. At present, the largest and thinnest IN718 alloy castings have a diameter of 1.93 m, a wall thickness of only 1.5 mm, a casting weight of 534 kg, and a pouring weight of 1900 kg, representing the highest level of nickel-base superalloy investment casting in the world.

A big gap still exists between China and the leading countries in the field of precision forming of large-scale castings for aeroengines, and due to a lack of enough technical resources and support, for many years Chinese aviation enterprises were unable to produce high-quality and large-scale complex thin-walled load-bearing structural castings. Until the end of the twentieth century, Beijing Institute of Aeronautical Materials began to develop and master the integral precision casting technology of the turbine case with a diameter of 570–800 mm. After 2000, China began to develop the integral precision casting of the turbine case with a diameter of about 1 m. With fundamental research and technology breakthroughs, some domestic universities, scientific research institutes, and enterprises had preliminarily developed the precision casting technology of the large-scale high-temperature alloy castings with a diameter of more than 1 m and a wall thickness of about 2 mm. However, the overall domestic mass production level of superalloy turbine cases is still about 20 years behind the international top level. The backwardness of the whole precision casting technology of large castings has become one of the bottlenecks in the development of China's commercial aeroengines.

With the continuous progress in structural design of advanced aeroengines, the load-bearing structural parts will develop towards the direction of larger size, thinner wall thickness and more complex structure in the future. Taking a turbine case developed for a commercial aircraft engine as an example, the outer flange size is more than 1.5 m, while the minimum wall thickness is less than 1.8 mm; moreover, there are a large number of positions with variable cross-sections. The inner quality and reliability requirements of the part are also very demanding. The precision forming of such a casting is quite difficult, which poses great challenges for the existing processing technology. Consequently, it is urgent to carry out the construction of relevant infrastructure as well as the fundamental research and key technology development of precision forming of large castings, in order to solve the problem in manufacturing superalloy turbine cases for large airliner engines.

1.2 Classification of Precision Casting

Precision casting is mainly divided into investment casting(lost wax casting), ceramic mold casting, metal mold casting, pressure casting, and lost foam casting. Ceramic mold casting is a metal casting process that uses refractories, binders, and catalysts to form the surface layer of ceramic mold. It can produce castings with surface roughness of Ra = 3.2–12.5 μm, and dimensional accuracy of CT5–CT8. It has advantages of high refractoriness of ceramic mold, stable high-temperature performance, simple processing, low investment, and high efficiency, and is thus widely used in the production of castings with complex surface shapes, such as plastic mold, glass mold, rubber mold, die casting mold, forging die, stamping die, metal mold, core box, and cutting tool.

Metal mold casting is a method of casting that uses metal mold, which is characterized by high mechanical properties, high precision, and surface finish of castings, high yield of process, and simple processing. It is easy to realize mechanization and automation, but has high cost in mold, poor fallback, and is prone to cause casting cracks. Metal mold casting is commonly used for aluminum alloys, zinc alloys, and magnesium alloys.

Pressure casting, also known as "die casting", is a process where the liquid metal is injected into the die at high a speed under high specific pressure and solidified into a casting. It is characterized by close dimensional accuracy, good surface finish, thin-walled castings, and superior mechanical properties, but the castings are lacking in internal integrity. It is applicable to low melting point zinc, aluminum, and magnesium alloys, and the resulting products cover automobile, motorcycle, communication, household appliances, hardware products, electric tools, and many other fields.

Lost foam casting is a casting process in which the model (usually made of polystyrene foam) does not need to be removed from the mold cavity. It has the characteristics of molding sand without adding binder, absence of parting surface and draft angle, high dimensional accuracy, and good surface finish. It is suitable for copper alloys, aluminum alloys, magnesium alloys, and cast-iron alloys, to produce components such as automobile engine cylinder heads, exhaust pipes, intake pipes, crankshafts, hubs and brake discs.

Investment casting, also known as lost wax casting, plays an important role in nurturing human civilization and promoting social progress. The history of investment casting can be traced back to more than 4000 years ago, originated from Ancient Egypt, China and India. For example, various kinds of bells and vessels with fine patterns and characters were cast in ancient China, such as Wang Zi Wu Tripod and bronze Jin in the late Spring and Autumn period, bronze Zenghou Yi zun-pan in the Warring States period, bronze Poshan incense burner inlaid with gold decorations in the Han Dynasty, etc. The main steps of investment casting includes wax pattern making, tree making, shell making, dewaxing, sintering, and casting [4]. The practical application of modern investment casting technology in industry began in 1940s when it was used to manufacture Aeroengine Blades in the United States. At that time, developing jet engines required the manufacture of heat-resistant alloy components with complex shapes, accurate sizes, and smooth surfaces, such as blades, impellers, and nozzles. However, since it was difficult to machine heat-resistant alloys and since the shape of components were complex, it was nearly impossible or extremely difficult to manufacture these components by other methods. Therefore, it is necessary to find a new precision forming method. With reference to the ancient wax loss precision casting, through a series of improvements on materials and manufacturing, the investment casting method has received important development on the basis of ancient technology. As an advanced near-net forming technology, its main advantages include: (1) it can manufacture metal castings with complex inner cavity structures, high melting temperatures, and high chemical activity [5], (2) the dimensional accuracy can reach 5% of the nominal size, and the roughness level is Ra = 0.8–3.2 μm, (3) the cost is relatively low, and (4) it has a wide range of material adaptability and excellent production flexibility. The continuous improvement of the investment casting technology has enabled the development and production of key parts in the aviation industry, and in return the development of the aviation industry has also promoted the further development and widespread application of investment casting, which has been used in automobiles, machine tools, ships, internal combustion engines, steam gas turbines, telecommunication instruments, weapons, medical devices, arts and crafts, cutting tools, and other industries. In the past decade, investment casting has been developed rapidly. The consumption of investment castings in major industrial countries in the world is increasing at an average rate of 7–12%, and that of specified investment castings is increasing at an amazing rate of 30%. For example, the export value of high-value-added castings used in the fields of aviation, aerospace, and industrial gas turbines in the developed countries such as Europe and the United States has accounted for more than 70% of the total export value [6].

1.3 Difficulties in Investment Casting of Large Castings

From the perspective of the overall development trend domestically and abroad, large-scale structural parts for aeroengines are moving to the direction of larger sizes, thinner wall thicknesses, and more complex structures. Taking the K4169 TRF and

ZTC4 intermediate case designed by one type of verification engine of a large airliner as examples, their design sizes are 1.4 and 1.2 m in diameter, respectively, most of the casting thicknesses are less than 2 mm, the wall thickness difference is more than 10 times, and the thin-walled area accounts for more than 80% of the overall component surface. There are a large number of special-shaped parts such as hollow support plates, mounting bosses, assembly seats, and stiffeners. The number of dimensions of the TRF that should be controlled is more than 300, while that of the intermediate case is 570. These structural characteristics make the large-scale castings different from the small and medium-sized ones in the process of filling and solidification, such as the "large size effect" due to the large-scale size, the "variable cross-section effect" due to the complexity of the structure, and the "thin wall effect" due to the thin-walled surface. At present, the manufacturing yield of K4169 large castings in China is low, and casting defects such as porosity, deformation, and dimensionally being out of tolerance have become the main issues of large castings. Generally speaking, there are two major difficulties in the precision forming technology of large-scale nickel-base superalloy castings: prediction and control of solidification defects and dimensional accuracy control of castings.

(1) Prediction and control of solidification defects

The large size, thin wall, and complex structure of aeroengine case castings make it difficult to predict and avoid solidification defects in the precision forming process. The large size of the casting makes the flow path of melt filling greatly increased during the pouring process and thus the filling process becomes difficult to control, which easily leads to defects such as misrun and cold shut. The temperature distribution of the whole casting system will also become nonuniform, and the hot spot (i.e., isolated pools of liquid form inside solidified metal) exhibits the characteristics of large quantity and wide distribution, which leads to the difficulty of feeding and many porosity defects. The complexity of the casting structure further highlights the hot cracking problem caused by a sudden change of wall thickness. Moreover, the thin-walled structure of the casting results in the increase of resistance during filling such thin-walled channels for superalloys that typically have a wide crystallization temperature interval. At the same time, the solidification sequence of the thick and large parts connected with thin-walled parts is difficult to adjust, in particular, the latent heat release of the thick and large parts of the casting may result in the remelting of the thin-walled parts as well as other special phenomena, thus leading to the formation of the solidification defects, i.e., thin-walled porosity and harmful brittle phases. Because of the dispersion of hot spots caused by the complexity of the structure, the grain size distribution of large-scale complex thin-walled castings is nonuniform, and the mechanical properties of each part are obviously different, which has potential danger for the subsequent service safety of castings. Therefore, "large size effect", "variable cross-section effect", and "thin wall effect" in the process of casting forming become the main causes of solidification defects.

(2) Dimensional accuracy control of castings

The investment casting is characterized by long process flow, complex influencing factors, and difficult control of casting dimensions. For large castings, from mold manufacturing, wax pattern assembly, shell preparation, and baking to alloy casting, it has undertaken dozens of processes, involving thousands of process parameters. The process parameters contained in each process may have an impact on the dimensional accuracy of final castings and the accumulation and transfer of errors cannot be ignored, which is a critical reason why the accuracy of large castings is difficult to control. With the trend into large-scale size, thin-walled surface, and complex structure of castings, the dimensional accuracy is more difficult to control. First, the deformation degree of large wax pattern caused by the movement and nonuniform stress in the process of shell coating and sand spraying is increased due to the increase of self-weight and size, which is one of the reasons for the excessive size accuracy of castings; Secondly, the deformation and oversize caused by high temperature creep of the ceramic shell greatly affect the final size accuracy of castings, because the shell weighs hundreds of kilograms and there is a phase change during the baking process. In addition, the shell needs to bear the hydrostatic pressure in the pouring and solidification process. Thirdly, because of the structural characteristics of large castings, namely, "large size", "variable cross-section", and "thin wall", they will cause nonlinear deformation of castings under multiple constraints during the cooling process, making it difficult to control the shape and size accuracy of castings. The traditional "trial and error" method cannot meet the requirements of dimensional accuracy control of large castings. Therefore, it is necessary to combine the characteristics of large castings with simulation and typical casting experiments to study the formation mechanism of casting errors, in order to realize the robust control of tolerance in investment casting of large castings.

The characteristics of investment casting technology manifest itself as more suitable for the formation of large batch and small and medium size precision parts. Therefore, in the past, most of the relevant process design theories and methods focused on the manufacturing of small and medium-sized parts. However, for the investment casting of large-scale structural parts, there has long been lacking in the process design theory, the formation mechanism and control method of solidification defects, and the quantitative control method of manufacturing errors for the whole process, which will be described in detail as follows.

(1) Process design theory for large castings

Due to the requirements of structural design and lightweight, the shape and size of the characteristic structure of large-scale complex thin-walled castings for aeroengines have exceeded the traditional design concept of castings. For example, in the classical design theory, the wall thickness difference of the casting at the variable cross-sections should not be more than 3 times; otherwise the deformation and cracking of the casting may be induced by stress concentration. However, different from the traditional castings, the maximum wall thickness difference of the intermediate engine case is 12 times, and that of the TRF casting is more than 10 times. Moreover, the area of the thin wall parts in the intermediate case and the TRF accounts for more than 80% of the castings. The large area of the thin wall will lead to the defects of filling difficulty, porosity, harmful brittle phases, etc., and therefore needs to be avoided in the classical design theory. In addition, the large-scale and complex of castings limit the shrinkage of castings during solidification and cooling, and thus nonlinear deformation becomes a new problem. Besides the pouring system, it may be necessary to add integral stiffeners to restrain deformation in the process design. Therefore, the traditional process design theory is no longer suitable for the process design of large castings such as engine cases. As a result, it is urgent to study and explore the new process design theory of large castings.

(2) Formation mechanisms and control methods of solidification defects

The special structural characteristics of large castings bring about "large size effect", "variable cross-section effect", and "thin wall effect", which lead to the change of the solidification sequence, temperature field, and solute field distribution. The traditional defect formation mechanism cannot be directly used to describe the formation process of defects in special structures. For example, the formation of porosity and negative segregation in the thin wall parts is not as usual, which needs to be described by establishing the kinetic equations of solidification. In particular, for the porosity defect, which is the most difficult to control for large castings, filling and feeding under pressure is the best way to solve this problem. The essence of counter gravity or adjusted pressure casting is that feeding is not only realized by fluid flow under gravity, but also by pressure to keep the feeding channel unblocked. In this sense, adjusted pressure casting is more suitable for the preparation of large complex thin-walled castings. Nevertheless, the fluid flow and freezing behavior of superalloy during filling and solidification under pressure, the mechanism of micro flow between dendrites to eliminate porosity, and the optimization of process parameters are all rarely reported in the literature, which should be an important subject to be studied systematically.

(3) Quantitative control methods of manufacturing errors for the entire process

The law of formation, accumulation, and transfer of various error sources in the entire manufacturing process of large-scale complex thin-walled superalloy castings, such as die design, manufacturing, pressing and assembly of wax patterns, shell mold preparation, and baking are very complex. Traditionally, the dimension accuracy control and tolerance design of castings mainly rely on qualitative experience and error correction, which is of high cost, long cycle, and poor stability and makes it difficult to meet the urgent need of rapid research and development of key components of new generation aeroengines. Due to the three structural effects of large castings, nonlinear deformation under multiple constraints occurs during the cooling process, which makes the shape and size of castings difficult to control. There is an urgent need to (1) study the formation mechanism, transfer rule, and control method of manufacturing errors in the complete casting process, (2) establish the systematic model for numerical simulation of manufacturing errors, and (3) realize the quantitative analysis of thousands of process parameters in the complete casting process and accurate allocation of tolerance. Only in this way can the technical bottleneck of traditional qualitative analysis and "trial and error" experience be overcome, enabling the rapid improvement in the dimension accuracy of large-scale aeroengine castings and robust control of casting quality.

1.4 Research Status of Precision Forming Technology for Large Castings

1.4.1 Virtual Manufacturing of Large Complex Thin Wall Castings

Virtual manufacturing is the virtual realization of the actual manufacturing process on the computer, which includes the integrated modeling of product design, process planning, processing, and manufacturing. The use of computer simulation and virtual reality technology to realize the essential manufacturing processes of product development, manufacturing, management and control on the computer, allows to enhance the decision-making and control capabilities at all levels of the manufacturing process. Prior to the physical realization of manufacturing, virtual manufacturing technology allows us to know about the performance or operation state of future products, so that we can make forward-looking decisions and optimization schemes.

Investment casting is a type of near-net-shape manufacturing technologies with long process, many working procedures, and high requirements for production environment and product quality. Because of the inevitable time and material cost, the operational cost of investment casting is significantly high. Owing to the further research on the flow and solidification behavior of high temperature melt, as well as the development of computer performance and virtual reality technology, the modeling research on investment casting has made great progress. There have been some examples of using virtual design and virtual manufacturing to replace lengthy and expensive experimental verification.

Soon after the concept of "virtual manufacturing" was proposed, it draw great attention of the world. In recent years, almost all industrially developed countries focus on the research and application of virtual manufacturing. In the United States, the National Institute of Standards and Technology (NIST) established a virtual manufacturing environment (known as the National Advanced Manufacturing Test Bed); in Europe, many universities and research institutes cooperated with each other and with enterprises for developing virtual manufacturing technology. The historic development of virtual manufacturing technology in foundry process has experienced three stages. In 1960s, the numerical simulation of temperature fields during the solidification of castings was initiated based on the partial differential equation of heat conduction. In the 1970s and early 1980s, many researches focused on numerical simulations of the temperature field, involving thermophysical property parameters of materials, interface conditions, latent heat treatment, criteria and control of shrinkage cavity, shrinkage porosity, and other defects. In the late 1980s and 1990s, the practical and research work was further developed. The research focused on casting residual stress, grain structure simulation, and calculation of material properties. In the mid-1990s, a large number of powerful and full-featured commercial softwares were introduced in the market. Among many casting process simulation softwares, ProCAST is one of the representatives that can be used for investment casting. Based on some investigation and evaluation, NASA recommended this software as the first choice of CAE software in the field of aerospace. World-famous precision casting companies, such as Howmet, PCC, Pratt Whitney, GE Aviation, and Rolls Royce, have chosen ProCAST as an important tool for casting process analysis and new product development.

At present, virtual casting technology is mainly used in the field of casting design, numerical simulation of casting filling or molding process, visualization of results, and simulation optimization of the casting production process.

(1) Casting design and rapid prototyping

A virtual casting system has been developed in the laboratory of Integrated Computer Aided Research on Virtual Engineering Design and Prototyping (ICARVE) at the University of Wisconsin. The system uses stereo glasses to observe 3D images, language to build various geometric models, and data gloves to determine the size and position of geometry. At present, ICARVE laboratory has successfully completed the design of injection and die-casting parts by using this system. The goal of this system is to achieve the design efficiency of 10–30 times faster than the traditional CAD method.

Additive manufacturing (AM) is a high-tech achievement newly developed in the 1990s. It is an advanced manufacturing technology that integrates modern disciplines such as computer, optics, electricity, precision instruments, and materials. It is applicable to the asymmetric, irregular curved or complicated parts and dies with fine structures that cannot be processed by traditional methods in practice. Rapid prototyping is a new concept of manufacturing technology. It abandons the traditional machining method and controls the 3D NC forming system according to the geometric information of parts generated by CAD. Materials are piled up by laser or other methods to form a component that is completely consistent with the geometry in the computer. At present, the main molding methods adopted are stereolithograhy (SLA), selective laser sintering (SLS), fused deposition modeling (FDM), laminated object manufacturing (LOM), and solid ground curing (SGC).

(2) Numerical simulation of mold filling and visualization of results

Many commercial simulation programs of pouring process have the visualization module, which is developed using 2D image processing technology and allows users to visualize the simulation results more intuitively and to analyze the forming process of castings. Researchers at Tsinghua University [7] have made much progress in numerical simulation of the temperature field in casting process, prediction of shrinkage cavity and porosity, simulation of filling process, 3D stress analysis, and microstructure prediction. It can predict the temperature field, flow field, and stress field of cast steel, cast iron, and nonferrous alloy castings, and predict the possible occurrence of casting defects (the position and size), such as shrinkage cavity, shrinkage porosity, hot cracking, etc.

Yan et al. [8] established a finite element numerical analysis model for the aero-engine case. Using the MSC.Marc software, they simulated the temperature field, stress field, and deformation of the thin-walled parts of the casting. Meanwhile, the influence of the temperature field and stress field on the deformation was analyzed qualitatively, providing the basis for reducing the deformation of the actual case parts.

Based on the results of numerical simulation, Zhang et al. [9] applied the shape optimization method in the field of structural optimization to the design of pouring and riser system of castings, and established the integrated system of design, analysis, and optimization. The optimization model of casting process was established, and the sensitivity analysis formula of design variables was carried out. For the first time, the global convergence method (GCM) was used to automatically obtain the search direction and search step length of the optimization process. Finally, an example of riser optimization design was given to prove the reliability of the optimization model and algorithm, which could ensure the quality of casting and achieve the purpose of minimum material consumption.

Based on the momentum equation, continuity equation, volume of fluid equation and energy conservation equation, Xue et al. [10] developed a 3D simulation program coupling fluid flow with heat transfer in the filling process of casting. They performed the simulation of temperature fields in the filling process of nickel-base alloy blades, and used the criterion to judge the position of shrinkage cavity and porosity in the solidification process of nickel-base alloy blades. Li et al. [11] compared different criterion functions of shrinkage cavity and porosity, and predicted the possibility of shrinkage cavity and porosity of IN738 shell mold thin-walled plate casting using the Niyama criterion function.

In order to solve the issue that the computer takes a lot of time to repeat the "trial and error" process, the Finite Solutions company of the United States proposes to link the multivariable optimization operation software to AF Solid solidification simulation software, so that the simulation and optimization can be completed in a single step, thus greatly improving the efficiency of computer simulations.

(3) Simulation and optimization of casting process

This includes simulation analysis of productivity, production cycle, equipment utilization, logistics, and other information, such as workshop and section production process simulation, allowing to analyze the production capacity and optimize the allocation of equipment resources. Through computer simulation analysis, virtual operation can be carried out for various situations of production process. The bottleneck that restricts productivity in existing production system can also be analyzed, and the feasibility of new production scheduling scheme can be predicted. These procedures can help companies optimize the production process, reduce the blindness of investment, improve production efficiency, and fully realize the company's potential.

Although the research in this field is still in progress, it has shown broad application prospects, high application value, and great development potential in the foundry industry both domestically and abroad. LAEMPE, a German foundry equipment manufacturer, used discrete event simulation and robot simulation technology to build a core production line for Waupaca company. The Swedish Foundry Association also provided virtual production analysis for large-scale enterprises and completed a series of production simulation works, such as a complete virtual production analysis from melting to product delivery in an aluminum foundry. FSC company in the United States has successfully used virtual manufacturing technology to complete the upgrading of production system and avoid unnecessary investment in the melting equipment.

1.4.2 *Precision Manufacturing Technology of Large Complex Thin Wall Castings*

In the first half of the twentieth century, foreign countries began to study the vacuum melting and investment casting technology of nickel-base superalloys for aircraft engines. In the 1970s–1980s, breakthroughs were made in alloy design and investment casting technology for large castings. A series of high-performance nickel-base superalloy castings were rapidly industrialized. British AE company has applied a series of investment castings to new engines. The overall dimensions of these castings are generally 300–600 mm, and the minimum wall thickness is 0.8–1.5 mm. In addition, the grain size can be controlled, and the fatigue life of castings can be greatly improved with the help of hot isostatic pressing technology. Hitchiner, PCC, and Howmet companies have cast a series of large-scale thin-walled integral castings for the engine of General Electric company in the United States, among which IN718 alloy is used as the casting material of TRF, with the thinnest section size of 1.25–1.75 mm, the outer diameter of up to 1.93 m, and the casting weight of up to 2040 kg. At present, the smelting process and equipment of large-scale casting mostly use a vacuum induction degassing and pouring furnace (VIDP). Through the pattern design, the system has great flexibility, and it is easy to realize the special pouring process such as refining, degassing, vacuum slag removal, instant chemical composition control, bottom pouring, and so on.

The research institutes of high temperature structural materials and its precision casting technology in China mainly include Beijing Institute of Aeronautical Materials, Shenyang Foundry Research Institute, Institute of Metal Research Chinese

Academy of Sciences (CAS), Central Iron and Steel Research Institute, Northwestern Polytechnic University, Harbin Institute of Technology, Shanghai Jiao Tong University, etc.

Focusing on the precision manufacturing technology of large-scale complex thin-walled castings, three aspects of investigations have been mainly carried out: the filling behavior and solidification structure control of large-scale complex thin-walled castings, the formation mechanism and control of solidification defects, and the dimensional accuracy control of castings.

(1) Filling behavior and solidification structure control of large complex thin wall castings

The fluid flow and growth kinetics of dendrites in the filling and solidification process of castings are always the frontier issues in the field of solidification, and also the key factors affecting the complete filling of thin-walled structures. Because of the high temperature and opacity of metal alloys, the solidification process always appears as a "black box". People have been expecting to observe and analyze the melt filling and solidification process directly, to better control the casting process parameters and hence to obtain the castings with complete shape and excellent quality.

The permeability of X-ray renders X-ray imaging as the tool that metallurgists have expected for. The early X-ray intensity is low, which could not meet the needs of scientific research and engineering application. As early as 1974, Japanese researchers began to use high-energy X-ray to study the melting and solidification process of Si crystal. Because the X-ray energy was not high enough at that time, and because the resolution of picture tube-type video camera used was also very low, the change of crystal could only be identified from the rough outline.

Since its discovery in 1947, synchrotron radiation has been used as a major experimental method in the field of physics because of its intensive photon flux, high energy, high spatial and temporal resolution, high collimation, and other characteristics and advantages. By the 1980s, the second generation of synchrotron radiation light source was created, and synchrotron radiation had gradually become a general interdisciplinary research tool. Japanese and French researchers published the research results on Sn and Al–Cu alloys, respectively, both using synchrotron radiation imaging, which initiated the use of synchrotron radiation in the solidification field.

In the early stage, limited by synchrotron radiation performance and ability of signal receiving and analysis, it was difficult to get clear real-time in-situ X-ray images, which was not helpful to the study of solidification process and not even compared with the contribution of solidification simulation using transparent organic material. With the construction and operation of the third-generation synchrotron radiation light sources, together with the rapid development of electronics, optics, machinery, computers, and other disciplines, the capability of synchrotron radiation has been greatly improved since 2000. Meanwhile, the application of synchrotron radiation to the study of solidification process of metal alloys has been widely spread all over the world.

Yasuda et al. studied the problem of microsegregation in the solidification process of Al–Si alloy at the Spring-8 light source. Utilizing the European Synchrotron Radiation Facility (ESRF), Faraji et al. [12] carried out research on the nucleation kinetics and grain refinement of Al–Si alloy during solidification; Terzi et al. analyzed the mechanism of dendrite coarsening of Al-10wt.% Cu[1] during solidification; Bogno et al. [13] studied the process of equiaxed dendrite structure of Al–Cu alloy; Shuleshova et al. [14] observed the in-situ solidification process of Ti–Al alloy; Lengsdorf et al. studied the dendrite growth characteristics of Al-Ni alloy under different gravity conditions. These studies showed a more and more in-depth use of synchrotron radiation in the study of solidification process of molten metals.

Based on the study of the solidification process of a low melting point model alloy, Yasuda et al. [15] investigated the dendrite structure growth change of iron-based alloy at SPring-8. Husseini et al. [16] conducted the crystal structure analysis and defect monitoring of nickel-base superalloy at the beam line of Advanced Photon Source (APS) at Argonne National Laboratory by in-situ imaging. The above works established an experimental foundation for the synchrotron radiation imaging of the formation and evolution of the solidified structure, as well as the defect analysis of high melting point alloys.

With the development of the third-generation synchrotron radiation light source in Shanghai, the research on materials by means of synchrotron radiation has been carried out rapidly in China. Since 2010, researchers from Shanghai Jiao Tong University have carried out a series of studies on the solidification structure evolution and melt structure of Al–Cu alloys at the imaging and small angle scattering beamline of Shanghai Synchrotron Radiation Facility (SSRF) [17, 18], and have made preliminary progress in the imaging research of solidification process of superalloys, as shown in Fig. 1.1.

Owing to the development of computing method and capability, numerical simulation of filling and solidification process based on finite element and finite difference method, together with the microstructural simulation of dendrite growth dynamics

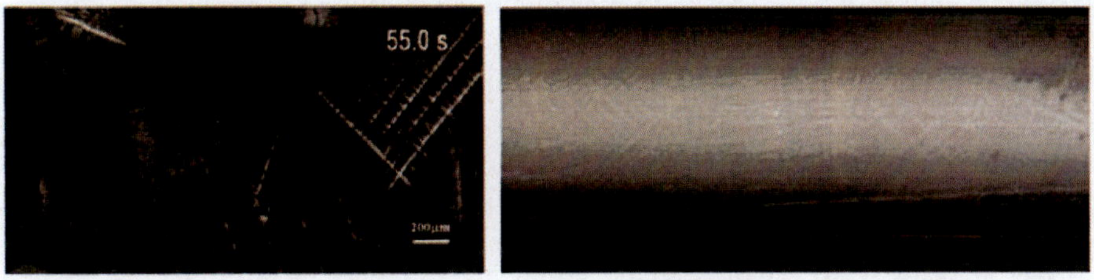

(a) (b)

Fig. 1.1 Synchrotron radiation imaging of solidification process of metal alloys: **a** Al-10wt%Cu [18] and **b** K4169 superalloy

[1]Mass percentage.

based on phase field method, have become a new focal point in the field of solidification. The simulation of filling and solidification process mainly focuses on the influence of melt static pressure, pouring temperature, filling speed, alloy composition, solidification sequence, and gravity direction on the temperature field distribution at the metal/mold interface. The microstructural simulation of dendrite growth dynamics mainly focuses on the simulation of dendrite tip shape, crystal orientation selection, and growth rate in simple alloy systems.

While the above researches have obtained fruitful results in terms of the basic theory of filling and solidification, they still suffer from some remaining issues that make them not directly applicable to the precision forming process of large castings. In general, the solidification systems studied above were relatively simple, and the influence of multiple components was not considered. In addition, the influence of casting structural characteristics on filling and solidification behavior was neglected. In particular, the effect of pressure on macroscopic and microscopic fluid flow and the formation mechanism of the dendrite arm spacing, dendrite tilt, and solidification structure are all lacking in the relevant mathematical model.

In the 1980s–1990s, research institutes in China, such as Northwestern Polytechnic University, Institute of Metal Research CAS, Tsinghua University, Harbin Institute of Technology, Dalian University of Technology, etc. had carried out in-depth and systematic research on some fundamental scientific issues of solidification. The research on filling and solidification process had also focused on the simulation of transparent organic matter and thermodynamic calculation. Aiming at the core scientific problem of solidification disciplines, substantial progress has been made on the solid–liquid interface stability, the formation principle of solidification structures (in particular, for the solidification process far from equilibrium), the nucleation, and interface morphology. Zhuangqi Hu and his group [19] have investigated and developed the preparation of hollow turbine blades and DZ38 alloy series materials in the early stage. The influence of Al, P, B, Ti, Re, C, Co, W, S, Si, Zr, and other elements on the structure and properties of nickel-base single crystal superalloys (DD8, K17G, DZ38, GH761, DD32, etc.) had been systematically studied. The creep behavior of single crystal alloy under the condition of high-cycle fatigue and low-cycle fatigue were both analyzed. The effect of coarsening of γ' phase on the high temperature properties was studied and the effects of heat treatment and pre-pressing on the rupture property of superalloys were also examined. In addition, the influence of melt refining, structure, and superheat on the alloy properties were explored. In the field of casting alloys, Zhou and Zeng [20] used mathematical modeling and hydraulic simulation experiments to study the change of free surface of liquid during filling. They also used the method to examine the influence of external pressure field on the solidification process and the influence of casting process parameters on the temperature field in the counter gravity casting and the adjusted pressure casting of aluminum alloys. Li et al. [21] have carried out a lot of research in the field of counter gravity casting of large aluminum alloy and magnesium alloy castings. They have made important progress in liquid and semi-solid forming, solidification theory of multi-component and multi-phase alloys, counter gravity casting equipment, and the technology of large complex thin-walled structure parts. Liu et al.

[22] have performed considerable research in the field of non-equilibrium solidification, directionally solidified structures, and ultra-fine structures of superalloys, established the theoretical framework of dendrite transformation, and proposed the idea of controlling casting porosity using fine-grained casting.

Numerous research works on phase field simulation of dendrite growth have also been carried out in China. Zhao et al. [23] studied the quantitative influence of undercooling and temperature on the coupling coefficient, anisotropy coefficient, and other parameters in the phase field model. They also discussed the evolution of dendrite morphology and studied the dependence of grain morphology on the growth rate and phase field parameters. The criteria of the dendrite branching during growth and the initiation of the lateral branching were also analyzed. Following the phase field model of the coupled flow field originally proposed by Chen et al. [24] used the finite difference method to implement the model numerically and was able to simulate the solidification process of pure metals in 2D, through which the influence of different convection velocities on the dendrite growth had been shown. They also used the phase field model coupled with the flow field developed by Lan and Shih to simulate the dendrite growth process and the solute field of a Ni–Cu binary alloy under isothermal conditions; the SIMPLE algorithm was employed in the simulation to calculate the pressure field, showing the asymmetric growth of dendrites under convection. Chen et al. [25] used an adaptive finite element method to solve the governing equations of the phase field model, which allowed for simulating the evolution of a single equiaxed crystal in undercooled nickel melt under the condition of large calculation domain and thin interface layer, rendering the simulation result of the phase field model more relevant to the real process. Wang et al. [26] used phase field simulation coupled with solute field to study the influence of solute diffusion coefficient of the solid phase D_s on the dendrite morphology and microsegregation during dendrite growth of a Ni–Cu binary alloy.

(2) Formation mechanism and control of solidification defects

In the mushy zone at the front of solid-liquid interface, the interdendritic microscopic flow has a great influence on the dendrite growth, solidification microstructure, and composition distribution. Research has focused on the relationship between the microscopic melt flow and the dendrite arm spacings as well as the liquid phase fraction, targeting at establishing a finite element model to predict the location, size, and quantity of interdendritic micro shrinkage [27]. In addition to using transparent alloys to simulate the interdendritic flow condition and the interaction with dendrites, external physical fields are usually used to control the flow characteristics in order to reduce segregation, break high-order dendrites, and refine the solidification structure.

Nastac and Stefanescu [28] studied the solidification kinetics of IN718 alloy casting and proposed a predictive model of the formation and evolution of harmful phases such as NbC/Laves. Their results showed that the solidification process has a significant impact on the distribution and microsegregation of carbon in Laves phase, element redistribution, and high-temperature mechanical properties. Knorovsky et al. [29] found that the element segregation during the solidification of IN718 could promote the γ/Laves and γ/MC eutectic structure. The work of Whitesell and Overfelt [30] showed that the insufficient filling in the solidification process would result in a large tensile stress in the mushy zone, which was released through the plastic deformation of the dendrite network and the nucleation of microporosities in the interdendritic liquid. According to the tensile test and thermodynamic calculation of IN718 alloy in the mushy zone, it has been shown that the non-uniform deformation and stress concentration caused by high strain rate can lead to fracture at a level of low strain. The formation mechanism of porosity caused by stress in liquid phase is also identified and the corresponding stress threshold values under different strain rates are obtained as well. Penya et al. [31] used a Bayesian network method to establish a model for the formation of microporosity and product quality control in the process of investment casting.

Studies on solidification defects in superalloys have been carried out by Northwestern Polytechnic University, Institute of Metal Research CAS, Nanjing University of Science and Technology, Shanghai Jiao Tong University, and other research institutes in China, mainly focusing on the metallurgical quality control of master alloys, optimization of process parameters such as cooling rate and pouring temperature, application of electromagnetic stirring, and addition of refiners for grain refinement. In terms of structure control of high-temperature materials, the group led by Zhuangqi Hu from the Institute of Metal Research CAS has carried out research on the optimization of microstructure and properties of nickel-base and cobalt-base superalloys since 1980s. Minggao Yan and Chunxiao Cao at Beijing Institute of Aeronautical Materials have carried out intensive studies on the high temperature structural materials, in particular, with regard to the microstructure, strengthening mechanism, and superplasticity theory of titanium alloys; nevertheless, there are very few of works on the solidification structure and defect control of large-sized superalloy and titanium alloy castings.

The study of the group led by Zhuangqi Hu focused on the microstructure and properties of K465, IN718, and other alloys, as well as on the influence of casting process on the microstructure and properties. In particular, for a given cast superalloy of K417G, the influence of the size effect on the microstructure and properties of castings was documented, and important progress had been made in the turbocharger and other parts for aeroengines. Chen et al. [32] studied the influence of casting process parameters such as solidification rate on the formation tendency of microporosity, composition segregation, microstructure, and property stability of Inconel 718C alloy. The results showed that as the solidification cooling rate was decreasing, the tendency of forming microporosity decreased and the stability of alloy rupture life increased, whereas the grain size increased and the average rupture life decreased. Xiong et al. [33] studied the effect of microstructure refinement of K4169 alloy ingot on shrinkage porosity and element segregation. Their results suggested that the lower the casting temperature was, the smaller the length of primary dendrite axis and the secondary dendrite arm spacing were; meanwhile, the fraction of equiaxed grains in the casting was increased up to 90%, and the shrinkage porosity and the segregation of major elements in the casting were reduced, with, however, the size, quantity, and morphology of MC carbides and Laves phase showing little changes before and after grain refinement. Wang et al. [34] showed that the lower the cooling rate was, the more significant the degree of segregation of Nb and Mo in the solidified structure of IN718C were. Tang et al. [35] studied the fine grain casting process of superalloys and its effect on fatigue life, of which the results showed that 100% fine grain castings could be obtained by properly controlling the shell and pouring temperatures, with the low-cycle fatigue life of castings being increased by 2–3 times.

In conclusion, although some researches have been carried out in the field of solidification structure control and defect suppression for high temperature structural materials in China, there is little systematic studies on the mechanism of defect formation of large-scale superalloy castings, in particular, in terms of the formation mechanism and the method of suppressing defects due to the distinctive "large size effect", "thin wall effect" and "variable cross-section effect" of large-scale castings.

(3) Dimensional accuracy control of castings

The tolerance analysis and dimension control method of complex systems in precision casting are becoming a new trend of academic research both domestically and abroad. In general, the precision forming process of large castings is very complicated. There are certain errors at each individual step of the entire process, such as mold design and processing, wax pattern pressing and assembly, shell coating and baking, melt filling and solidification, and mold removal and casting post-treatment. The traditional "trial and error" method to achieve the dimensional accuracy and deformation control of large-scale castings can only ensure that the accuracy of a certain step and/or a specific local structure may meet the requirements, but cannot guarantee the accuracy of the entire casting, which inevitably leads to a high cost and long processing time. Therefore, it is urgent to establish relevant theories to predict and control the size deviation and deformation in large castings, such that a system of accurate processing and dimension control can be established for the complete casting process of large castings.

The deformation and size deviation of wax pattern are the primary factors that cause the precision out of casting tolerance. The dimension control of wax pattern is the first step in both theoretical research and production practice. Bonilla et al. [36] reported the numerical simulation results of wax pattern pressing process based on the solid expansion/contraction theory and heat conduction theory and was able to establish a model of wax pattern accuracy according to the thermophysical properties of wax pattern materials, molding characteristics, and wax pressing process parameters. Sabau et al. [37], on the basis of the thermophysical and thermomechanical properties of the wax itself, focused on the rheological properties of the wax during the pressing process and constructed the constitutive equation of wax pattern deformation. These studies represent some of the pioneering works on the pressing process and quality control of wax pattern; nevertheless, most of the cases in question are wax patterns with relatively small sizes and simple shapes. For the large wax pattern system with large areas, thin walls, abrupt sections, and other structural features, the mechanism of deformation and dimensional deviation, together with the corresponding tolerance design and prediction, have not been reported so far.

The manufacturing process of ceramic shell is another factor that affects the dimension accuracy of castings. Although there are many new materials and technologies for ceramic shell, there are few reports on the study of shell size accuracy control. The influence of the shell on the casting size and other errors is mostly coupled with the melt casting process. While most of the previous studies had focused on the simulation of the heat transfer between the shell and the liquid metal, they relied only on the heat transfer coefficient between the shell and the metal material and thus could not simulate the final dimensional accuracy of the casting. Song et al. [38] used a 3D nonlinearly coupled thermo-mechanical model to analyze the influence of the shell on the dimensional accuracy of the casting during the casting process, and the simulation and analysis results were well verified by experiments. Following the existing studies, Rafique et al. [39] established the heat conduction model of shell and melt in the process of filling and solidification using the standard heat transfer equation containing the heat transfer coefficients of all materials, which was previously used to predict the dimensional error of casting in the process of pouring. However, reports are still missing on systematic theoretical studies of the interaction between ceramic shell and melt as well as the relationship between the ceramic shell and the special structure of large-scale castings in investment casting.

With regard to controlling the dimension accuracy of castings, the domestic factory mainly relies on the empirical data of small and medium-sized castings. The relevant researches mainly focus on the accuracy control of wax pattern and shell, including the analysis of the influence of pressing mold, pattern material and pattern making process parameters on the quality of wax pattern, putting forward measures to improve the quality of wax pattern, and optimizing the binder performance, refractory performance, and shell making process parameters in the manufacturing process of mold shell. Lin et al. [40] proposed a nonlinear error analysis model considering the coupling effect of the part shape error and material error, developed a complex system error analysis software, and applied it in the fields of automobile, aircraft, and train. Generally speaking, however, the research on the error analysis method of complex system and the principle of casting dimension control have not been widely carried out in the field of investment casting.

(4) Patent analysis of investment casting of superalloys in China

In total, 172 patents were identified in the database of China from 1990 to 2020 when using "superalloy" and "casting" as the key words (Fig. 1.2).

According to these search results, the total number of patents in the field of superalloy investment casting is relatively small in China. Nevertheless, the number of patent applications has been increasing since 2005. From the perspective of distribution, they are mainly concentrated in metallurgy, chemical industry, and other relevant industries. The patent holders include Shanghai Jiao Tong University, Institute of Metal Research CAS, Central Iron and Steel Research Institute, etc. Among these patents, there are a number of patents for "process" and "method" of superalloy investment casting, such as "a vacuum traveling wave electromagnetic refining

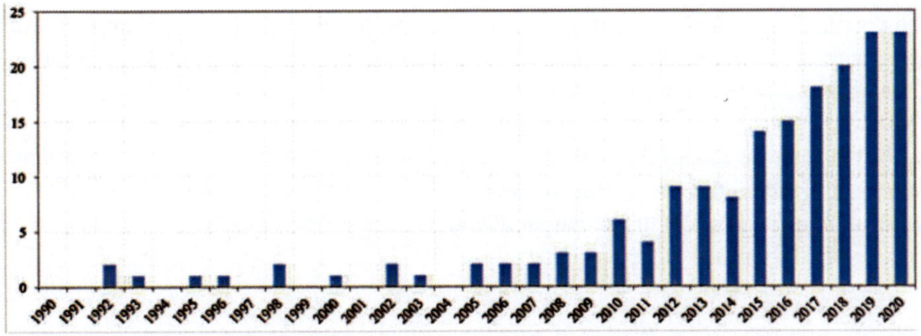

Fig. 1.2 Distribution of patent applications for superalloy casting in China from 1985 to 2012

device for superalloy investment casting", "a grain refiner for casting superalloy and a method to use it in superalloy investment casting", "a pressure regulating precision casting device for complex thin wall superalloy castings", and "a precision casting method for complex thin wall superalloy castings", etc.

(5) Patent analysis of investment casting of superalloys in the world

Using the SooPAT search tool, the patents of 99 countries were searched using "nickel, superalloy, casting, engine" as the key words, with a result showing a total of 1580 patents found. The top 10 applicants are shown in Table 1.1, which shows that General Electric (GE, US) has applied for 453 patents, accounting for 35.98% of the total number of patents. GE is the world leader in the aviation engine industry, and other top companies include United Technologies Corp., Howmet Res Corp., Siemens, PCC, etc.

Table 1.1 indicates that most of the relevant patents are from the United States, Japan, Russia, and China. In particular, the number of patents is large in the United States and Russia, both of which are leading countries in aircraft engines. In terms of

Table 1.1 Top ten applicants of patents related to superalloy casting

Applicant name	Application number	Percentage (%)
1. GEN ELECTRIC	453	35.98
2. UNITED TECHNOLOGIES CORP	201	15.97
3. Howmet RES CORP	64	5.08
4. HONEYWELL INT INC	35	2.78
5. SIEMENS WESTINGHOUSE POWER	28	2.22
6. Howmet CORP	22	1.75
7. PCC AIRFOILS INC	17	1.35
8. ROLLS ROYCE PLC	15	1.19
9. SIEMENS AG	15	1.19
10. SIEMENS POWER GENERATION INC	14	1.11

the content, most of the patents are focused on welding, coating, single crystal, and alloy smelting, whereas no patents for the development technology of large complex thin-walled castings have been found.

In other words, in order to achieve breakthroughs in the precision forming of high-temperature material castings for large passenger aircraft engines, China must rely on independent innovation to establish its own research platform and form a development strategy with independent intellectual property rights.

1.4.3 Quality and Reliability Inspection of Large Complex Thin-Walled Castings

Usually, the conventional inspection method for castings can ensure that the appearance, shape, surface quality, alloy composition, and mechanical properties of castings can meet the requirements. However, there are many factors during the casting process, e.g. the uncertainty of internal quality and the formation of defects (e.g., blowhole, inclusion, precipitated phase, porosity, and hot cracking), which can seriously affect the performance and service safety of castings and is especially important for large passenger aircraft engine castings. At present, the internal quality analysis methods of castings mainly include ultrasonic inspection methods, fluorescent magnetic powder detection methods, X-ray photography detection methods, and industrial CT. Among them, the X-ray photography method has relatively high accuracy and thus has been widely used.

With the development of high-energy X-ray industrial CT technology, industrial CT has gradually become an effective way of non-destructive testing of internal defects for large-scale components, and is currently being widely used in aerospace, aviation, weapons, nuclear power, machinery manufacturing, and so on. Nevertheless, the spatial resolution of the aforementioned inspection methods are quite limited (0.1 mm for ultrasound, 0.2 mm for X-ray photography, 0.03 mm for CT), and hence the composition and type of bulk defects cannot be determined accurately.

With the progress in material science and the development of research techniques, it gradually becomes clear that the microporosity, inclusions, and composition of fine precipitates in castings are in fact closely related to the mechanical properties of materials, such as the high-temperature creep, stress rupture, and fatigue. The difficulty is that casting defects cannot be accurately detected by the non-destructive testing methods mentioned previously. Apparently, there is an urgent need to understand the size and spatial distribution of porosities and inclusions, as well as the composition and properties of precipitates, in order to determine the metallurgical quality of the casting and the resulting impact on service. Furthermore, most of the alloys used in aircraft engine castings (such as Ti-6Al-4 V and nickel-base superalloys) are polycrystalline multiphase materials, which implies that the properties of different phases should be significantly different due to the various crystal orientations. In addition, the deformation of polycrystalline multiphase materials depends on the interaction of grains with different orientations, as well as the anisotropy associated with the different phases. Moreover, the strain distribution and tension–compression asymmetry in hardening, which are due to differences in intergranular or interphase elastic modulus and ductility, can have a major impact on the high-temperature creep and stress rupture properties of alloys. Therefore, clarifying the strain process of polycrystalline multiphase materials is another key to ensuring the metallurgy of aircraft engine castings in order to improve the service safety.

Recently, the rapid development of synchrotron radiation technique has driven the fast development of the relevant researches. For example, Puncreobutr et al. [41] from the UK have thoroughly studied the morphology and evolution of defects in bulk aluminum alloy specimens using high energy X-ray, as shown in Fig. 1.3.

It can be seen that high energy X-ray imaging and diffraction make it possible to accurately and effectively analyze the 3D distribution of casting defects before and after failure and to better evaluate the performance of aircraft engine components.

In addition, Korsunsky et al. [42] from Oxford University utilized the energy-dispersive X-ray diffraction setup at the high energy white-beam synchrotron beamline I12-JEEP at DIAMOND Light Source to study the four-point bent beam Ti-6Al-4 V samples. Diffraction patterns from the bent polycrystalline Ti-6Al-4 V samples were collected using a newly designed 23-cell "horseshoe" detector and then interpreted using Pawley refinement to determine elastic strain. According to this important report, the quantitative data of the tension–compression strain asymmetry of the polycrystalline Ti-6Al-4 V relative to the grain orientation was finally obtained, which provided key data for analyzing the service performance. Hofmann et al. [43], also from Oxford University, used the high-energy transmission polychromatic beam Laue diffraction setup developed at beamline ID15A at ESRF to analyze the crystal structure of pure Ni. This novel finding demonstrated two methods based on the high-energy X-ray diffraction for analyzing the lattice orientation and strain distribution in the grain size range of bulk polycrystalline alloys. Using the beamline ID15B at ESRF, Leo Prakash D. G. et al. from the University of Manchester studied the variation of strain of bulk Ti-6Al-4 V alloy under the tension–compression load with the sample size of 40 mm × 40 mm × 22 mm. French scientist le Graverend J. B. studied the effects of fine precipitation on non-isothermal creep and creep-fatigue behavior, with the sample sizes of 14 mm × 6 mm × 2 mm.

As one of the top three aircraft engine manufacturers in the world, Rolls-Royce plays a very important role in the development and manufacture of aircraft engines. Since the establishment of the British national light source, i.e. DIAMOND Light Source in 2007, Rolls-Royce has cooperated with it to build a time-resolved beamline, namely I12-JEEP, for *in-situ* real-time imaging of defects and diffraction analysis of stress–strain in key aircraft engine components. After the beamline was put into use at the end of 2010, it first analyzed and tested the engine of the Boeing 787 Dreamliner.

With the construction and application of Shanghai Synchrotron Radiation Facility (SSRF), the related research on high-energy synchrotron radiation has been gradually carried out in China. Taking the construction of SSRF Phase-II beamline project as an opportunity, the development of China's own high spatial resolution and

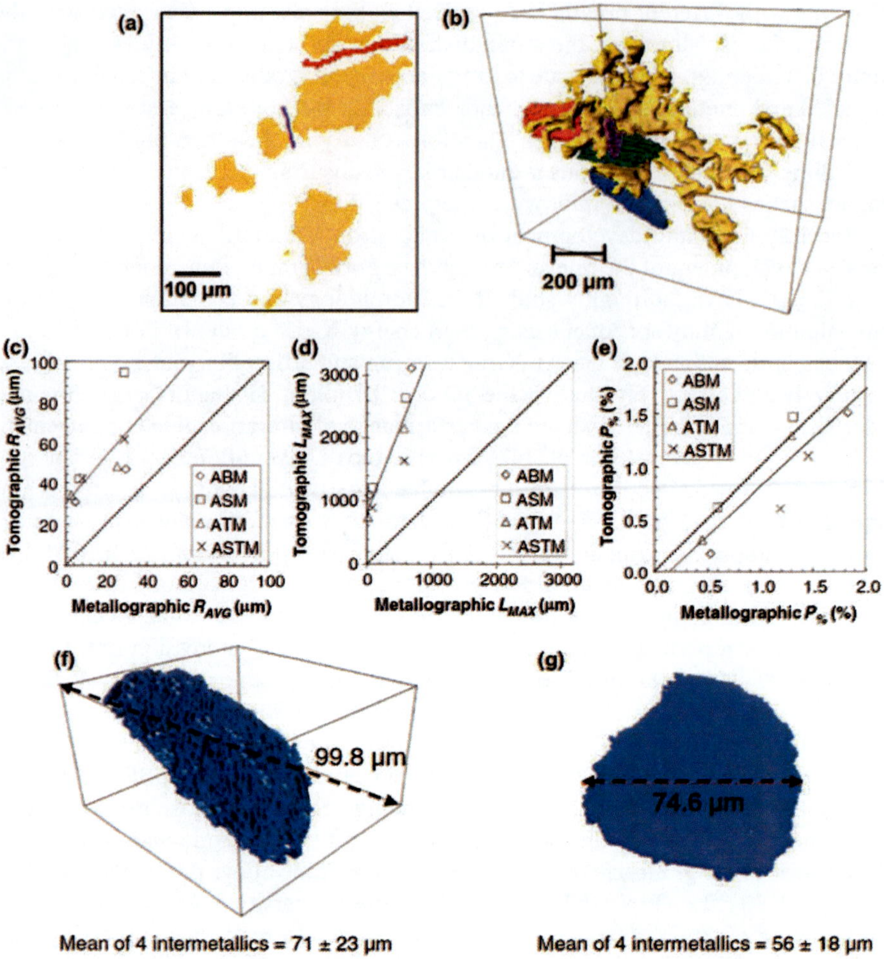

Fig. 1.3 Comparison of 2D and 3D solidification defects (porosity and intermetallic compounds) in a bulk aluminum alloy [41]: **a** digitized 2D optical micrograph of porosity and intermetallic compounds in an A319 aluminum alloy; **b** true 3D morphology of the same sample; **c** equivalent radius, R_{avg}; **d** maximum length, L_{max}; **e** percentage porosity, $P_{\%}$ comparison between metallographic (2D) and tomographic (3D) analysis for the same alloy; (f) x–y–z bounding box measurement of the intermetallic from (**b**); **g** average size of intermetallic compounds

time-resolved *in-situ* real-time analyses on defects, stress–strain relationships, and high-temperature creep performance have become an important goal of this major infrastructure construction.

1.5 Technical Development Trend of Domestic and Abroad

1.5.1 Virtual Manufacturing of Large Complex Thin-Walled Castings

Nowadays, the Virtual Manufacturing technology has allowed that the actual castings can be produced more quickly and economically and hence made companies much more competitive in the market. In recent years, the Virtual Manufacturing technology has been increasingly chased by users, resulting in a much rapid development. When being used in mould manufacturing and rapid model preparation, Virtual Manufacturing has currently evolved into the following trends.

(1) Integration

The design of the mould directly using the 3D feature modeling system has enabled virtual assembly and automatic inspection of the various parts of the mould. In addition, it is possible to complete the planning and machining simulation of the pass and automatically generate computer programs to quickly produce casting moulds of high quality. Adopting this method, the design speed of mould development can be greatly accelerated, which allows the designer to be freed largely from the heavy computational work and focus more on more creative works.

(2) Rapid Prototyping

Rapid Prototyping has promoted rapid creation of casting entities, which can significantly shorten the new product development cycles and hence reduce the corresponding budget. Together with CNC machining, casting, metal cold spray, and silicone moulds, Rapid Prototyping has become a powerful tool for manufacturing modern models, moulds, and parts. The effective combination of Investment Casting and Rapid Prototyping has achieved lower costs with higher efficiencies of production and has delivered the purpose of individualization, diversity, and rapidity of casting production.

(3) Concurrent Engineering

Concurrent Engineering is an effective method to minimize design errors at the beginning of mould design, taking into account all aspects of mould processing, i.e., assembly, usage, and waste disposal. Owing to the continuous development of remote design and manufacturing technologies, off-site design and manufacture of casting parts has become more and more feasible and convenient. Remote design and manufacturing technologies can give full play to the resources of different countries and regions to optimize design and manufacturing conditions.

(4) Computer Detection and Control of Casting Forming Process

By means of computer detection and control technology, the fluidity of molten metal, casting performance, and the working state of the main and auxiliary machines of the moulding line can be effectively detected. In addition, on-line monitoring, integration, and intelligence can also be realized in the foundry industry.

1.5.2 Manufacturing Technology of Large and Complex Thin-Walled Investment Castings

As the aircraft engine performance and reliability continue to improve, the structure of critical hot-end components has also changed drastically. Significant changes are developing towards not only the entirety, thin wall, and hollow direction, but also gradually towards the structural function integration of structural bearing and gas guiding. All these require that the components should possess higher stress-bearing capacity and dimensional precision, better dimensional stability, as well as surface roughness, better fatigue performance, and service life. Undoubtedly, these new requirements proposed higher requirements for near-net-shape investment casting system of key materials, e.g., nickel-base superalloys.

The comparison of the capability in manufacturing large investment castings of superalloys between China and other countries is shown in Table 1.2, indicating a huge gap between China and other leading countries. Taking the selection of casting process as an example, in order to avoid misrun defects in thin-walled parts of castings, centrifugal casting technology has typically been adopted with large-scale vacuum centrifugal casting equipment being constructed in China. In contrast, the gravity casting has been generally solved and used in other leading countries

Table 1.2 Comparison of manufacturing capability of large investment castings of nickel-base superalloys between China and abroad

	China	Abroad
Casting size	~1 m	~2 m
Minimum wall thickness	1.8 mm	1.5 mm
Casting precision	Tolerance: CT6; surface roughness: $R_a = 3.2$	Tolerance: CT5; surface roughness: $R_a = 1.6$
Single casting weight	<500 kg	2000 kg
Casting process	Centrifugal and gravity casting	Gravity casting
Alloy selection	K4169	IN718, GTD222, RENE 220, Inconel 939
Service temperature	650 °C	650–850 °C

through the fundamental research of filling behavior and solidification characteristics of superalloys. The latter method avoids not only technical problems such as alloy segregation, uneven organization, and precise positioning of the centrifugal casting, but also the large investment associated with the centrifugal casting equipment.

Since 2000, in order to further reduce the weight of castings and improve their long-term service performance, researches have been performed outside of China on the complete filling of ultra-large-area thin-walled castings and the control of micro-defects in large-scale castings. In 2010, the world's top large-scale aircraft engine investment casting manufacturer, the American PCC company, has begun to explore the investment process of ultra-thin large-scale TRF casting with a minimum wall thickness down to 1 mm. Up to now, the current technology maturity has reached 6–7 grades, providing a technical basis for lightweight design and manufacturing of the next generation of high performance aircraft engines.

Over the years, foreign countries have always being imposing strict technical blockades on China's investment casting technology and equipment for large-scale structural components, and the corresponding technical information has rarely been publicly reported. With the support of the National Key Basic Research Project, Shanghai Jiao Tong University has systematically studied the basic issues such as the filling behavior and solidification structure control of large castings under multi-field coupling, the formation mechanism and control method of casting defects of large castings, the formation mechanism of errors in complex casting systems, and the precision control of castings. In addition, the effects of alloy composition and temperature on the melt structure and solidification behavior of high-temperature alloys were preliminarily revealed. The technology of melt over-temperature treatment and solidification structure control was developed, and the optimal design scheme for casting system of large-scale gravity casting process was proposed. In terms of material research, the criteria for microporosity and harmful brittle phase formation of K4169 large castings were established and the typical defect map of the TRF casting was preliminarily established, revealing the relationship among the characteristic structure of large castings, defects, and mechanical properties. Moreover, the error formation mechanism in the wax mould pressing process was revealed, and the new high-performance back shell material and its preparation technology were developed. In 2009, based on the above technical preparation, the TRF castings with diameter of about 1.1 m and minimum wall thickness of about 2.2 mm were successfully manufactured, which made a breakthrough in the development of large turbine rear casing castings in China. Since 2010, domestic manufacturers have been able to produce integral precision castings of rear casing, which provides an important technical basis for lightweight design of aircraft engines in China. However, compared with other leading countries, the domestic rear casing castings are still far behind on the lightweight and long-term service performance. In particular, the homogeneity of the microstructure in the whole casing is not good enough and the fatigue life of different parts in the main bodies vary significantly, which can seriously affect the stability and reliability. These problems imply that the overall casting process technology is in urgent need of significant improvements. Meanwhile, the research results also show that the gravity casting process can only guarantee the complete filling of the area of 100 mm × 100 mm for the thin wall with a thickness of 1 mm. Therefore, in order to manufacture larger and thinner casing castings, it is necessary to explore new methods and processes.

The research institutes such as the University of Birmingham in the UK have conducted in-depth studies on the new casting process. They compared the casting process such as gravity casting and vacuum-assisted counter-gravity casting, and found that these casting processes can significantly improve the material utilization rate and the performance of castings. Anon of the British Investment Casting Trade Association (BICTA) pointed out that counter-gravity casting used in the production of aviation components has the potential to improve production efficiency and reduce production costs, which is a technology worthy of promotion.

A foundry company, named American Metal Casting Technology, LLC., invented a new counter-gravity low-pressure inert gas protective casting process (CLI) for manufacturing nickel-base superalloy castings. The relevant patent was licensed to the famous Hitchiner Company for aviation component production. Compared with vacuum-melted and gravity-melted parts, the castings produced using this process have a slight increase in oxygen content, but can still be controlled within 50 ppm, and, more importantly, the tensile and fracture properties are in line with customer requirements. Hitchiner has utilized this technology to produce castings of nickel-base superalloys such as the IN718 turbine diffuser housing with a diameter of 889 mm and a height of 625 mm. In addition, Howmet produced titanium alloy castings with a maximum weight of 47 kg and a maximum size of nearly 900 mm using a similar technology. The US has also formed a research center for counter-gravity casting and control technologies, led by the University of Iowa and the University of Tennessee, and has achieved technological leadership in the automatic control of the casting process.

Compared with foreign countries, the research and application of superalloy counter-gravity casting in China are still far behind. Figure 1.4 clearly shows the size and weight range of large-scale superalloy castings that have been produced abroad using counter-gravity casting. However, in China, the research on such technologies has not been carried out effectively.

Furthermore, the introduction of the tolerance design concept into the error analysis and control of the investment casting process has shown great potential to promote the development of large-scale casting dimensional precision control theory and technology. In the basic research of tolerance design of complex systems, foreign researches mainly focus on modeling and predicting dimensional deviation of complex manufacturing systems. In the past decade, the Oak Ridge National Laboratory and the Edison Materials Technology Center have listed casting precision control as a special research project.

In the early stage, error analysis of complex products was generally based on multi-parameter linear system modeling. Among these, Gilbert used a linearized dimensional deviation analysis model to study the cumulative effect of dimensional tolerances during assembly. By the mid-1990s, the theory of error modeling based on nonlinear systems began to receive widespread attention. Inoue et al. studied the coupling effect between temperature, phase transition, and thermal stress in

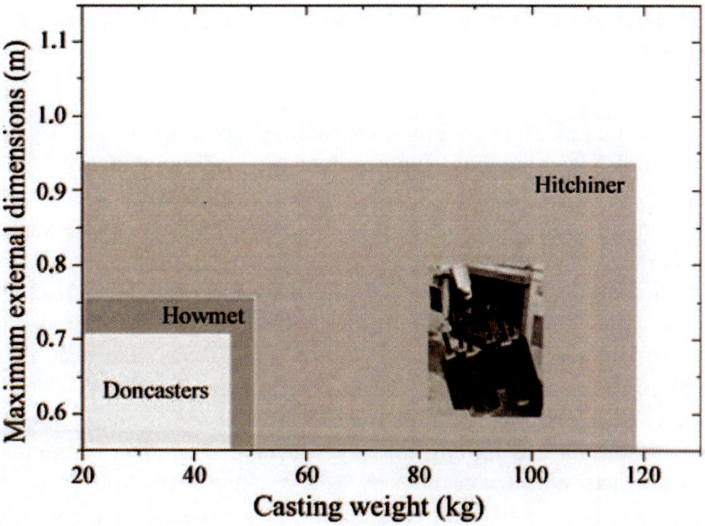

Fig. 1.4 Comparison of superalloy counter-gravity casting technology

the temperature varying process together with phase transition, and established the constitutive equation. In 1997, Hu first proposed the deviation flow theory to analyze and predict the transmission mechanism of multiple deviation inputs throughout the assembly process. Shi and Jin used the state space method in cybernetics to establish a state equation describing the multi-process deviation transfer, which significantly improved the control level of manufacturing precision.

At present, the analysis of the formation causes of solidification defects of large castings ignores the formation dynamic mechanism of solidification defects based on the structural characteristics of castings and lacks in the necessary research methods and in-depth studies on the mechanism and influence of the external force field on solidification defects suppression. In fact, the structural characteristics would cause large changes in the flow field, temperature field, and solute field. And there is an interaction between the parameters of the solidification process and the formation of defects, which cannot be ignored. In addition, a variety of force field effects cause the solidification process under multi-field coupling to proceed away from equilibrium conditions, which may have a favorable effect on the regulation of solidification structure and defects. It is expected that the solidification defects such as microporosity may be suppressed effectively through vacuum pressurization or adjusted pressure forming process and it is currently of great significance to carry out relevant research in-depth.

1.5.3 Quality and Reliability Inspection of Large-Scale Castings

With the advancement of science and technology, some countries began to study the use of digital X-ray technology instead of the traditional X-ray film exposure technology. In early 2008, the US Air Force organized a number of companies including aircraft manufacturers Boeing, Lockheed, aircraft engine suppliers GE, R&R, PW, Honeywell, aviation castings manufacturers Howmet, PCC, etc., to officially study the application of digital X-ray in aviation castings. After more than a year of hard work, the preparation of reference standards for digital images was completed. Simultaneously, high-energy industrial CT has also become the main technique of non-destructive testing of large castings.

In addition, the research platform based on the fast imaging and *in-situ* analysis of the synchrotron radiation technology has been developed rapidly in order to reveal the stress and strain distribution of castings over the past few years. Many research results have been achieved in the study of bulk engineering materials and components, the detection of internal information of components, and basic research fields in metallurgy/crystallography. Four internationally renowned synchrotron radiation sources, i.e., SPring-8 of Japan, ESRF of Europe, APS and SSRL of the US, have all built high-energy engineering beamlines for engineering applications. For example, the ESRF ID15 beamline is designed for industrial materials and components research, using superconducting Wiggler and Undulator composite light sources, X-ray photon energy up to 750 keV. This beamline includes two experimental stations: ID15A can realize energy dispersive diffraction, high energy micro-diffraction, and *in-situ* dynamic imaging, mainly applied to engineering materials, material surfaces, and interfaces hidden in materials, and is used for material engineering, metallurgy, and aerospace engineering; ID15B can realize high-resolution angular diffraction and high-energy scattering research, and is equipped with a high-temperature and high-pressure gas loading system, which can provide *in-situ* research such as gas–solid and liquid–solid catalytic reactions. The I12 beamline station under construction in the UK DIAMOND Light Source is also specially designed for engineering materials, which uses a superconducting Wiggler light source with an X-ray photon energy range of 50–150 keV. It contains two experimental stations: EH1 and EH2. Opened in September 2009, EH1 is located in the experiment hall and mainly used for diffraction and imaging research. EH2 is located outside the experimental hall and provides space for large-scale engineering and processing experiments. It is very suitable for analyzing high-density materials such as the *in-situ* studies of large industrial materials and aerospace engine components. Except the 5 beam lines dedicated to industrial applications, the industrial sections have built more than 4 beamlines at SPring-8 light source in Japan. It is mainly used in the development of new medicines, photoelectric materials, structure of engineering materials, stress measurement, and so on. APS, NSLS, and other light sources in the US also have a number of high-energy beamlines for material research. The NSLS X17 beamline consists of five experimental stations. The experimental methods include high-energy diffraction, scattering, and so on, mainly used for material science.

Considering the rapid development of synchrotron radiation technology abroad as described previously, it is thusly imperative to build a high-energy, high spatial resolution, and high time resolution engineering harness with high-precision support platform and several analytical means to facilitate the research on aircraft engine materials, component preparation technology, and service safety evaluation platform in China.

1.6 Focuses of Each Chapter and Main Issues to Be Solved

Chapter 2 focuses mainly on the design of investment casting process for large castings, including the gating system design and optimization. Chapter 3 is mainly about the wax pattern preparation process for large complex thin-walled castings via the physical properties test of wax and finite element simulation. Chapter 4 introduces the selection, manufacturing, and mechanic properties testing methods of shell mould systems. Chapter 5 discusses the method of composition optimization based on the effect of alloying elements on the melt structure and thermophysical properties, explores the filling law of castings using hydraulic simulation, and reproduces the evolution law of microstructures in thin-walled and variable cross-section positions of castings using synchrotron radiation technology. In Chap. 6, aiming at the defect issue of large castings, the non-destructive testing methods and related optimization measures are first introduced. Then the structural effects of defects are constructed based on the anatomy of large castings and the prediction criteria for the formation of microporosity and Laves phase are established as well, together with the optimization principle of gating system being proposed based on microscopic defect suppression. Finally, the repair method and repair process of large castings are introduced. In Chap. 7, targeting at the problem of deformation and oversize of large castings, the transmission laws of comprehensive errors in the whole process of wax mould preparation, shell mould preparation, and filling and solidification are investigated. Meanwhile, the robust control method of multi-process parameters and the technology of error source detection and traceability are studied. The technology has formed the tolerance design and process parameter control system method for dimensional precision control. It has also constructed the technical system of error detection, evaluation, diagnosis, and control in precision casting process, and has established a new theory of error simulation of large castings. New technologies and methods for quantitative analysis, process optimization, accurate detection, and rapid diagnosis of large casting errors have been proposed as well. In Chap. 8, to tackle the formability problem of thin-walled complex castings, a new adjusted pressure casting equipment for superalloy has been developed, and the influence of adjusted pressure casting process on thin-wall filling and solidification microstructure is studied. Finally, the ultra-thin-walled complex castings represented by floating wall tiles are trial-produced by adjusted pressure casting method.

References

1. B.Y. Tu, M. Zhao, T.S. Shang et al., Advanced structure and key manufacturing technology of aeroengine. Aeronaut. Manuf. Technol. **7**, 53–56 (2014)
2. X.S. Xie, The development and application of high temperature materials in China. Mater. Mech. Eng. **1**, 2–8 (2004)
3. W.D. Huang, *Challenge and Countermeasures for Materials Forming Technology Induced by Development of Advanced Aircraft and Aeroengine* (CSAA Science and Technique Youth Forum, Discovery, Innovation and Communication, 2004).
4. C.Z. Li, Investment casting. Res. Stud. Foundry Equip. **2**, 73–75 (1984)
5. Z.G. Lv, History and development trend of investment casting industry in China. Foundry **4**, 347–356 (2012)
6. R. Williams, H. Richard, Global investment casting market conditions beginning to stabilize after 2009 lows. Incast **1**, 8–9 (2011)
7. D. Pan, Q.Y. Xu, B.C. Liu, Multi-scale modeling of solidification process of Ni-based superalloy investment castings. Chin. J. Nonferrous Metals **20**(2), 329–338 (2010)
8. H. Yan, Y.D. Zhang, B.K. Wang et al., Numerical analysis of casting deformation in case shell on aeroengine. Dual Use Technol. Prod. **4**, 41–43 (2009)
9. D. Zhang, W.H. Zhang, F.G. Li, Application of shape optimization based on numerical simulation in feeder design. Foundry **53**(2), 129–132 (2004)
10. X. Xue, C.M. Yin, B.D. Zhou et al., Prediction of shrinkage defects in a third class moving turbine blade by invested casting. Foundry **51**(4), 228–231 (2002)
11. D.Z. Li, S.F. Su, J.Q. Wang et al., The shrinkage porosity functions and its application in IN738 castings. Foundry **6**, 13–17 (1998)
12. M. Faraji, J.P. Wright, L. Katgerman, In-situ observation of the nucleation kinetics and the mechanism of grain refinement in Al–Si alloys (part I). Mater. Lett. **64**(9), 1016–1018 (2010)
13. A. Bogno, H. Nguyen-Thi, G. Reinhart et al., Growth and interaction of dendritic equiaxed grains: In situ characterization by synchrotron X-ray radiography. Acta Mater. **61**(4), 1303–1315 (2013)
14. O. Shuleshova, W. Loser, D. Holland-Moritz et al., Solidification and melting of high temperature materials: in situ observations by synchrotron radiation J. Mater. Sci. **47**(11), 4497–4513 (2012)
15. H. Yasuda, T. Nagira, M. Yoshiya et al., In-situ observation of peritectic solidification in Sn–Cd and Fe–C alloys, in *3rd International Conference on Advanced in Solidification Processes*, 27, 2012
16. N.S. Husseini, D.P. Kumah, J.Z. Yi et al., Mapping single-crystal dendritic microstructure and defects in nickel-base superalloys with synchrotron radiation. Acta Mater. **56**(17), 4715–4723 (2008)
17. Q. Dong, J. Zhang, J. Dong et al., Anaxial columnar dendrites in directional solidificationofanAl-15wt.% Cualloy.Mater. Lett. **65**(21–22), 3295–3297 (2011)
18. Q. Dong, J. Zhang, J.F. Dong et al., In situ observation of columnar-to-equiaxed transition in directional solidification using synchrotron X-radiation imaging technique. Mater. Sci. Eng. A **530**, 271–276 (2011)
19. Z.Q. Hu, Y.J. Kang, J.S. Zhang et al., Effect of trace elements on directionally solidified IN738 alloy (development of DZ38G alloy). J. Mater. Eng. **S1**, 102–108 (1992)
20. J.M. Zeng, Y.H. Zhou, A new forming technology for aeronautical castings: adjusted pressure precision casting process. Aviat. Maintenance Eng. **10**, 18–19 (1997)
21. X.L. Li, Q.T. Hao, X.Q. Jie, Equipment technology in low pressure casting for large magnesium alloy components. Rare Metal Mater. Eng. **43**(3), 703–706 (2014)
22. L. Liu, J. Zhang, J. Shen et al., Advances in directional solidification techniques of superalloys. Mater. China **29**(7), 1–9 (2010)
23. D.P. Zhao, T. Jing, B.C. Liu, Influence of phase-field parameters on the dendrite morphology. Acta Metallrugica Sinica **39**(8), 813–816 (2003)

24. Z. Chen, C.L. Chen, L.M. Hao, Effect of phase field parameters on the dendritic growth in a forced convection. Rare Metal Mater. Eng. **7**, 1209–1213 (2009)
25. Y. Chen, X.H. Kang, D.Z. Li, Phase-field modeling of free dendritic growth with adaptive finite element method. Acta Physica Sinica **58**(1), 390–398 (2009)
26. Z.P. Wang, R.Z. Xiao, C.S. Zhu et al., Phase-field simulation of binary alloy during directional solidification. Rare Metal Mater. Eng. **S2**, 369–373 (2006)
27. K.D. Carlson, C. Beckermann, Prediction of shrinkage pore volume fraction using a dimensionless *Niyama* criterion. Metall. Mater. Trans. A **40**(1), 163–175 (2009)
28. L. Nastac, D.M. Stefanescu, Computational modeling of NbC/laves formation in INCONEL 718 equiaxed castings. Metall. Mater. Trans. A **28**(7), 1582–1587 (1997)
29. G.A. Knorovsky, M.J. Cieslak, T.J. Headley et al., INCONEL 718: a solidification diagram. Metall. Trans. A **20**(10), 2149–2158 (1989)
30. H.S. Whitesell, R.A. Overfelt, Influence of solidification variables on the microstructure, macrosegregation, and porosity of directionally solidified Mar-M247. Mater. Sci. Eng. A **318**(1–2), 264–276 (2001)
31. Y.K. Penya, P.G. Bringas, A. Zabala, Advanced fault prediction in high-precision foundry production, in *6th IEEE International Conference on Industrial Informatics* (2008), pp. 1570–1575
32. W. Chen, C.C. Li, H Li et al., Effect of casting parameters on mechanical properties of a superalloy. Foundry (9), 871–874, 2005
33. Y.H. Xiong, A.M. Yang, P.J. Li et al., Effects of pouring temperature and refiners on microstructures of superalloy K4169. J. Aeronaut. Mater. **21**(4), 5–8 (2001)
34. L. Wang, J.X. Dong, C.Q. Li, Effect of cooling rate on the segregation and Rayleigh number of IN718 during solidification. J. Univ. Sci. Technol. Beijing **29**(12), 1222–1227 (2007)
35. X. Tang, F.X. Liu, A.D. Yang et al., Investigation of K417 fine grain casting thermal parameters. J. Mater. Eng. **7**, 23–25 (1995)
36. W. Bonilla, S.H. Masood, P. Iovenitti, An investigation of wax patterns for accuracy improvement in investment cast parts. Int. J. Adv. Manuf. Technol. **18**(5), 348–356 (2001)
37. A.S. Sabau, S. Viswanathan, Material properties for predicting wax pattern dimensions in investment casting. Mater. Sci. Eng. A **362**(1–2), 125–134 (2003)
38. Y.H. Song, Y.N. Yan, R.J. Zhang et al., Three dimensional non-linear coupled thermo-mechanical FEM analysis of the dimensional accuracy for casting dies in rapid tooling. Finite Elem. Anal. Des. **38**(1), 79–91 (2001)
39. M.M.A. Rafique, J. Iqbal, Modeling and simulation of heat transfer phenomena during investment casting. Int. J. Heat Mass Transf. **52**(7–8), 2132–2139 (2009)
40. Z.Q. Lin, X.M. Lai, S. Jin et al., Digital methods of complex product manufacturing precision control and its development trend. J. Mech. Eng. **49**(6), 103–113 (2013)
41. C. Puncreobutr, P.D. Lee, R.W. Hamilton et al., Quantitative 3D characterization of solidification structure and defect evolution in Al alloys. JOM **64**(1), 89–95 (2012)
42. A.M. Korsunsky, X. Song, F. Hofmann et al., Polycrystal deformation analysis by high energy synchrotron X-ray diffraction on the I12 JEEP beamline at Diamond Light Source. Mater. Lett. **64**(15), 1724–1727 (2010)
43. F. Hofmann, X. Song, T.S. Jun et al., High energy transmission micro-beam Laue synchrotron X-ray diffraction. Mater. Lett. **64**(11), 1302–1305 (2010)

Chapter 2
Investment Casting Process Design of Large-Size Superalloy Castings

Guoxiang Wang

2.1 Outline

2.1.1 The Applications of Large-Size Superalloy Investment Castings

The performance of aeroengines is built on the delicate continuous rotatory structure, on which the main difficulty in production also centers. The advancing tactical and technical targets of modern aircrafts require better aircraft engines, of which the basic service characteristics includes high temperature, high pressure, high speed and high reliability, durability, environment-friendly, and easy to maintenance. With these harsh requirements, the aircraft engine has become the most complex and sophisticated industrial product in human history [1]. Aeroengines are the determining factor of flexibility, reliability, economy, and environment-friendliness of the aircraft, whose performance is the fundamental of aircraft development.

Superalloys are widely used in aerospace and aviation fields. They can be made into turbine disks, turbine blades, and various other structural parts of the engine. In addition, most large structural castings are made of superalloys. For example, more than 30% (mass percentage) of the key components in modern aircraft engines are made of IN718 [2] (similar to K4169). Nowadays, the most widely used superalloy castings are isometric superalloy castings under medium temperature. With the help of new technology, the aviation industry aims to higher performance and lower cost. Large thin-walled integrated castings have the advantages in terms of cost and performance and have more applications other than assembly parts [3]. As the aeroengine tends to be delicate and complex, the components will have more and more multilayer thin-walls, narrow cavities and deep holes, and narrow slots structures.

In aeroengines, large-size investment castings such as nickel-base superalloy rear case frames should withstand the driving force and all vibration loads of the

B. Sun et al., *Precision Forming Technology of Large Superalloy Castings for Aircraft Engines*, https://doi.org/10.1007/978-981-33-6220-8_2

engine. As a result, their quality directly affects the performance, life, and reliability of aeroengines. Such castings usually have large sizes, complex structures, and abrupt changes of cross-section, big thin-wall areas, and strict tolerances. With the increasing performance of the aircraft, the thrust-to-weight ratio also should be increased. The structure and materials of Key Hot End Components also have great changes. For instance, integrated thin-walled and hollow structures are usually used; the materials should have higher bearing capacity, better mechanical properties, dimensional stability, durability, and lower cost as well. These new demands push a rapid development of large superalloy castings.

2.1.2 The Development of the Domestic and Abroad

Due to the big advantage of high precision and surface finish, the investment castings require little machining.Only when some parts of the castings have very high requirements, they will need a little machining allowance, with some castings need certain grinding or polishing allowance as well; in general, no machining is needed.Since investment casting is a process with little or no allowance, it can save a lot of machine tools and time and save raw materials. The other advantage is that this process can produce complex castings with various alloys, especially superalloy castings. It is very difficult to produce the thin wall hollow struts with streamline outline of the turbine rear frame (TRF) of the jet engine. But these support plates can be quantity producted with consistency use the investment casting process, even there is no stress concentration of the tool marks after machining.

A series of technologies have been developed to make large complex thin-walled integral superalloy castings [4–7]. The American companies, such as Hitchiner, PCC and Howmet have produced a series of large thin-walled integral castings for GE engines. The TRF castings for turbine were made of alloy IN718, with a thickness 1.25 ~ 1.75 mm and a size of 1.93 m and pouring metal weight of 2040 kg. In addition to the large size and thin wall, the general linear tolerance of investment castings is CT4-6, and some special linear tolerance can be up to CT3, while the surface roughness is getting smaller and smaller down to Ra 0.8 μm.

Since the beginning of the twenty-first century, China's aviation industry has been developing rapidly, with various advanced aircrafts having been developed continuously. However, in sharp contrast, China's aeroengines have seriously dragged the development of aircrafts. For a long period of time, a significant dependence on purchasing foreign aircraft engines has posed a huge threat to China's national security, which has become one of the most urgent problems to solve.

Casting technology is one of the important bases of aeroengines. However, the design, control, and casting process often rely on the experiential knowledge, which is difficult to ensure casting quality, and the rejection ratio is high. Due to the technology blockade, we have to develop these advance technologies ourselves. With the steady accumulation of technology and continuous innovation and improvement, we have to establish the investment casting technology system of the large complex thin-wall castings with our own independent intellectual property rights.

2.1.3 The Process of Investment Casting

The process of investment casting includes pattern made of wax and assembled into a tree, then coated with multiple layers of refractory material on the surface of the pattern. After drying and curing, the mold is autoclaved and the pattern melt off, then pours liquid metal into the roasted shell and obtains investment mold casting [8–10].

Normal process of investment casting follows the steps: mold preparation → pattern → shell preparation → pattern meltout → roasting → pouring → cleaning → repair & grinding → qualified castings.

Figure 2.1 shows the process of large complex thin-wall investment castings, virtually the same as the normal investment casting except for some special steps in the process. According to different workplaces, the casting process can be roughly divided into the following four sections:

1. pattern making (injection wax, pattern cleaning, pattern measurement, and pattern assembly);
2. shell preparation (coating, autoclaved, roasting);

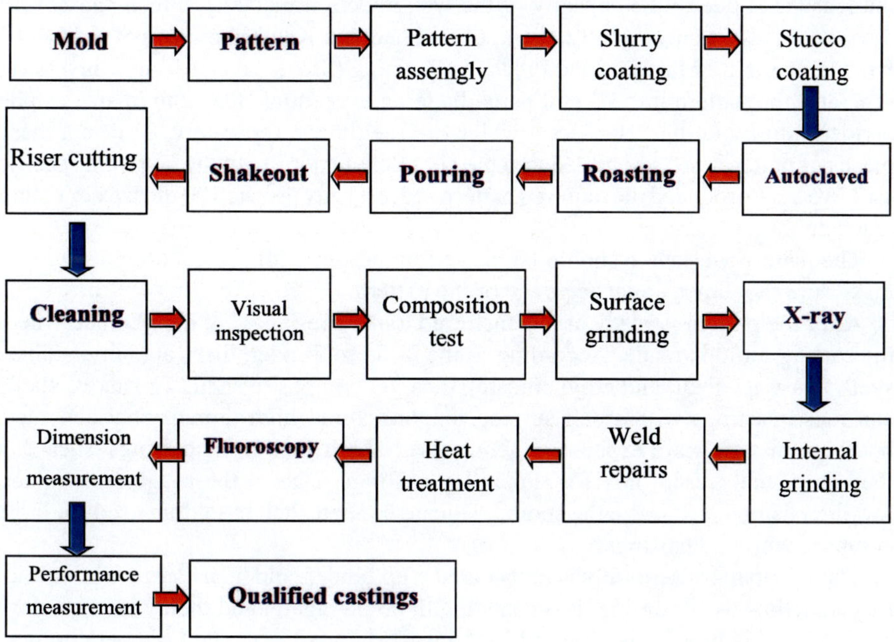

Fig. 2.1 The process of investment casting

3. pouring (metal melting, pouring);
4. cleaning and measurement (shake out, riser cutting, blasting, grinding, measurement).

The investment casting should start with the wax pattern that is exactly the same as the required casting, it is necessary to prepare the mold to make pattern by injecting wax. Normally the mold is made of steel or aluminum alloys, and sometimes the mold can be made of easily fused gold, epoxy resin, gypsum, etc. with a small batch of production. The cavity of the mold forms the shape of the pattern. The mold can be made into horizontal parting, vertical parting, etc., according to the complexity of the casting. The cores should be used if the pattern has internal cavity or external geometric characteristics that are difficult to reject. The cores should be used repeatedly with high strength, not easy to deformation, and easy to ensure the size accuracy of pattern, they can be made of steel, aluminum alloys, and other metal. Targeting at the lightweight, a large number of hollow and thin-walled castings are produced, most of which have internal cavities with complex structure that cannot be ejected from the mold. Thus, the only choice is the water-soluble core, which can dissolute or collapse in the presence of water or acidified water. Ceramic cores can be used in case of more complex and smaller cavities, such as hollow blades. Before injecting wax, the core is put into the mold in advance, and the melted wax is injected into the cavity of the mold. Then the pressure is maintained until the wax is solidfied and the pattern can be ejected from the mold after cooling. It is difficult to cast large-sized castings using the traditional investment casting. One of the reasons is due to the capacity of the wax injection machine, which can not fill the whole pattern successfully at one time. Shanghai Key Laboratory of Advanced High temperature Materials and Precision Forming (Shanghai Jiao Tong University) seperates the pattern into several parts. Patterns exceeding 1000 mm in size can be divided into more than 10 pieces, which are then made separately. Later on, these pieces of patterns are assembled into the complete pattern with the help of a specific tool. We have produced the required pattern successfully use the 100 ton wax injection machine.

The pattern workshop should be located in a room with constant temperature to make sure the dimensional accuracy of the pattern.

After the pattern is ready, it is transferred to the shell making department, where the casting mold is made. According to the different binder, there are ethyl silicate shell, silica gel shell, and ethyl silicate-silica gel composite shell. The above shells can make castings with small surface roughness and high dimensional accuracy, but these materials are expensive, giving rise to high cost of the castings, such that they are usually used only for superalloy castings. Due to the rough surface and low dimensional accuracy, the sodium silicate bonded shell is seldom used for large complex thin-wall castings.

Then the pattern assembly is first coated with binder, and then coated with refractory materials. After drying, it is coated with binder again, and then with refractory materials. This process is repeated in the workshop with constant temperature and humidity to ensure that the shell is of a high quality. The shell will gain certain strength after more than ten times of repeatitions. Then the shell is sealed and dried, after which the shell is put into the high-pressure steam dewaxing tank. The wax pattern melts off under high temperature and high pressure, leaving the shell without wax inside. The shell is then sent into a roasting furnace. The strength of the shell is further improved under a high temperature of 1000°C for a long time. After these steps, the shell can withstand the pressure and scouring of metal liquid during pouring and the metal can then be poured into the shell.

Before pouring, proper amount of superalloy ingots is put into the crucible of the vacuum induction furnace, with the furnace being vacuumed and the alloy being heated up. After the melting is completed, the high temperature shell is removed from the roasting furnace and transferred to the vacuum furnace quickly. The high temperature molten metal is then poured into the cavity of the shell under the action of gravity, and then the vacuum is removed. The casting is taken out of the furnace together with the shell and is cooled down at room temperature.

After pouring, the cleaning and inspection of large thin-wall superalloy investment castings have additional steps, which are rather complex. After shakeout and cutting of the riser, visual inspection shall be carried out, followed by holes welding repair, grinding, and X-ray and fluorescence inspection. If there are any defects, welding shall be carried out after removing the defects, and then repeat heat treatment and X-ray and fluorescence inspection to ensure that there are no long any casting defects. Note that some castings also require hot isostatic pressing.

The hot isostatic pressing (HIP) is to put the castings into a sealed tank filled with high purity argon gas and then apply equal static high pressure along all directions to the castings at high temperatures according to Pascal's principle, which will close up the holes in the castings and hence obtain highly dense castings. The experimental results show that the mechanical properties, in particular, the fatigue properties of the superalloy castings have been improved significantly after HIP.

To sum up, the large thin-wall investment castings have a time-consuming and complex process, wherein all the processing factors can affect the final quality. It is a huge challenge to control the defects and deformation, in particular, the shape control. In addition, tolerance accumulates in mold, pattern, shell, and castings and the mold, wax pattern, shell, and the master alloy will affect the final shrinkage rate. As a result, the materials and process must be strictly controlled.

2.2 Function and Design Principle of Gating System for Large Complex Thin-Wall Investment Castings

For the purpose of high efficiency, low cost, and reliable quality of castings, the filling and solidification process must be effectively controlled. The clean molten metal can be solidified sequentially from casting to riser while no turbulent flow is allowed in the casting cavity. For the larege castings such as TRF, the main problem is to design the process to control the filling mode and solidification sequence according to the complex structure. The complex structures in TRF castings can be seperated into two types, i.e., large-area thin-wall structures and varying cross-section structures. During the filling, the former, due to the narrow cavity, may misrun and the abrupt change of cross-section of the latter may cause the disturbance of melt flow, leading to bubble involvement and other problems and affecting the stability of filling and the density of castings. During the solidification, the large area thin-wall structure has a higher cooling rate than the thick-wall structure, and the abrupt change of cross-section also has an important influence on the growth direction and stability of the dendrites. These factors make it difficult to obtain a uniform solidification structure of the large casting.

2.2.1 The Threshold Value of Large Thin-Wall Castings

Friction and surface tension are the main factors of structure resistance during the filling.

Surface tension is the result of molecular attraction in the fluid. It is the property of the substance itself, which is related to the temperature and the properties of the substance at and near the phase interfaces. The additional pressure caused by the surface tension must be considered when dealing with the flow between thin walls.

According to the Young-Laplace equation [11],

$$\Delta p = \sigma(\frac{1}{R_1} + \frac{1}{R_2}) \tag{2.1}$$

where σ is the surface tension and R_1 and R_2 are the principal radii of the curvature of the interface. The relationship between the curvature radius and cavity size is shown in Fig. 2.2.

It is assumed that (a) the wetting angle of liquid flow and the top plate is equal to that of the liquid flow and the side of the plate and (b) during the whole filling process, the wetting angle is a constant, which, in fact, is related to the filling rate and the pressure of the riser tube.

From above:

$$R_1 = -\frac{W}{2\cos\theta} \quad R_2 = -\frac{H}{2\cos\theta} \tag{2.2}$$

where, W and H refers to the width and wall thickness of the runner, respectively, and θ is the wetting angle (Fig. 2.2).

Combining Eqs. (2.1) and (2.2), we get

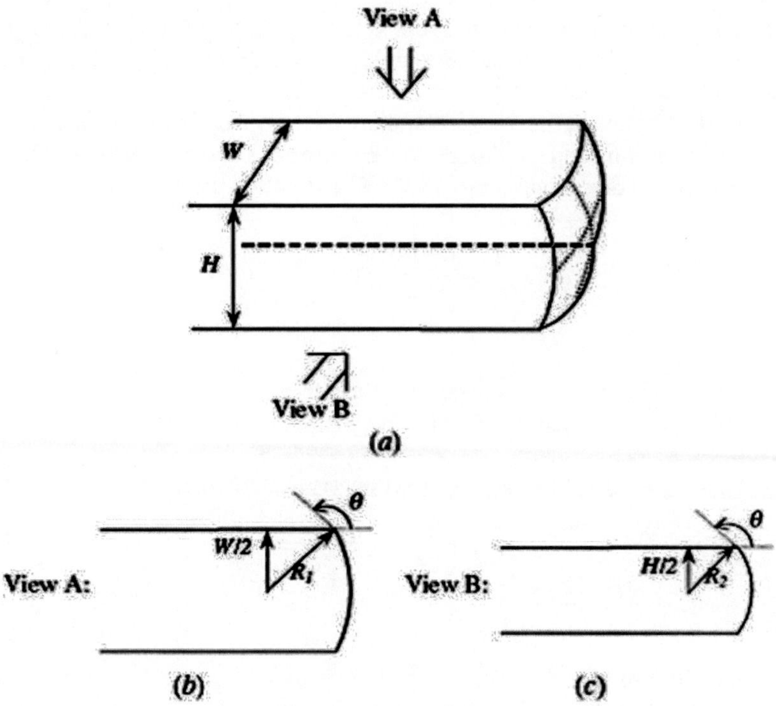

Fig. 2.2 The relationship between curvature radius and cavity size

$$p = -2\sigma \cos\theta \left(\frac{1}{W} + \frac{1}{H} \right) \tag{2.3}$$

If it is assumed that the alloy liquid does not wet at all with the cavity wall, the Laplace force can be simplified as,

$$p = -2\sigma \left(\frac{1}{W} + \frac{1}{H} \right) \tag{2.4}$$

The resistance of fluids to shear deformation is called viscosity. When the fluid flows uniformly, the velocity distribution on the flow section is not uniform due to the viscous action, and there is a relative motion between the two adjacent flow layers. As a result, there is shear stress (friction force) between the fluid flow layers as well as between the fluid and the boundary, exhibiting certain flow resistance. The flow resistance generated in the uniform flow section is called the along resistance or friction resistance.

For the runner with the parallel plate, the formula of flow friction pressure drop is:

$$\Delta p = 12L\mu\bar{v}/H^2 \tag{2.5}$$

where, Δp is flow friction pressure drop, L is the length of the runner, μ is hydrodynamic viscosity, $\bar{v}$ is the average velocity in the parallel runner, and H is the plate spacing.

For thin-wall castings, the Weber Number '*We*' criterion (surface tension) is used as the primary criterion and the Froude '*Fr*' criterion as the secondary one. Therefore, the wall thickness corresponding to $We/Fr = 1$ is called the critical wall thickness, given as

$$\delta_c = \sqrt{\frac{\sigma}{\rho g}} \tag{2.6}$$

where, ρ is the density, and g is the gravity acceleration.

At 1500°C, these properties of nickel-base superalloy K4169 are:

$$\sigma = 1.75\,\mathrm{N/m},\ g = 9.81\,\mathrm{m/s^2},\ \rho = 7210\,\mathrm{kg/m^3}$$

and we get

$$\delta_c \approx 4.9mm \tag{2.7}$$

As to nickel-base superalloy castings, if the wall thickness is less than 4.9 mm, it can be considered as thin-wall castings.

For large thin-walled castings, 'Re' criterion is the main criterion, that is, the viscous force becomes the dominant resistance to fluid, leading to

$$\frac{\text{viscous force}}{\text{surface tension}} = \frac{12\mathrm{L}\mu\,\bar{v}/H^2}{2\sigma\left(\frac{1}{W}+\frac{1}{H}\right)} = \frac{6\mathrm{L}\mu\,\bar{v}\,W}{\sigma H(W+H)} > 1 \tag{2.8}$$

$$\frac{LW}{H(W+H)} > \frac{\sigma}{6\mu\,\bar{v}} \tag{2.9}$$

According to Eq. (2.9), when the thickness of the plate meets the condition of thin-wall castings, the width of the plate has little influence to the flow. Therefore, the large thin-wall plate can be regarded as the ratio of the filled path (length of the plate) and the characteristic structure size (thickness of the plate).

As to nickel based superalloys, $\mu = 6.75$ mPa·s, $\sigma = 1.75\,\mathrm{N/m}$ at 1500°C, when $\bar{v} = 0.2\,\mathrm{m/s}$, $H = 1$ mm, $W = 200$ mm, we get

$$\mathrm{L} > 0.217\,\mathrm{m} \tag{2.10}$$

For nickel-base superalloys, a plate with a width of 200 mm, a length of 217 mm, and a thickness of 1 mm can be regarded as a large thin-wall casting when the wall thickness meets the thin-wall condition. See Table 2.1.

Table 2.1 Characteristic Dimensions of Large Area Thin Wall of Nickel-based Superalloy

Thickness of the plate (mm)	Width of the plate (mm)	Filling velocity (m/s)	Critical length (mm)
H = 1	W = 200	0.2	217
H = 2	W = 200	0.2	436

2.2.2 Casting Design and Concurrent Engineering

Nowadays, the computer has become an indispensable advanced tool for manufacturing industry with its outstanding advantages such as powerful computing ability, graphics processing power, and storage capacity. CAD (computer aided design), CAE (computer aided engineering), CAM (computer aided machining), and other technologies have been extensively used in foundry industry.

Investment casting has many advantages, as well as many disadvantages. It has a long chain of processes, which are always complicated with long production cycles and many factors affecting the quality of castings. This has influenced the application and development of investment casting. With the rapid development of computer technology, the application of computer technology has brought great changes to the production of investment castings, such as the structural design, process design, pattern mold design and manufacture, wax pattern forming, and the preparation of shell and core.

2.2.2.1 Application of Numerical Simulation in Structure Design and Process Design of Investment Castings

The investment castings are becoming lighter, thinner and net shaping. The net castings or near net castings has been put forward in recent years, in order to meet the demand of modern industry for high quality parts. This requires the structure of investment casting to be more reasonable and the process be more optimized.

The traditional investment casting process consists of the following 5 steps:

(1) The user issues the design blueprint to the foundry;
(2) The foundry makes a budget. For the purpose of easy production and low cost, the foundry gives some advises to the designer;
(3) The foundry designs necessary equipment;
(4) The foundry issues drawings to the mold workshop or pattern workshop;
(5) Metal casting and castings inspection.

In the process of casting structure design, pattern mold design, wax injection process parameters design, gating system design, etc., the traditional production mainly relies on the actual experience of engineers, they seldom use scientific theory. Especially for complex parts and important parts, it is often necessary to modify the casting structure, pattern mold and gating system repeatedly to meet the technical requirements.

The application of CAE simulation software has partially replaced the physical tryout. CAE technology is gradually used in product development, performance simulation, manufacturing process, and production management, etc. It not only provides effective guarantee for the whole life cycle of product development such as structural strength, lightweight design, kinematics check, fatigue life, and collision safety, but also greatly reduces the tryout cost and product developing time and thus saves the cost of new product development.

Computer has powerful computing and graphics processing ability. It can combine numerical analysis, database, and visualization with the classical theory of heat transfer, flow, and solidification. In particular, through modeling the casting filling, cooling, and solidification, we can obtain the flow field, temperature field, and stress field of the investment castings, and predict the organization and the defects, such as cold shut, porosity, cracking, and deformation, etc. Therefore, computer simulation can be used to study the casting structure and casting process, and provide useful information on castings structure and process design, and to avoid the blindness of traditional structure design and process design which is sbased on experiences. It can shorten production preparation time and save trial production cost.

2.2.2.2 Application of Concurrent Engineering in Investment Casting

With the continuous development and popularization of computer technology, concurrent engineering will be widely used in the investment casting industry in the future.

Concurrent engineering is to establish the electronic data communication mechanism which is closely related to the investment castings user and the factory, and to carry out the parallel product and process design of the user and the foundry. Users give the 3D geometry of castings to the foundry online. The foundry engineers can easily check the 3D model before the computer screen and then show the users and designers possible problems under different process conditions after CAE simulation, such as hot crack, porosity, etc., in order to make improvement and obtain high quality castings. Similarly, the manufacturing process of pattern mold, pattern, and the shell can also be realized simultaneously. In this way, the earlier the foundry gets involved in product development, the shorter the development time and the lower the cost, which increases the market competitiveness of the product [12].

2.2.2.3 Application of Computer Technology in Investment Casting

The application of computer modeling and simulation in the investment castings overcomes the shortcomings of the investment castings process and makes this technology more flexible and more adaptable to meet the requirements of fast development, high quality, and complexity.

(1) The CAE technology helps to get the reasonable casting structure and the most optimal process design of investment castings.
(2) The application of rapid prototyping in the pattern and the mold has greatly shortened the time to prepare the pattern and the mold.
(3) Compared with the traditional method, 3D printing technology can reduce the time of shell making.
(4) Complex ceramic cores can be produced by computer-controlled laser fabrication.

With the continuous development and application of computation technology, concurrent engineering and integrated manufacturing will be widely used in the investment casting industry in the future.

2.2.3 Design of Gating System

For the large complex thin-wall superalloy castings, centrifugal casting has some disadvantages, such as propensity to component segregation, which gradually reduces its application. Engineers are more likely to use the gravity casting method. This section mainly discusses the basic rules of the gating system under the condition of gravity casting.

2.2.3.1 The Determination of Pouring Position

The rules of pouring positions are as follows: ① Important parts of the casting should be placed at the bottom as far as possible. ② Important machined surfaces should be facing down or upright. ③ Make the large plane of casting face down to avoid scarring defects. For large plate castings, tilt casting can be used to speed up the rise of metal surface and prevent scarring defects. During tilting casting, the total gating system pressure head should be in the range of 200 ~ 400 mm according to the size of the mold. ④ The casting shall be fully filled. For castings with thin-walled parts, the thin-walled parts should be placed in the lower half or below the ingate to avoid defects such as misrun and cold shut. ⑤ Easy to feed the casting. ⑥ Chillers should be avoied, which are very difficult to place in the process of investment castings. ⑦ The pouring position and the cooling position of the casting should be the same, in order to avoid turning over the mold after metal pouring.

2.2.3.2 The Parting Line

For investment casting, the metal pours into a whole shell mold without parting lines. However, the pattern molds have parting lines. The parting surfaces are the places where the two halves of the pattern mold meet. The parting surface is usually selected after the location of wax injection is determined. The parting surface affects the pattern accuracy, cost, and productivity to a great extent. It should be carefully analyzed, compared, and design. The parting line design should conform to the following rules: ① All or most of the wax pattern should be placed in the same half. ② The number of parting lines should be minimized; the wax pattern should be more accurate with less parting lines, as well as a smaller number of mold components. ③ The parting surface should be a plane. It is easier to manufacture the flat parting surface and then ensure the accuracy of the wax pattern. ④ The mold components, two halves and cores, should be easy to assemble and easy to inspect as well. ⑤ Parting surface is usually selected on the largest section of the wax pattern, so that the mold is not too high.

2.2.3.3 Rules of the Gating System

The general rules of the gating system design are as follows: ① Help the melt metal fill the cavity steadily and smoothly, avoiding air entrapped due to the turbulence, and scour the core, avoiding the oxide inclusion and shell debris. ② During the filling,the flow direction and speed can be controlled to ensure clear and complete shape. ③ The cavity should be filled at the right time. The gating system should have the ability to block slag and overflow slag, and purify the melt metal. ④ Control the temperature distribution of the metal to benefit the feeding, reducing the casting stress, and preventing the casting from deformation and cracking.

2.2.3.4 Types of Gating System [13]

The right gating system type is one of the important parts in the process design. It is related to the alloy composition, structure, size, technical requirements, and production requirements of the casting. The gating system can be classified by the position of the ingate or by the section area ratio of the gatings.

(1) The position of the ingate

a. Top filling system. The melt metal is introduced from the top of the cavity.

The advantage of top filling system is: (i) the temperature of the upper part of the casting is higher than that of the lower part during the filling and solidification, which is beneficial to feeding the casting; (ii) it can have large metal flow, the cavity can be filled in a short time with strong filling capacity; (iii) the gating system and riser need less metal and they are easy to be cut after casting.

The biggest disadvantage of the top filling system is that the melt metal falls down from the high place, which has a great impact on the shell. In addition, it is easy to cause the splash, oxidation, and gas entrapment, as well as to lead to the oxidation slag and porosity defects.

The top filling system is suitable for small and medium sized castings with small quality, low height, and simple shape.

b. Bottom filling system. The ingate is located at the bottom of the casting.

The advantage is: (i) the metal fills the cavity smoothly; (ii) the runner is mostly filled with metal during the filling, regardless of the ratio of runner area and ingate area, which keeps the slag out of the cavity.

The disadvantage is that, after filling, the temperature of the lower part of the casting is higher than that of the upper part, which is on the contrary to the sequential solidification from bottom to top and thus weakens the feeding effect of the top riser. The bottom of the casting, especially near the ingate, is easy to overheat, which results in shrinkage porosity, coarse grain, and other defects. The filling capacity is poor, so it is easy to have misrun and cold shut for large thin-walled castings, even though it consumes more metal.

The disadvantages of the bottom filling system can be solved through some technical methods, such as fast pouring, dispersing multi-runner, exothermic riser, and so on.

Bottom filling system is widely used in aluminum and magnesium alloys and other non-ferrous castings, which is also suitable for all kinds of ferrous castings with complex shapes and high requirements.

c. Medium filling system

The ingate of medium pouring system is between the top pouring system and the bottom pouring system, whose advantages and disadvantages are also between these two pouring systems. It is widely used in small and medium-sized castings with a small height and large horizontal size. When the casting quality is high, the falling height of the melt metal, that is, the depth of the lower half cavity, should be limited when high quality castings are desired.

d. Multi-step gating system for setting up multi-layer ingates at different heights of casting.

A well-designed multi-step gating system has the following advantages: the metal filling the cavity smoothly from bottom to top; the gas escapes the cavity smoothly; the temperature of the upper casting is higher than that of the lower part after filling, which is benificial to sequential solidification and help the riser feed the casting. The strong filling ability can avoid defects such as cold shut and misrun. In addition, the local overheating near the ingate can be reduced with multiple ingates. The main disadvantages are as follows. Correct calculation and structural design are needed. Otherwise, the metal will enter the cavity from the upper and lower ingates at the same time in a mess; the temperature of the lower part is also higher than the other part of the casting due to too much metal entered the cavity from the bottom ingates. Multi-step gating system is widely used for pouring large and medium-sized steel castings and iron castings.

(2) The section area ratio of the gatings

The gating system can be categorized into closed, open, and semi-closed types according to the proportion of the cross-section area of the sprue, runner, and ingate.

a. Closed type

The cross-section area of the spure, runner, and ingate decreases sequentially ($A_{sprue} > A_{runner} > A_{ingate}$). As the cross-section area of the runner is getting smaller and smaller, the flow velocity is higher and higher in this gating system. Since the velocity at the ingate will be significantly high, the metal will impact the shell with splashing and violent oxidation. However, during the entire filling from the begining, the sprue is kept in full state, and the slags in the liquid metal is easy to float up to the top of the sprue and avoid entering the cavity. In addition, the volume of this gating system is small and need less melt metal.

b. Open type

The cross-section area of the spure, runner, and ingate enlarges sequently ($A_{sprue} < A_{runner} < A_{ingate}$). The characteristics of the open gating system is opposite to the closed one. The main advantage of the open gating system is that the metal flow in the runner and the ingate is slower and the metal enters the cavity smoothly. The disadvantage is that the slags can hardly be kept out from the cavity while the runner is half filled up during the initial stage of filling.

c. Semi-closed type

When ($A_{sprue} < A_{ingate} < A_{runner}$), it is called the semi-closed gating system. Its characteristics is between the open type and the closed type, that is, the melt flow is relatively stable and the filling capacity and slag rejection capacity is better. It is suitable for small and simple structure castings.

The section area ratio of the gating system has a big effect on the casting quality. As a result, it is important to choose the correct gating ratio during the design of gating system.

In practice, the engineers have accumulated a lot of experience to choose the right ratio, leading to many relevant data in the literature. However, it is difficult to give out an effective, simple, and easy method due to the different casting structures, processes, and other specific conditions.

2.2.3.5 The Dimension and Structure Design of the Gating System

After the type of the gating system and the ingates have been determined, the size and structure of the gating system can be further determined. The minimum section area of the gating system (that is, the choke section) is calculated by hydraulic or empirical formula. And the gating ratio is determined according to the structural characteristics and geometry of the castings. Finally, the size and structure of each part are determined and the size and structure of each part are then checked and optimized using computer simulations.

(1) The minimum section area calculation by hydraulic formula

Based on fluid mechanics, the melt metal fluid is regarded as a common fluid and the gating system as a channel. For the semi-closed gating system, the cross-section at the bottom of the sprue has the minimum section area. If the metal at the pouring basin is one end and the outlet of the sprue is the other end, between the two ends, according to the Bernoulli equation, Eq. (2.11) is used to calculate the minimum section area.

$$A_{\text{min}} = G/0.0433\gamma\tau\mu\sqrt{H_P} \quad (2.11)$$

Where, G is the total mass of liquid metal (kg), A is the cross section area at the outlet of the sprue (cm^2), μ is the flow consumption coefficient, τ is the pouring time(s), γ is the density of liquid metal (g/cm^3), and H_P is the average calculated static pressure head (cm)

For a closed gating system, the ingate has the minimum cross-section area. The Bernoulli equation can also be used to derive the above formula. According to this formula and the value of each factor, the minimum cross section area of the gating system can be calculated.

According to the type of gating system, we can know where the minimum cross-section area is located, in the sprue, runner, or ingate. Then we can get all the cross-section areas of the sprue, runner, and ingate according to the appropriate ratio ($A_{sprue} : A_{runner} : A_{ingate}$), which allows us to obtain the whole dimensions of the gating system in a preliminary way. Since the predetermined values of '*G*' and '*r*' in the initial calculation are estimated values, and since the actual ratio of the section area of each part is different from the one selected, the calculation results still need to be calibrated against experiments.

(2) Section ratio design method

Experiments shows that the position, quantity, shape, and size of ingate have a great influence on the quality of the castings when the alloys are easy to oxidize. Sometimes the method above do not work. Therefore, "backward induction" is used to calculate the size of the gating system on the basis of practice.

The "backward induction" is to determine the number and cross-section area of the ingates according to the specific casting process at first, and then to determine the size and structure of the runner and sprue according to the total cross section area of the ingates and the selected gating ratio. The specific steps are as follows.

a. According to the characteristics of casting structure, select the type and structure of gating system;

b. According to the alloy, casting structure characteristics, and process, the ingate number and total cross-section area are determined by experience. Normally the engineer makes tables through induction and summary according to the workshop experience, to design the similar gating system.
c. Select the thickness, width, and length of the ingate according to the casting wall thickness connected to the ingate.
d. Select the gating ratio according to the casting characteristics and determine the dimensions of the runner and sprue.

2.2.3.6 Riser Design

The riser size is calculated using the modulus method.

The casting solidification time depends on the ratio of the volume of the casting to the heat transfer surface area. The ratio is called the solidification coefficient, or modulus for short. Let us write it this way:

$$M = V/A \tag{2.12}$$

where, *M*—modulus; *V*—volume; *A*—heat transfer surface area.

In order for the casting and riser to solidify sequentially, the riser modulus *M* must be greater than the modulus of the part to be fed. In the vacuum condition, due to the lack of atmospheric pressure, the feeding depends on the gravity of metal only. Therefore, the modulus of the riser should be more than 1.5 times of the modulus of the part to be fed under vacuum conditions.

2.3 Casting Equipment

According to the characteristics of superalloy, it needs to be melted under vacuum condition. As a result, the vacuum furnace is necessary. Normally, the pouring weight of the melt metal is more than 4 times the weight of the casting, so the melting furnace should have a volume to hold above 400 kg for the casting of 100 kg. According to the integral large complex thin-walled investment casting, the diameter of castings should exceed 1000 mm, considering the thickness of the shell and appropriate operating clearance, the diameter of the slurry bucket and dewaxing kettle should exceed 1500 mm. During the slurry coating and stucco coating, the manipulator grasps the pattern assembly to dip in the slurry bucket and then place it into the stucco coating machine. According to the size of the casting, the payload of the manipulator should be more than 500 kg, in particular, for the manipulator for the back layer shell. If the sand is scattered by hand, the labor intensity of the workers is high. In addition, there is still uneven shell and the repeat-ability is difficult to ensure.

Thus, it is difficult to ensure the shell quality of large investment castings. For the partitioning wax pattern, 100T wax injection machine is needed, such as MPI51660 of MPI Company, a larger wax injection machine is needed if we want to inject the integral wax pattern.

To sum up, the necessary equipments for large complex thin-wall integral investment castings are as follows. The furnace: 500 kg and above vacuum induction furnace. The slurry bucket:the diameter should be 1500 mm or above, at least one bucket for surface layer and one bucket for back layer. The dewaxing kettle: the diameter should exceed 1500 mm. The wax injection machine: the locking force should be 100T or above. The Roasting furnace: the width of the furnace should be 1700 mm or above.

2.4 Process Design

The characteristics of large size thin-walled complex castings should be fully considered in the process design. Thin-walled castings have the characteristics of fast melt heat transfer, rapid rise of liquid level, and large resistance caused by surface tension during the filling. During the solidification process, the melt solidification time is short, the time windows of dendrite formation and feeding are narrow, and the temperature of the melt metal drops significantly after a long distance of flow.

For the complete filling and effective feeding of thin-wall structures at the most edge, the correspondence between solidification characteristics and process parameters at the characteristic structure has been studied in real time through synchrotron radiation in situ. Based on the original process based on gate design, the process parameters suitable for thin-walled casting have been designed.

It is assumed that after the melt enters the cavity, it flows laminar away from the ingate, and the whole shell has the same temperature before cavity filling. After the melt flows through the shell, heat is transferred between the high-temperature melt and the shell. Since the thermal conductivity of the shell is low and the filling speed is rather high, it is assumed that the shell keeps the original temperature unchanged before it contacts with the melt.

Solidification sequence at the thin wall is as follows. As the melt begins to solidify after filling the cavity, there is no directional solidification and the temperature at the bottom is lower than that ad the top.

During the filling process, the temperature change of melt along the flow route can be used to analyze the melt temperature at the thin wall of the farthest end.

Let the flow distance from the gate to the farthest thin wall be L and the filling height be H (Fig. 2.3). The radius of the castings is R_g. If the casting is a rotor, then $L^3 \geq R_g$. For simplicity, let $L = R_g$. K_M is the thermal conductivity between the shell and the melt, and K_m is the thermal conductivity of the melt itself.

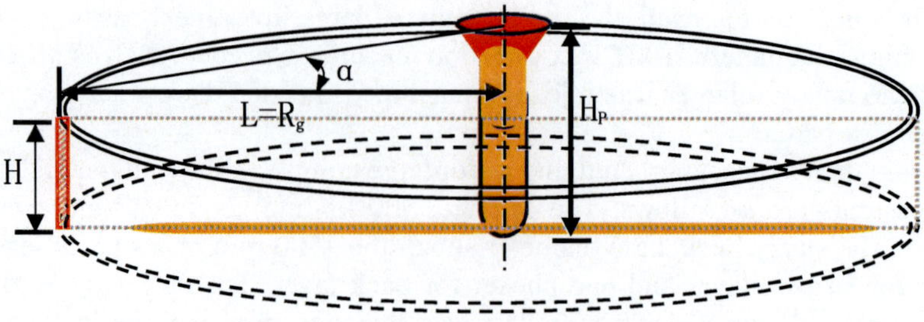

Fig. 2.3 Solidification process of large size thin-walled castings

Assume that the initial temperature of the shell is T_M, and after contact with the melt, the shell temperature is the same as the melt temperature *T(x)*.

Then the unit mass melt flows from the gate to the final thin-wall position, and the temperature at different positions is as follows:

$$T_0 - x\frac{dT}{dx} = T(x), x = L, T(x) = T_t \tag{2.13}$$

The heat dissipation during the flow to the final position is as follows

$$(T_0 - T_t) \cdot C = \frac{L}{v} \cdot K(T_0 - T_M) \tag{2.14}$$

where L is the flow distance of the melt and v is the flow velocity of melt under the condition of free liquid level in the cavity. S_g is the total ingate section area, The pouring quality per unit time is m, then:

$$\mathrm{v} = \sqrt{2gH_p} \tag{2.15}$$

At the same time, $\mathrm{v} = \frac{m}{S_g \cdot g}$.

Assuming that the melt flow always keeps laminar,during the flow process, the heat conservation state per unit melt per unit time is:

$$C\frac{dt}{dx} = K_M(T_0 - T_{(x)} - K_m\frac{dt}{dx} \tag{2.16}$$

The left-hand side above is the loss of melt heat per unit volume per unit mass. The first term on the right-hand side is the heat transfer from melt to shell, and the second term is the heat transfer from unit volume and unit mass melt with higher temperature to this melt unit. By combining the above 3 equations, we can obtain T_t.

When the melt flows to the final position of the cavity, the temperature of the melt must be guaranteed to fullfill. The fitting formula is obtained according to the experimental data,

$$t_{2mm} = 3.14312 \times 10^{-9} \times T_t^{3.30319} \tag{2.17}$$

According to Eq. (2.17), the dendrite network formation time can be set as the time since melt fully filled to dendrite network formation. According to the size and shape of the casting, t_{2mm} can be set. To get the smaller grain size of the casting, the pouring temperature should be as low as possible.

The melt temperature T_t can be derived according to the dendrite network formation time. According to this temperature, we can get the pouring speed, the shell temperature, and pouring temperature, as well as the height of the sprue.

The accuracy of the above design is based on the in situ and real-time monitoring data of the dendrite network formation, as well as relevant assumptions. Actual casting data and in situ analysis has shown that this design parameter can be used as a reference.

2.4.1 *Design of Gravity Casting Process for Large Complex Thin-Walled Castings*

Gravity casting is the most widely used casting method of investment casting for superalloys. It has two options, either directly pouring from melting furnace or pouring from converter. At present, most superalloy castings are directly poured in vacuum induction furnace. The quality of superalloy castings, besides that the master alloy must meet the quality requirements, depends mainly on casting process parameters.

2.4.1.1 Design of Gating System for Gravity Casting of Large Complex Thin-Walled Superalloy Castings

Turbine rear frame casting, as shown in Fig. 2.4, are of large size, complex structure, and thin wall, which are typical to large complex thin-walled castings. According to the structural characteristics of rear case frames, the section changes sharply and it is a large-area thin-walled piece, which makes it difficult to fill. Therefore, the bottom filling and open gating type is selected. At the same time, for the better feeding, the upper runner is used to let the melt metal flow into the riser directly at the final stage of solidification.

Fig. 2.4 Turbine rear frame castings

(1) The rising speed of liquid metal

a. Pouring time

The filling up time of cavity is the pouring time, which is an important parameter of the gating system. The longer the pouring time, the higher the required pouring temperature. However, if the pouring time is too short, there will be casting defects due to turbulence flow. Therefore, there is an optimal pouring time for a given casting.

Pouring time depends on the type of the alloy, the complexity of the casting, the wall thickness, and section size, etc. The pouring time is usually calculated by empirical formula, which is based on the experience of the engineers. It is hard to derive from theory.

Since the thickness of the casting is largely affected by the ratio of volume to surface area of the casting, the wall thickness is also an important parameter apart from the weight of the casting when calculating the correct pouring time. Usually, the casting weight is considered when the gating system is not fully designed. In addition, it is considered that the gating system is fully filled before the metal liquid begins to enter the cavity, so the weight of the gating system is generally not considered. However, if the size of the gating system is almost the same as the casting size, the weight of the gating system should be considered in the calculation of pouring time.

For complex thin-walled castings with the pouring mass less than 450 kg:

$$t = K_1 \sqrt[3]{W} \tag{2.18}$$

where t is the pouring time (s), W is the total weight of the casting, the gating system and the risers (kg), and K_1 is a constant number (Table 2.2).

Table 2.2 Recommended values of K_1 for castings with different wall thickness

δ/mm	K_1
1.5–2.5	1.62
2.5–3.5	1.68
3.5–8.0	1.85
8.0–15.0	2.20

b. The rising speed of liquid metal in the cavity

The filling speed of liquid metal has a close relationship with casting quality. If the filling speed is too fast, the melt metal will impact the shell with splashing and turbulent flow, resulting in the oxidation inclusion and porosity defects of the castings. On the other hand, if the filling speed is too slow, it is easy to have defects such as cold shut and misrun, especially for the large thin-walled castings that are difficult to fill. Therefore, reasonable filling speed and sufficient liquid metal flow are required during the cavity filling. The velocity V is,

$$V = c/t \tag{2.19}$$

where, c is the height to fill(mm), t is the filling time (s).

Normally the filling time of large castings should conform to Table 2.3.

(2) Determination of average pressure head H_p

The average pressure head is determined by Eq. (2.20):

$$\mathrm{H_p = H_o - \frac{p^2}{2C}} \tag{2.20}$$

where, H_p is the average pressure head (mm);H_o is the pressure head above choke section (mm); P is the height of the cavity above the choke section (mm); C is the total height of the casting at the casting position (mm).

This is easily obtained from Fig. 2.5:

For bottom filling system where $P = C$, we have $H_p = H_o - \frac{P}{2}$;
For top filling system where $P = 0$, we have $H_p = H_o$.

Practical experience shows that if the sprue is too low, the filling pressure and feeding pressure may not be high enough. It is then easy to have defects such as the edges and

Table 2.3 The relation between minimum metal rising speed and wall thickness

Wall thickness mm	<4	4–10	10–40	>40
Minimum speed mm/s	30–100	20–30	10–20	8–10

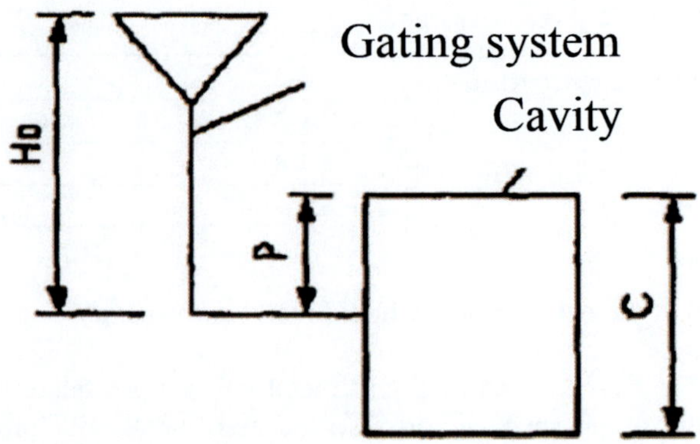

Fig. 2.5 Diagram of each variable in the calculation formula of average pressure head

contours not being clear, misrun, concave on the upper surface, and other defects. In particular, for large superalloy castings, the melt metal should maintain enough residual pressure head H_M such that the liquid from the pouring basin and the sprue can effectively feed the casting during the solidification. When the superalloy casting solidifies in a vacuum furnace, the residual pressure head should be of sufficient height. The following equation is the relation between residual pressure head and pressure angle of melt metal:

$$H_M = L \times tg\alpha \tag{2.21}$$

where, as shown in Fig. 2.3, L is the horizontal distance from the furthest point of the casting to the center line of the sprue (in units of mm) and α is the pressure angle (in units of °).

In the vacuum state, the feeding pressure needed during the solidification mainly comes from the gravitational potential of the liquid metal. Therefore, the pressure angle in Table 2.4 should be appropriately enlarged. In the vacuum state, 1.5 times of the pressure angle in the table should be taken.

Table 2.4 The relationship between pressure angle and casting size

Wall thickness (mm)	3–5	5–8
L (mm)	Pressure angle	
600	13–14	9–10
800	12–13	9–10
1000	11–12	9–10

2.4.1.2 Influence of Pouring Parameters of Superalloy

(1) Pouring temperature

Pouring temperature has a significant effect on the dendrite, grain size, porosity, and the size and number of $\gamma + \gamma'$ eutectic structure of the superalloy castings. The pouring temperature should be 100–150 °C above the liquidus temperature. If the pouring temperature is too low, the cavity will be insufficiently filled and there will be cold shut, misrun, and porosity as well. However, if the pouring temperature is too high, it will lead to dendrite thickening and significant grain coarsening, together with significantly increased $\gamma + \gamma'$ eutectic size and serious porosity issue. For large complex thin-walled rear case frames, the influence of pouring temperature is more complicated. The rear case frames have many structures where the cross-section changes abruptly. With high pouring temperature, thick sections tend to form porosity and severe segregation, whereas with low pouring temperature, the parts with thin wall thickness will have serious porosity. Therefore, it is very important to choose the appropriate pouring temperature during the experiments.

(2) Filling speed

Filling speed has an obvious effect on casting quality. If the speed is too slow, there will be defects like misrun, cold shut, and quenched grain. If the speed is too fast, the melt is easy to rush out of the pouring basin. It is generally beneficial to the cavity filling when the speed is faster, but it is also easy to entrap air and inclusions. The appropriate filling speed should be: for thick large castings or bottom filling system, the filling speed should be fast at first and then slow down, such that there is a certain amount of high-temperature melt in the pouring basin that can reduce shrinkage cavity and porosity in the casting. For small thin-walled castings, it is better to pour slowly at first and then quickly, which can reduce the defects such as air entrapment and inclusions. As to superalloys, the filling rate at the ingate should be about 0.9 kg/s if there is ceramic filter. If no filter is present, the filling speed can be slower, generally about 0.6 kg/s.

(3) The shell temperature

The shell temperature is also an important parameter affecting the quality of supper-alloy castings. The shell should be preheated for superalloy investment castings and the preheating temperature of the shell should not be too low. Otherwise the casting is not easy to be filled and there will thus be cold shut, porosity, and inclusions. In addition, the shell temperature should not be too high. If not, the degree of under cooling of the melt will be too small such that the nucleation rate becomes lower and the grain size becomes larger, thus reducing the plasticity and toughness of the alloy. The appropriate roasting temperature should be selected according to the shell condition and alloy requirements.

(4) Effect of cooling rate

The solidification rate of superalloy casting is mainly affected by the pouring temperature, shell temperature, and the condition whether or not there is sand surrounding the shell. The microstructure and mechanical properties of the alloy as cast are affected accordingly. The influence of pouring temperature and shell roasting temperature is introduced in the previous sections. Usually, the lower the pouring temperature or the lower the shell temperature, the faster the solidification proceeds. However, if the shell temperature is too low, it may result in too large melt temperature gradient and thus lead to the columnar crystals.

When the shell is put into the sand-filled box, the solidification rate is different in vacuum and atmosphere environments. Compared with single shell casting, sand filling casting in atmospheric environment has good insulation effect, resulting in slow cooling rate and coarse grain. The solidification rate of single shell casting is faster and the grain size is smaller due to the good effect of convection and heat dissipation in the atmosphere. In constrast, under the vacuum condition, the solidification rate of single shell casting is slow while the sand surrounding the shell absorbs the heat from the casting. Consequently, the cooling speed is faster and the grain becomes smaller.

2.4.2 Simulation of Casting Process

On the basis of the traditional design, numerical simulation technology is used to quantify the gating system and riser and evaluate whether or not the process design goal can be achieved. Based on the simulation results, the optimal pouring system and riser is selected. This process should start at the casting design stage, which will effectively reduce the process difficulty and shorten the product development cycle. This section describes a more detailed simulation of the actual casting process.

The first step of numerical simulation is to acquire the casting model, 3D model of pouring system, and boundary conditions. After the 3D modeling, provide the software with the composition and material type of the casting metal and the shell, together with the pouring temperature, shell temperature, environment temperature, pouring speed, heat transfer coefficient, etc. The software will divide the casting, riser, and the shell into small elements, combined with the discretization in time, and then calculate the temperature and velocity of every small cell. Finally, we obtain the result of the liquid metal flow field and temperature field during the filling and solidification process es, together with some criteria. The engineer then analyzes the results and further optimize the gating system and riser.

2.4.2.1 Model of Casting, Gating System and Riser

Using 3D software, such as UG NX, Pro/E, CATIA, SolidWorks, Solidedge, CAXA, and other CAD software, we first construct the 3D geometry model of casting, gating system, and riser, and import the result into CAE software through STL or other interfaces.

2.4.2.2 Input of Casting Parameters and Boundary Conditions

Metal: IN718 (Liquidus:1337°C; Solidus:1255°C).
Shell: Corundum.
Casting process: investment casting (gravity).
Casting parameters: pouring temperature 1500 °C; shell temperature 1000 °C; the pouring time is 12 s.

2.4.2.3 Analysis of Simulation Results

(1) Information from filling simulation

Check the mold filling sequence, flow field, velocities, and temperature in the cavity, loss of superheat and temperature distribution in the melt. Pay attention to whether the temperature of the melt is lower than the liquidus during the pouring process. If the melt temperature is lower than the liquidus, there may be cold shuts and misruns. If there is turbulence in the flow, slags and inclusions can not float up. If two melt flows collide in the cavity, there may be weld marks or cold shuts. Finally, check the temperature distribution of the melt after pouring, in particular, whether the melt temperature is high at the top of the cavity and low at the bottom, which can help the feeding.

(2) Information from solidification simulation

Check the solidification paths and feeding patterns. The metal in the riser should remain in liquid until the casting is solidified and the feeding paths should keep connected.

(3) The criteria of casting defects

a. Niyama criterion

Niyama et al. evaluated the influence of flow resistance on shrinkage porosity, which was a comprehensive prediction method for determination of center line shrinkage in cast steels. The method proposed by Niyama, namely Niyama criterion, represents the relationship between the gradient (G) and cooling rate (R) [14]. It is calculated according to the following formula:

$$Niyama = G/\sqrt{R} \tag{2.22}$$

where, *Niyama* is dimensionless, *G* is the gradient, and *R* is the cooling rate.

The volume contraction between the liquid state and liquid-solid state during the solidification is the main cause of shrinkage and porosity. When the feeding path is unobstructed, the dendrite has not formed a skeleton, the volume shrinkage results in shrinkage cavity (primary or secondary), which is located in the upper part of the casting. However, when the dendrite forms a skeleton and the feeding path is blocked, the volume shrinkage of the hot melt surrounded by dendrite results in micro porosity (within the dendrite range). Solid shrinkage has little effect on the formation of shrinkage cavity and porosity. As a resutl, solid shrinkage is not considered. During the solidification, the fraction of solid in the cell increases and the fraction liquid decreases. When the fraction of solid in the cell is greater than a critical value, the cell will become an immobile cell, such that it cannot feed the others. In every calculation step, it should be determined which cell is flowable and which cell is immobile. Then by substracting the flow cell from the top down and making the reduce number of the flow cell equal to the increase number of shrinkage cavity cell, a collection of all the shrinkage cavity cell becomes the primary or secondary porosity, while the coordinate of the cell is the position of the shrinkage cavity. When the solidified dendrite forms a skeleton, the melt surrounded by dendrite will shrink and form porosity. The criterion $G/\sqrt{R}$ proposed by Niyama can be used to predict the location where shrinkage occurs. When the $G/\sqrt{R}$ value of the cell is less than a critical value, the cell may contain porosity. The porosity distribution range is the collection of all possible porosity cells. For 2D temperature field, the *G* and *R* values of 26 neighboring cells of each node should be calculated, wherein the maximum value of $G/\sqrt{R}$ represents the interdendrite feed ability. Points that reach or remain below the defined $G/\sqrt{R}$ value are considered as porosity and otherwise not. Niyama considered that the critical value was 1.0, but Lee Wenzhen et al. found that the critical value was related to the size of the casting, with the range of the critical value is from 0.8 to 1.10, with the upper limit corresponding to large castings and the lower limit to small castings [15].

We recommend that the critical value is 0.8 for large castings. When it is larger than this value, it will be gray on the scales, that is, there is no shrinkage in this region. If it is less than this value, it means that the region cannot be feeded effectively, resulting in possible shrinkage.

b. Porosity criterion

Porosity criterion is an effective way to visualize porosities in the casting. The portion of porosities at the end of solidification simulation is displayed. An analysis of porosity problems is more reliable when combining Niyama criterion and Porosity criterion together, which guarantees the optimizing casting process and reducing casting rejection rate.

Porosity criterion is set as 0.001 ~ 10. The color scale can be used to correspond to the porosity of 0.001% ~ 10%. For example, the blue area indicates that the porosity is zero, that is, there is no shrinkage problem, whereas the white area indicates that the porosity is very high and the value is close to 10%. The corresponding porosity can be found on the color scale.

c. FSTIME criterion

The FSTIME criterion means the time that the casting needs to reach the critical portion of solidified melt. It is used to determine the causes of porosity and solutions. It is generally believed that after the melt reaches the critical "*Fraction Solid*", it is no longer flowable due to dendrite lap joint. If the FSTIME of a part of the casting is larger than that of any other points nearby, there will be porosity tendency. The critical "*Fraction Solid*" varies according to the composition of the liquid metal.

d. HOTSPOT criterion

The HOTSPOT criterion shows the solidification time of hot spot during solidification, with a unit of "*s*". This criterion shows porosities in these residual melt regions.

The HOTSPOT criterion shows the solidification time in the local area of the casting by the color scale. The solidification time is longer in these residual metal regions that are called hot spots, and porosity problem will occur during solidification. Through the comprehensive analysis of the solidification time criterion and the HOTSPOT criterion, it can be seen that, for large castings, if no riser is added, the hot spot part of the casting will have defects.

2.5 Shape and Size Control of Large Castings

Quality control of large castings have two aspects: casting defects and size defects. The shrinkage rate of wax pattern and casting are different depending on whether it is free or restrained. In addition, different restrained conditions can result in different shrinkage. For small parts, the size does not change much for different shrinkage rates; but for large castings, the change is large enough to exceed the tolerance and make scrap. Casting deformation is also easy to cause shape and position deviation and lead to scrap.

2.5.1 Factors Affecting Deformation [16]

The process of investment casting is complicated, and there are many factors that affect the dimensional accuracy. The dimensional accuracy of casting will be affected by the casting structure, material, pattern making, shell making, roasting, pouring, and other factors.

(1) Effect of pouring temperature: The higher the pouring temperature, the greater the shrinkage; the lower the pouring temperature, the smaller the shrinkage.
(2) The influence of casting structure: ① The shrinkage is greater with a higher wall thickness, and smaller with a thinner wall thickness; ② The free shrinkage is bigger, whereas the shrinkage becomes smaller when restrained.
(3) The influence of shell materials: Since the expansion coefficient of zircon sand or powder is only 4.6×10^{-6}/°C, the resulting expansion can be ignored.
(4) The effect of shell roasting: the shell expansion coefficient is small, the expansion is only 0.053% when the shell temperature is up to 1150°C; In addition, some data show that there may be partial collapse due to phase transition after contact with high-temperature melt; however, experience shows that the effect can be ignored.
(5) The influence of material: The higher the carbon content in the material, the smaller the line shrinkage, and vice versa.
(6) Effect of pattern on line shrinkage: ① Radial (restrained) shrinkage of pattern is only 30–40% of the longitudinal (free) shrinkage, and the influence of wax injection temperature on free shrinkage is far greater than that on the restrained shrinkage (the best wax injection temperature is 57–59 °C. The higher the temperature, the greater the shrinkage). ② The line shrinkage of wax is about 0.9–1.1%. ③ When the pattern is stored, it will further shrink, with a shrinkage value about 10% of the total shrinkage. However, after 12 h, the size of the pattern is almost stable. ④The wax injection temperature, injection pressure, and pressure hold time. Studies have shown that the size deviation of pattern is more than 40% of the casting size deviation [17, 18].

To sum up, there are three key processes in investment casting that can have great influence on the size and deformation of the casting, which include pattern preparation, shell preparation, and casting solidification and cooling.

2.5.2 *Control of the Shape and Size of Large Castings*

The factors influencing the casting shape and size can be classified into two aspects: material and process. The raw materials and process should be strictly controlled during the manufacturing. The raw materials such as wax, shell, and master alloy must be stable and controllable. Regarding the process, in order to optimize process parameters and reduce deformation, the traditional experience design has been followed and information technology has also been used, together with numerical simulation being carried out for pattern making and casting processes. After the process parameters are set down, strict control is required, including the environmental conditions of the wax injection and shell making workshop, which should be controlled within a certain range.

2.5.2.1 Numerical Simulation of Deformation

(1) Numerical simulation of casting stress

During the solidfication, the cooling rate in the casting is different everywhere due to the interaction of the geometric structure and boundary conditions and the cooling shrinkage is also different, resulting in deformations caused by stresses in the casting system. Numerical simulation of stress and strain can help better understand the dynamic changes of stress and deformation during the casting solidification. As a result, research and prediction of casting deformation and residual stress analysis can guide the castings at the shop floor scientifically [19–21]. Nowadays the casting simulation softwares can successfully combine the flow field, temperature field, and stress field together to simulate the casting process of large castings [22–31]. The subsequent stress, crack, and deformation can be well predicted as well. Based on these simulation results, the process parameters are modified to make the cooling of the casting as evenly as possible, thus reducing the deformation of the casting.

(2) Numerical simulation of wax injection

The dimensional accuracy of wax pattern directly affects the dimension accuracy of final casting. As a result, it is necessary to simulate the injection of wax pattern. To this end, first measure the theological and compressible properties of wax and then construct the mathematical model and carry on the numerical simulation of wax injection. Based on the simulation results, the runner system for the wax pattern is optimized and the displacement field of the wax pattern deformation is predicted. Combined with experiments data, the shrinkage rule and shrinkage mechanism in different directions of wax pattern have been studied. In addition, the mathematical shrinkage model of large thin-wall pattern has also been established to optimize injection parameters and reduce wax pattern deformation.

2.5.2.2 Dimension Detection During the Investment Casting Process

There are many factors that affect the shape and size of the casting, some of which have not been fully studied. Consequently, the simulation can hardly predict the deformation of the casting exactly. Currently, inspection, rework, and rejection rates account for 35 percent of the cost of casting. Both process measurement and final product measurement are the key to realizing accurate process control and cost reduction. In the United States, the next generation manufacturing technology plan will carry out research on digital imaging detection technology of castings for air plane and aerospace [16]. Measurement is indispensable for casting development and production. It is necessary to carry out more studies on the size measurement technology for wax pattern, shell, and casting in the process of investment casting.

Wax pattern, especially the low temperature wax pattern, is soft. The traditional contact measurement will affect the dimension accuracy of the thin-wall wax pattern.

3D digital detection is a new advanced technology with broad application prospects, which is of great significance owing to its flexibility, automation measurement, and product design [32–36]. Non-contact techniques, such as knuckle-arm or hand-held laser scanner, are used to scan the full surface of wax pattern and casting rapidly and form a solid surface model. Then the 3D surface tolerance comparison, section deviation analysis, and rapid dimension measurement will be performed by some softwares, which provide a reference for the accurate study of casting deformation. The new scanning measurement technology can accurately record the process of deformation in the entire process of investment casting, quantitatively record the size deformation in each process, and help determine the main factor influencing the casting size and deformation deviation, which are then used to control the deformation of large complex thin-walled castings.

2.6 Wax Pattern Mold Design

According to the process of investment casting, casting is transformed from wax pattern, implying that the shape and size of wax pattern can directly affect the shape and size of casting. Researches have shown that the size fluctuation of casting is largely inherited from the size fluctuation of its corresponding wax pattern [17]. The wax pattern size deviation accounts for over 40% of the casting size deviation [18]. In order to obtain the correct shape and size of the wax pattern, the size of the pattern mold cavity should be controlled so that the casting with reliable dimension accuracy can be obtained. This means that it is necessary to correctly assign the shrinkage rate of each dimension of the casting when designing the wax pattern mold. The dimensional accuracy of the final casting is realized by setting the dimension of the wax pattern mold to compensate for the size change during the wax injection and the casting process. The principle of mold surface compensation design is to apply an appropriate amount of reverse deformation to the deformed parts to offset the shrinkage deformation of wax patterns and castings during solidification and cooling. Therefore, it is very important to calculate the size of mold cavity.

In the long-term study and practice, the three influential factors of wax pattern shrinkage, i.e., alloy shrinkage, shell expansion, and deformation are all taken into account comprehensively and the researchers have carried out a lot of shrinkage calculation solutions and achieved certain results [37]. However, casting structure, including the wall thickness and shape, has more a significant effect on the total shrinkage rate, but is relatively difficult to calculate. For complex castings, some parts are constrained, some are half constrained, and some are constraint-free. As a result, they should be analyzed carefully, with different shrinkage rates being assigned to each part, in order to achieve high dimension accuracy castings. Due to the lengthy process and many relevant factors, the total shrinkage rate is difficult to determine. Therefore, the shrinkage compensation rate of the mold obtained by the "formula calculation" also has a large uncertainty. Although it can be ignored for small castings, for large castings it cannot. Otherwise it may lead to out-of-tolerance.

The compensation shrinkage method of mold cavity design has been lacking in accurate theoretical guidance, and still remains in the design practice as "experience + experiment" [38], which mainly includes empirical data method, formula calculation method, and trial and error.

For important castings, specific parts should be tested under different processing conditions, in order to find out the appropriate total shrinkage under the shop floor conditions. Then the product shrinkage compensation and pouring system are adjusted accordingly, which is followed by designing and manufacturing the pattern mold, and finally move to production. When the traditional "experience + experiment" trial and error method is used to design the mold, several modifications are needed. Although the "trial-production method" can finally get the appropriate shrinkage compensation rate, its long cycle and high cost seriously restrict the development of investment casting. The pattern mold design of high temperature castings for aeroengines is always a bottleneck problem.

In order to shorten the R&D cycle time, reduce the mold repair, and mold cost, the method of split pattern mold is used in the R&D stage.

2.6.1 Split Pattern Mold

Due to the large, complex, and thin wall of large castings, if the wax injection is carried out using the traditional wax pattern mold, it is difficult to control the dimension accuracy of wax pattern. In addition, it is difficult to adjust the dimension inaccuracy caused by improper shrinkage. For circular castings (such as TRF, intermediate casing, etc.), since they are axisymmetric and geometrically meet the requirements of separate designs, split wax pattern molds can be applied.

As to the TRF, it has 10 hollow joint struts. The wax pattern is divided into 10 pieces along the center surface of the two adjacent struts. In this way, the wax pattern partition is 1/10 of the whole wax pattern. Prepare the 10 wax blocks by the wax injection machine, adjust the positioning size of special fixture tool, and then put the wax blocks into the fixture tool, weld the wax blocks, and finally form the whole wax pattern.

The advantages of split wax pattern are:

(1) Easy to adjust. When the incorrect shrinkage rate is used, we can also get the correct size by adjusting the wax fixture tool, such that re-manufacture the pattern mold is avoided.
(2) The size of the wax pattern is smaller and it is easier to get the correct shape and dimension accuracy of the wax pattern. The wax pattern rejects things such as short runs and hence low density are reduced and the wax pattern mold is easy to open and operate;

(3) Compared with the overall mold, the split mold is smaller in size, much simpler in terms of the structure, and easy to design and manufacture. Even with the addition of design and manufacture of special tools, the preparation cycle is greatly shortened and the cost of tooling can be greatly reduced.

2.6.2 *Overall Pattern Mold*

Although the split pattern mold has many advantages, it needs to weld the wax blocks together, which need time, and even more time for injecting the 10 wax blocks separately. For overall mold, one injection is just needed to get the wax pattern, and it does not need to assemble and weld the wax blocks. Clearly, the disadvantage of split pattern mold becomes more obvious for mass production. In addition, it is easy to leave welding traces with improper operation, and the wax pattern is easy to deform during the welding manually. Therefore, after trial production, the shrinkage problems of the casting have been controlled, the overall pattern mold can be used in mass production, which can easily ensure the dimension accuracy of the wax pattern and improve the production efficiency as well.

References

1. Longtengriyue. How hard is an airline jet engine, Aerospace knowledge. (12), 28–31, 2010
2. Q. Huan, Development and process of superalloy INCONEL 718 (GH4169). J. Mater. Eng. **40**, 8 (2012)
3. J. He-Fu, Development and manufacturing technology of gas turbine engine. Aeronaut. Manuf. Technol. **5**, 36–39 (2007)
4. H. Xiangping, Overview of foreign investment investment casting development. Spec.-Cast Non-Ferrous Alloy. **12**(3), 40–43 (1992)
5. N. Hai, X. Chengmu, Application and development of titanium alloy casting abroad. China Foundry Mach. Technol. **6**, 1–3 (2003)
6. L.-T. Zhang, C. Lamei, L. Guoli et al., *Theory and Practice of Near Net-shape Investment Casting* (National Defense Industry Press, Beijing, 2007)
7. C. Lamei, T. Xin, Z. Yong et al., Progress of advanced near net-shape investment casting technology of superalloys. J. Aeronaut. Mater. **26**(3), 238–243 (2006)
8. T. Tianfu, C. Bing, B. JIANG, *Investment Casting Technology* (China Machine Press, Beijing, 1991)
9. B. JIANG, *Practical Investment Casting Technology*, (Liaoning Science and Technology Press, Shenyang, 2008)
10. B. JIANG, *Investment Casting*, China Machine Press, 2004
11. Leiden; Boston: Brill, Bubble and drop interfaces Progress in colloid and interface science, vol. **2**, pp. 1877–8569, 2011
12. Z. Yuan, Product Design Guidelines for Manufacturing and Assembly, China Machine Press, 2011
13. Foundry Institution of Chinese Mechanical Engineering Society. Casting manual, vol. 5, casting process, 2nd edn, China Machine Press, 2003
14. P. Liwen, Z. Lijing, Z. Hu, Applicability of shrinkage porosity prediction for casting with Niyama criterion, J. Beijing Univ. Aeronaut. Astronauties, **37**(12), (2011)

15. W-Z. Li, *Numerical Simulation of Microstructure and Shrinkage Cavity Formation During Solidification of Castings* [D]. Tsinghua University, 1995
16. C. Bing, Dimensional stability and accuracy of investment casting. Spec.-Cast Non-Ferrous Alloy. **1**, 53–56 (2003)
17. H. Xiangping, The general situation of wax pattern precision in investment casting. Spec.-Cast Non-Ferrous Alloy. **24**(1), 156–159 (1999)
18. Z. Zhenyu, L. Bunv, Factors that affect the dimension accuracy of casting during mold making. Mech. Res. Appl. **15**, 15–16 (2002)
19. X. Yan, K. Jin-Wu, H. Tian-You, Application of contact element method in simulating thermal stress during solidification of casting. Comput. Appl. **27**(5), 506–510 (2006)
20. K. Jinwu, *Numerical Simulation Analysis of Thermal Stress Fields during Solidification of Steel Casting*. [D]. Tsinghua University, 1998
21. Y. Yi, J. Yuming, L. Liling et al., Numerical simulation of thermal stress field and thermal cracking of casting during solidification. Foundry Technol. **2**, 36–39 (2000)
22. L. Hui, S. Jian-song, Z. Ai-Qin, Numerical simulation and deformation analysis of thermal stresses in stress frame. China Foundry **59**(1), 38–41 (2010)
23. C. Jianguo, K. Jinwu, Z. Jiafeng et al., Extraction of deformation of castings from simulated displacement results. Foundry Technol. **29**(10), 1322–1326 (2008)
24. C. Bing, Numerical simulation of investment casting process. Spec-Cast Non-Ferrous Alloy. **25**, 683–686 (2005)
25. S. Liwen, H. Hua, X. Hong, Numerical simulation of thermal stress and thermal cracking of casting. Res. Stud. Foundry Equip. **3**, 20–22 (2006)
26. L. Meie, X. Jiandong, Research development of thermal-mechanical numerical modeling of casting stress field. Foundry **51**(3), 141–144 (2002)
27. Z. Hui, K. Jinwu, H. Tian-you, Study of effect factors on thermal stress during solidification process. Hot Working Technol. **35**(13), 69–72 (2006)
28. Z. Xianshu, Y. Shan, J. Jun-ze, Study on calculation model of resistance stress of cast mold. J. Dalian Univ. Technol., **36**(6), 687–691, (1996)
29. Z. Xianshu, J. Jun-ze, Thermo-elasto-plastic analysis of the dynamic stress in stress-frame specimen. J. Dalian Univ. Technol. **22**(3), 1–4 (1983)
30. S. Kun, Z. Dongbo, L. Bingheng, Numerical simulation of casting stress considering mold resisting stress. Acta Metall. Sinica **12**, 1258–1262 (2000)
31. F. Xianjun, L. Dunming, Z. Jianxin et al., Bidirectional coupled simulation for casting thermal stress based on ANSYS. China Foundry **60**(11), 1103–1106 (2011)
32. H. Xiaomei, L. Jiayan, C. Wei et al., Study on digieal design and check of turbine blades based on reverse engineering. Metrol. Meas. Technol. **29**(2), 8–10 (2009)
33. Z. Meng, C. Liliang, W. Junchang, Dimension measurement of castings based on 3d reconstruction technology with marking points. China Foundry **59**(4), 379–383 (2010)
34. Y. Yongquan, Z. Shichao, X. Zhenjia, Application of 3d scanning technology in new casting development. J. Stand. Qual. Mech. Ind. **478**, 37–41 (2013)
35. Z. Shichao, Y. Yongquan, Z. Shoushuang et al., Application of 3d laser scanning technology in casting dimension inspection. Automobile Technol. Mater. **12**, 61–64 (2012)
36. H.-S. Li, Qualify for CAE and CAI integration applications based on moldflow and geomagic. CAD/CAM Manuf. Inf. **8**, 51–53 (2009)
37. L. Jingguo, Z. Xiaofeng, L. Xintao, Research on the production technology in the process of investment casting. Mod. Cast Iron **6**, 67–69 (2006)
38. J. Liu, K. Bu, Y. Li et al., Casting shrinkage analysis of turbine blade. Mod. Manuf. Eng. **3**, 9–12 (2008)

Chapter 3
Dimensional Deviation and Defect Prediction of Wax Pattern

Donghong Wang

A large number of studies have shown that in many processes of investment casting there are three key processes that have a great influence on the dimension and deformation of castings, namely, wax pattern, shell preparation, and casting solidification [1–5]. Sabao et al. [3] analyzed the dimension errors that occurred in each process of investment casting. Figure 3.1 shows the fluctuation behavior of the casting dimension in the three key processes. The results show that the fluctuation in dimensions of castings is mostly inherited from the corresponding wax pattern fluctuation. Zhang et al. [6] also confirmed that the dimension deviation caused by wax pattern stage accounts for more than 40% of the total deviation from the design dimensions to the final dimensions of castings. It can be seen that the injection process of wax pattern has a decisive influence on the final dimensions for investment casting. Taking the wax material used in engineering practice as an example, this chapter expounds on the influencing factors of dimensional accuracy of wax pattern and the method for prediction and control of the wax injection process during investment casting.

3.1 Factors Affecting the Dimensional Deviation of Wax Pattern

The dimensional accuracy of wax pattern is affected by many factors. Firstly, the chemical composition affects the linear shrinkage caused by heating expansion and cooling shrinkage of the wax. Second, the linear shrinkage of the wax pattern is also related to the shape and dimension of the pattern, as well as to the position of the pouring system and the parameters of the injection process [7]. Therefore, the main factors that determine the final dimension of wax pattern include (1) the composition and properties of wax material, (2) wax injection process parameters, (3) the dimension and structure characteristics of pattern, and (4) injection channel system and die design.

B. Sun et al., *Precision Forming Technology of Large Superalloy Castings for Aircraft Engines*, https://doi.org/10.1007/978-981-33-6220-8_3

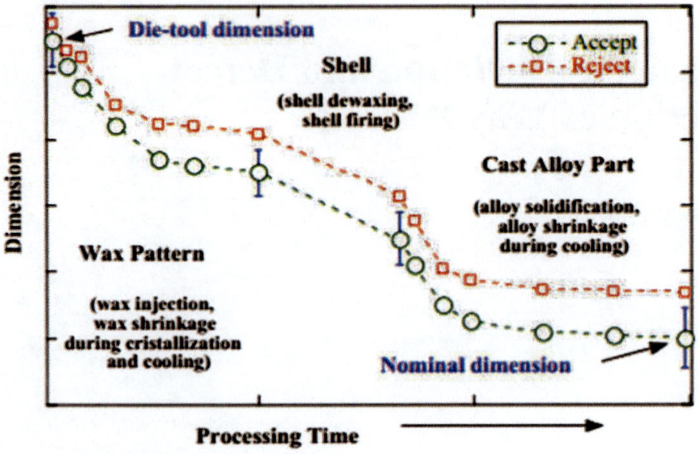

Fig. 3.1 Dimensions in the investment casting process [3]

3.1.1 Composition and Properties of Wax Materials

The investment casting wax is a complex mixture, which mainly includes synthetic wax, natural or synthetic resin, solid organic fillers, and water, as well as additives such as plastics, oil, and plasticizers. Its components also have their own uses. For example, the addition of resin can increase the strength of wax materials and the addition of fillers can improve the shrinkage performance of wax materials. Owing to the existence of these additives, the investment casting wax shows complex mechanical and thermal properties.

The composition of additives and fillers has a great influence on the properties of wax materials. At present, wax materials with fillers are widely used, such as powdered polyethylene, polystyrene, organic acid, fatty acids, and starch, etc., which accounts for 30–45% of the total amount. This type of wax has good heat preservation and liquidity and can be molded at relatively low temperatures. Wu et al. [8] found that the shrinkage of wax mold added to the wax material was more than 5% lower than that without filling materials. Liu et al. [9] improved the strength and shrinkage of wax by adding EVA (a polymer of ethylene and vinyl acetate). Hao [10] showed that the shrinkage properties of wax materials with similar chemical structures but different compositions were obviously different. These studies fully show that the filling material has a great influence on the properties of wax.

It should be pointed out that some performance indicators such as melting point, freezing point, penetration, hardness, and so on, which are usually used in engineering, only provide a general description of wax products in order to ensure their supply quality. These simple data cannot provide enough help for the selection of wax injection process parameters and the design of flow passage. However, some useful physical properties, such as shear viscosity, surface tension, the compressibility of wax, specific heat, thermal conductivity, and heat transfer coefficient between mold and wax, cannot be characterized and tested simply and efficiently. In order to understand the performance and behavior of wax, researchers have made great efforts. Harvey Feld et al. [11] studied the mechanical properties of investment casting wax. Han et al. [12] tested the rheological properties of wax materials and established the flow constitutive equation of wax materials for investment casting using the power-law viscosity model. Gebelin et al. [13] tested the viscosity state of wax materials and the viscosity of liquid injection, and fitted the rheological curve of wax materials using the Carreau and modified cross WLF viscosity model. These studies have established a critical relationship between the composition and properties of wax.

3.1.2 Wax Injection Process Parameters

The so-called wax pattern manufacturing process refers to the process of transferring wax material into the mold to form wax mold. The main parameters involved are wax temperature, injection temperature, mold temperature, injection pressure, holding time, and holding pressure. These parameters will have an impact on the dimensional accuracy of wax pattern. Taking the wax melting temperature and time as an example, the insulation tank should maintain a setting temperature slightly higher than the ejection temperature and be heated for at least 8 h. It cannot be used until the wax material melts evenly; otherwise it is easy to produce defects such as filling dissatisfaction, cold insulation, grain surface, flow pattern, and mesh grain.

The researchers studied the importance of the parameters of various wax injection stages, including wax injection temperature, wax injection time, injection pressure, holding pressure, and static time of wax mold after demoulding. Kelkar et al. [14] found that wax injection temperature has a great influence on the dimensions of the constrained shape. With increasing the injection cycle time, the size shrinkage of wax mold decreases, in particular, for the case of the constrained shape. However, injection pressure has little effect on the size shrinkage of wax mold. Horacek et al. [15, 16] suggested that injection pressure and rate had no significant effect on the shrinkage of wax pattern, but the holding pressure and time are the important factors affecting shrinkage. Bonilla et al. [17] suggested that most of the shrinkage of wax pattern occured within 2 h after demoulding and the pressing process had a little effect. Other studies have also given various conclusions, which, however, are even contradictory to each other and thus have not been widely accepted. In-depth and comprehensive studies on the influence of visible wax injection process parameters on the size deviation are currently insufficient and still lacking. Therefor it is necessary to systematically consider the process factors that affect the size deviation of wax mold and introduce the process robust optimization design, which can then help solve the problem of process optimization, as well as the optimization theory of wax injection molding from the perspective of basic research, and develop the large complex thin-wall wax mold with higher size accuracy and better stability.

3.1.3 Dimension and Structure Characteristics of Wax Pattern

(1) Variable cross-section structures
For the characteristics of variable section structures, the uneven shrinkage caused by the sharp change of interface thickness will lead to serious deformation. The researchers found that the use of wax mold core, wax mold drawing, and other techniques can effectively reduce the thickness of the part of the size deviation, with, however, the impact of the overall wax mold size deviation still unclear.

(2) Hollow thin-wall structures
In order to reduce the weight of castings, castings with hollow thin-walled structures are favored by designers, and technically, it is necessary to use a predetermined core during wax mold pressing. However, it is difficult to predict the dimensional deviation of the wax mold when preparing the hollow structure wax mold. Horacek et al. [16] suggested that the dimension deviation of the hollow structure castings was mainly due to the influence of injection parameters. The research based on injection molding showed that the deviation of preset water-soluble core in the wax pressing process may be the most dominant factor. In 2004, Bakharev et al. [18] confirmed the core deflection of injection molding process using simulation software. In 2009, Dong et al. [19] showed that melt temperature and injection time had little effect on core deflection size. In constrast, injection pressure was the most important process parameter of core deflection [20, 21]. Giacomin et al. [22] had focused on the study of core migration in the injection molding process. In their study, the viscoelastic model was used to describe the wax material and the predicted core shift was larger than the measured value. The results showed that the core shift predicted by the fifth-order nonlinear theory was very close to the measured value [22].

(3) Constraint and free shrinkage
Okhuysen et al. [23] designed feature parts to study the shrinkage of wax pattern. The features of ring shape were divided into three types of deformation: constrained, partially constrained, and free deformation. It was found that not all the dimensions of the characteristics changed along with the injection process parameters, and the shrinkage of free deformation wax pattern was larger than that of constrained deformation dimension. Yarlagadda et al. [24] studied the accuracy of H wax pattern injected using low-pressure injection molding of polyethyl carbamate (hard mold) and silicone rubber (soft mold). The results showed that the shrinkage rate of free deformation size was twice as much as that of a restricted deformation position.

3.1.4 Wax Injection Filling System and Mold Design

The effects of wax injection runner systems are often overlooked by researchers and process designers. For small castings, an extremely simple runner design can be met. But for wax molds of large complex thin-wall castings, it is necessary to design the runners into the runner itself and the gate. The gate is a very short and narrow-section flow passage connecting the splitter and the cavity. It can be placed at one or more parts of the workpiece and can be of various types, generally used according to the respective characteristics. The main function of the gate is to ensure faster flow velocity and better fluidity when filling the mold, prevent backflow of the melt, facilitate demolding of the product, and control the freezing time of the gate and the flow during the filling of the melt performance [25].

The wax injection gate design mainly includes the design of the number, position, shape, and size of the gates, wherein the number and position of the gates mainly affect the filling mode whereas the shape and size of the gates affect the melt flow properties. It has been reported that the uniformity of the size of injection molded articles can be improved by gate size optimization [26, 27]. The basic principles of the wax mold runner system design during investment casting are as follows [28–31].

(1) The gate location should be placed at the farthest point where a hysteresis effect may occur to eliminate or reduce the hysteresis. The so-called "hysteresis effect" here refers to the phenomenon that when the melt flows to the thick and thin junction, the flow resistance is large at the thin point. The phenomenon of "cutting near the road" is also caused by the blockage at the same place, which is also called "short shot".
(2) When determining the number and location of gates for a product, the flow ratio must be examined to ensure that the melt can fill the cavity. The flow ratio is determined by the ratio of the flow length to the thickness of the flow channel. If the calculated flow ratio is greater than the allowable value, it is necessary to increase the thickness of the product or change the gate position or use a multi-gate method to reduce the flow ratio.
(3) The cavity layout and gate opening position should be symmetrical to prevent the flash phenomenon caused by the eccentric load of the mold. The gate position should ensure that the flow length in all directions is equal to prevent resions of the part from experiencing overpressure.
(4) For gates with slender cores, the core will be deformed by the impact of the melt. In this case, the front side of the wax should be avoided to determine the position of the gate.

Based on the above analysis, it is seen that gate location is a key variable in injection mold design. An incorrect gate location will result in a series of defects such as overpressure, high shear rate, poor weld line properties, and warpage. For simple parts, we may get a satisfactory design by analyzing the relationship between the gate position and the product. For complex parts, however, these guidelines may help us judge the rationality of the gate location design, but cannot automatically provide us with a reasonable design. Therefore, the researchers [32, 33] have turned to numerical simulations to optimize the design and achieved good results.

3.2 Numerical Simulation of the Investment Molding Process

As mentioned earlier, the dimensional change in the wax mold preparation stage is the main cause of the oversize of the investment casting and needs to be predicted and controlled as much as possible. As precision castings continue to become larger, more integrated, and thinner, the pressing process of investment casting wax molds increasingly relies on numerical simulations as a powerful analytical tool.

A number of numerical simulation studies have been conducted in terms of the filling behavior of wax injection molding. For example, Gebelin et al. [13] used Moldflow software to simulate the flow process of the wax and analyzed the influence of different mesh types on the numerical simulation of the filling process and the influence of gate location on filling process. Sadegh et al. [34] performed numerical simulations of solidification time, cavitation, and temperature fields during wax filling. On the basis of the experiment, the numerical simulation of the mold filling process for investment casting was successfully realized by Han [12]. Sabau et al. [1] used ABAQUS software to simulate the viscoelastic deformation of a stepped wax pattern. Figure 3.2 shows the displacement in the length of the wax mold after the demolding and the displacement on the broadband after 3.5 h of demolding. The work of these researchers demonstrated the versatile applications of numerical simulation methods in the wax mold preparation process and the new methods of wax mold preparation research brought by the numerical simulation technology based on the flow and solidification theory of polymer materials.

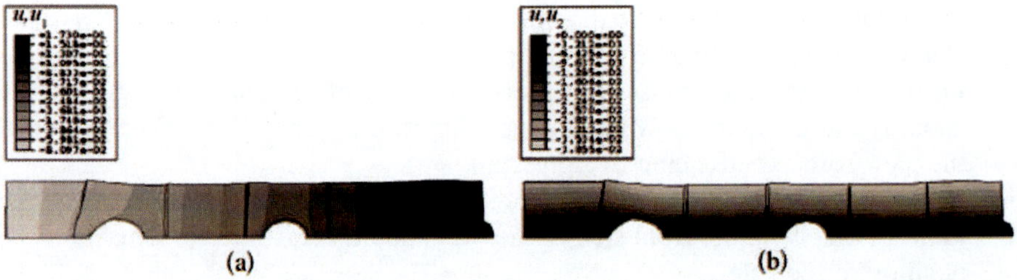

Fig. 3.2 Viscoelastic case: Displacement distribution (cm) in the wax pattern **a** along the length and **b** along the width at 3.5 h after the pattern is removed from the die. Displacements are magnified 20 times

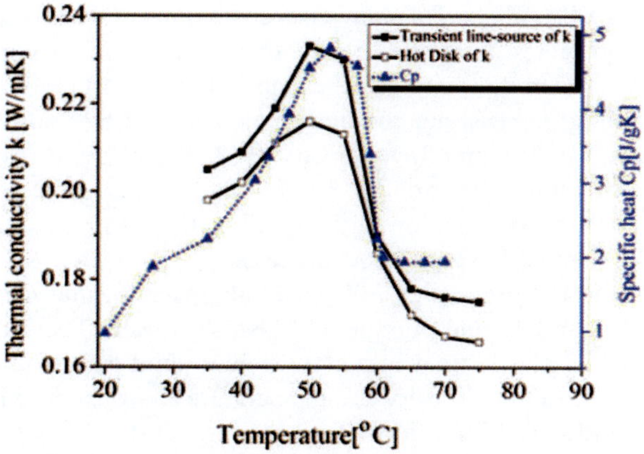

Fig. 3.3 Thermal conductivity and specific heat for the pattern wax

3.2.1 Performance and Theoretical Model of Wax

Before the establishment of the model, it is necessary to conduct an in-depth study on the properties of wax and measure the properties of wax, including thermal properties, rheological properties, and mechanical properties. At the same time, there is still a lot of work to be done for the constitutive model of wax. It must be pointed out that waxes vary in performance depending on the manufacturer. This chapter only uses a certain type of wax used in the engineering practice of the author as an example. The results in this paper are also related to the model, but the discussion is applicable to most waxes.

3.2.1.1 Thermal Performance

Figure 3.3 shows the relationship curve between the thermal performance of the tested wax and the temperature. As the temperature increases, the thermal conductivity first increases and then decreases, while the thermal conductivity of the liquid wax is small. It can be seen from the specific heat curve that near the softening point of the ring, the specific heat decreases rapidly, and it does not change substantially after 70 °C.

3.2.1.2 Rheological Properties of Wax

Quantitative analysis of the fluidity of wax provides a scientific approach to controlling the investment casting process and dimensional accuracy. It is considered that the shrinkage of casting is inversely proportional to the fluidity of wax [35]. In the standard for measuring the technological properties of mold, rheology is reflected by measuring the viscosity of mold at a predetermined temperature when it is in liquid state [2]. Figure 3.4 show viscosity curves of wax melt at different temperatures and shear rates. It can be seen that wax melt shows shear thinning behavior. However, the tested viscosity curve cannot reflect the true viscosity of the mold in the mold filling process, nor the dynamic change of the mold viscosity with the change of temperature and flow velocity. [36]. Therefore, it is necessary to carry out mathematical treatment on the basis of the measured rheological curve and establish the constitutive equation of the mold material to reflect the real dynamic performance of the mold material flow process.

In the process of filling and compaction, the viscosity is affected by not only temperature and pressure but also shear rate. Therefore, the Cross-WLF model of pseudoplastic fluid in non-newtonian fluid can be used to express the relationship between the viscosity and temperature and shear rate in the numerical simulation of mold melt filling flow [2, 13, 31, 36]. It describes the zero-shear viscosity, which is also sufficiently accurate at low shear rates. The flow process of wax melt in wax injection mold mainly considers the action of shear stress, ignoring viscoelastic behavior. The formula of cross-wlf viscosity model with shear thinning is shown as follows:

$$\eta = \frac{\eta_0}{1 + \left[\frac{\eta_0 \dot{\gamma}}{\tau^*}\right]^{1-n}} \tag{3.1}$$

$$\eta_0 = D_1 \exp\left[\frac{-A_1(T - T^*)}{A_2 + (T - T^*)}\right] \tag{3.2}$$

$$T_{au^*} = D_2 + D_3 p \tag{3.3}$$

$$\mathrm{A_2 = \bar{A}_2 + D_3 p} \tag{3.4}$$

Fig. 3.4 Viscosity versus shear rate at different temperatures of wax

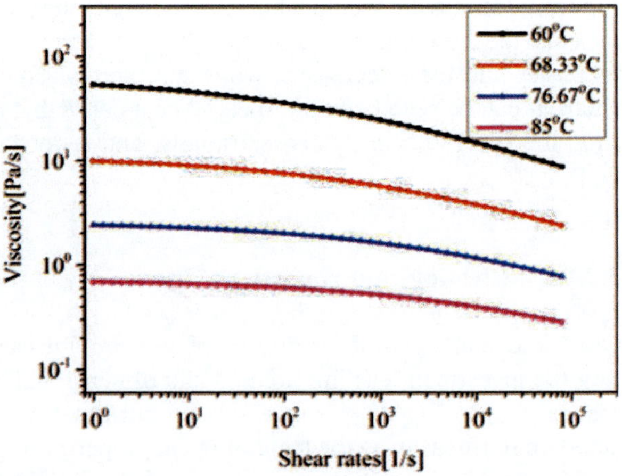

Table 3.2 Tait PVT model coefficients

Parameter	Value	Remark
b_{1m} (m^3/kg)	0.001078	Melted state
b_{2m} [m^3/(kg-K)]	6.271e-007	Melted state
b_{3m} (Pa)	1.69603e+008	Melted state
b_{4m} (1/K)	0.007912	Melted state
b_{1s} (m^3/kg)	0.001022	Solid state
b_{2s} [m^3/(kg-K)]	4.603e-007	Solid state
b_{3s} (Pa)	2.60746e+008	Solid state
b_{4s} (1/K)	0.005987	Solid state
b_7 (m^3/kg)	0.000056	
b_8 (1/K)	0.05867	
b_9 (1/Pa)	1.185e-008	
b_5 (K)	339.15	
b_6 (K/Pa)	1.490e-007	

where $\overline{A_2}$ is the average. Table 3.1 shows the viscosity model parameter values fitted using the least-squares method. The value of $\boldsymbol{D_3}$ is 0, indicating that the viscosity model does not take into account the effects of pressure.

3.2.1.3 Compression PVT Characteristics of Wax Melt

The pressure-specific volume-temperature (PVT) relation of polymer describes the change of specific volume of polymer with the change of temperature and pressure. As the basic property of polymer, it is the basis for flow analysis, mold design, injection molding process control, and process analysis. The PVT curve shows the influence of temperature and pressure on the specific volume of polymer. Figure 3.5 shows the PVT data of the tested wax material KC4017B.

PVT equation of state can also be used to study and calculate the specific volume properties of polymers, mixing and phase separation laws of polymers, and phase equilibrium calculation. In addition, the PVT equation of state is also used to describe the PVT relationship of the polymer, which provides the calculation formula and theoretical basis for the simulation and control of injection molding of the polymer. The modified two-domain Tait state equation is the most commonly used one to describe the PVT relationship of polymer in injection molding field. The formula of modified two-domain Tait state equation is as follows:

$$\rho = \left[\nu_0\left(1 - 0.0894\ln\left(1 + \frac{P}{B}\right)\right) + \nu_t\right]^{-1} \tag{3.5}$$

where for $T < b_5 + b_6 P$ we have

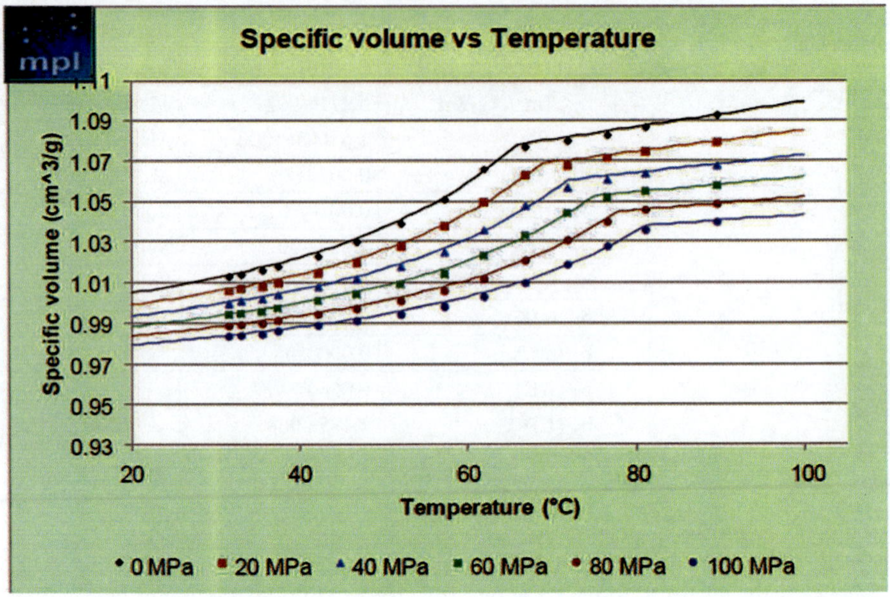

Fig. 3.5 PVT diagram of the wax

$$V_0 = b_{1s} + b_{2s}(T - b_5) \tag{3.6}$$

$$B(T) = b_{3s} \exp[-b_{4s}(T - b_5)] \tag{3.7}$$

$$V_1 = b_7 \exp[b_8(T - b_5) - b_9 P] \tag{3.8}$$

and for $T > b_5 + b_6 P$ we have

$$V_0 = b_{1m} + b_{2m}(T - b_5) \tag{3.9}$$

$$B(T) = b_{3m} \exp[-b_{4m}(T - b_5)] \tag{3.10}$$

$$V_1 = 0 \tag{3.11}$$

In the formula, $\boldsymbol{\rho}$ is the density of wax, i.e., the reciprocal of specific volume, $\boldsymbol{V_0}$ is a specific volume at zero pressure, and $\boldsymbol{B}$ is material of stress sensitivity. According to the thermodynamic property of polymer, the PVT relationship needs to be described in two temperature domains. The transition temperature of the measured volume at zero pressure is represented by b_s. The transformation temperature linearly increases with the increase of pressure, and the change is represented by $\boldsymbol{b_6}$. $\boldsymbol{b_1}$ represents the specific volume obtained by extrapolating the zero pressure isobaric curve to the transition temperature, and the value of this transition point is the same for both temperature domains. The relationship between the specific volume and temperature is contained in $\boldsymbol{b_2}$, while b_3 and b_4 represent the characteristics of the solid-state and molten state, respectively. When b_4 is determined, the specific volume becomes more sensitive to pressure with the increase of temperature. b_7, b_8, and b_9 characterize the specific volume vt in the solid-state. According to the experimental data, 11 parameters of the Tait equation are fitted using the multivariate nonlinear regression, as shown in Table 3.2.

3.2.1.4 Viscoelasticity of Wax Material

Under certain external conditions, the polymer from an equilibrium state can reach a new equilibrium state adapted to the external conditions through the thermal movement of molecules. This variable speed process is also known as relaxation process and the time required to complete the process is called the relaxation time. The relaxation process is related to the relative molecular weight of the polymer. Since the polymer has a certain molecular weight distribution, the relaxation time is not a fixed value but with a certain distribution, called the relaxation time spectrum. In general, the relaxation time spectrum has the most general functional form describing the viscoelasticity dependence on time or frequency. As can be seen from Bolzmann's superposition principle, all properties of materials are shown in all motion modes and contributions with different relaxation time, and all materials' functions measured by experiments are based on the same relaxation time spectrum. Therefore, the relaxation time spectrum is undoubtedly the core of all viscoelastic functions. Because the relaxation spectrum is central to describing the viscoelasticity of polymer fluid, it is a hot topic to obtain the relaxation spectrum.

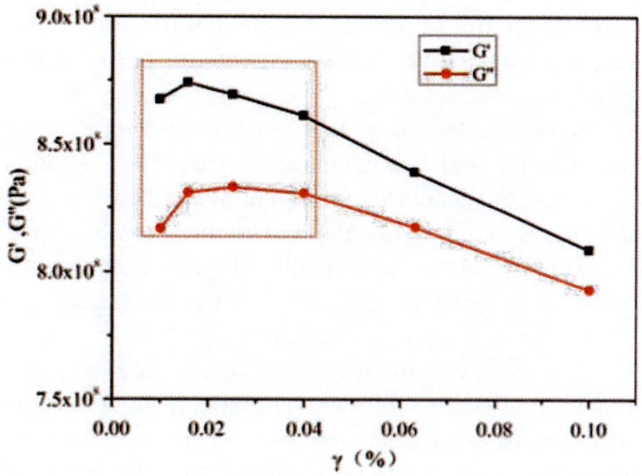

Fig. 3.6 Relationship between strain rate and storage modulus

It can be seen from Fig. 3.6 that the storage modulus G′ and the loss modulus G″ will change in a curve with the strain rate γ. When the curve enters the plateau range, the storage modulus G′ and the loss modulus G″ has nothing to do with the strain rate γ. That is, the storage modulus G′ and loss modulus G" at this time are only related to frequency, so that the relationship between wax frequency and modulus can be obtained. As shown in Fig. 3.6, it is most appropriate to set the strain rate of 0. 04% in the range of viscous elastic platforms.

According to the relaxation property of the polymer movement, the polymer chain can be made to have sufficient activity, so that the polymer can exhibit high elastic deformation, or it will take some time for the whole polymer to move and show viscous flow (measured by the relaxation time). With the increase of temperature, the relaxation time can be shortened. Therefore, the same mechanical relaxation phenomenon can be observed not only at a higher temperature in a shorter time but also at a lower temperature in a longer period of time. Therefore, according to the principle of time–temperature equivalence, the effects of increasing temperature and prolonging time on the molecular motion is equivalent. Reducing temperature and shortening time are equivalent to molecular motion and Viscoelastic behavior of polymers. With the help of conversion factor, the mechanical data measured at one temperature can be transformed into mechanical data at another temperature. If the experiment is carried out under an alternating force field, then the reduction in frequency is equivalent to prolonging the observation time, and the additional frequency is equivalent to shortening the observation time. Therefore, the practical dynamic mechanical data measured at different temperatures can also be superimposed by means of moving factors. Owing to the principle of time–temperature equivalence, the mechanical properties of the polymer measured at different temperatures or at different frequencies can be compared and converted, resulting in results that can not actually be measured directly from the experiment. For example, in order to obtain the stress relaxation behavior of natural rubber at a specified temperature, because the temperature is too low and the stress relaxation is too slow, it may take centuries or more to get the complete data, which is actually impossible. Therefore, under the principle of time–temperature equivalence, stress relaxation data can be measured at higher temperatures, which can be converted into the required low-temperature data.

According to the principle of time–temperature equivalence, the increase of temperature and the decrease of frequency are equivalent to the motion of the molecule; the decrease of temperature and the increase of frequency are equivalent to the motion of the molecule. As a result, each line spectrum is translated horizontally or vertically. After adjustment, two main curves of storage modulus (ω) and loss modulus (ω) can be obtained, and the master curve can be obtained by optimizing the main mode curve, as shown in Fig. 3.7.

The approximation method of calculating the relaxation spectrum from the material function is as follows. The steady-state relaxation modulus function is first obtained using the steady-state stress relaxation experiment or the dynamic shear test. Then the approximate solutions of the relaxation spectrum (λ) are obtained using these functions to obtain the steady-state relaxation modulus function (ω) or the

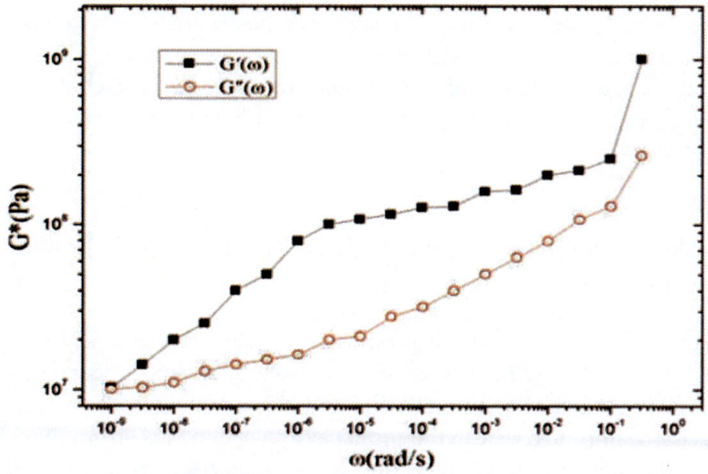

Fig. 3.7 Master curves for the storage (G′) and loss module (G")

dynamic modulus function (ω). Finally the approximate solutions of the relaxation spectrum (λ) are obtained using these functions [37]. In the process of calculating the discrete relaxation spectrum using linear least square method, in order to obtain more accurate results, the value range of the number of Maxwell motion elements is generally 5–9, and usually with a value of $N = 8$ [5]. Our calculation is $N = 5, 8$. It can be seen from Fig. 3.8 that the calculation results have the same changing trend

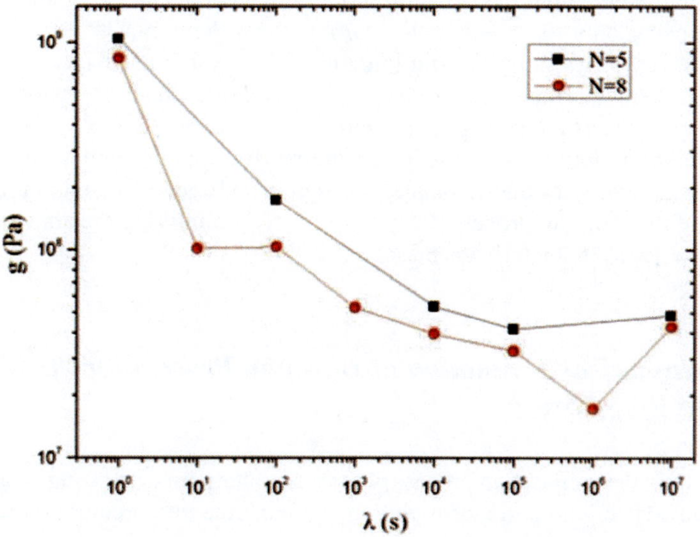

Fig. 3.8 The relaxation time spectrum of wax

for both $N = 5$ and $N = 8$, and it can be seen that the accuracy of $N = 5$ is obviously not as good as that of $N = 8$. With the relaxation time spectrum, the deformation caused by the stress relaxation of the residual stress can be predicted after the wax material is separated from the die, which is of great help to the correction of the die size.

3.2.1.5 Theoretical Basis of Numerical Simulation of Wax Injection Molding

In this chapter, the filling/backfilling process model of wax injection molding is based on the generalized Hele-Shaw flow of non-Newtonian fluid under non-isotherm, and the effects of material compressibility and phase transition are considered. The flow behavior of polymer wax in the mold cavity dependends strongly on the shear viscosity of the melt. In this chapter, the Cross-WLF viscosity model and the two-domain Tait empirical equation are used to describe the PVT relationship. The finite-element/finite-difference/controlled-volume methods have been used to solve the above unified mathematical model and realized the integrated numerical simulation of the injection filling/post-filling process.

In the filling/post filling stage of injection molding, the flow and heat transfer of melt in the cavity belong to the flow and heat transfer of a continuum. Therefore, the flow of melt should follow the law of conservation of mass, the law of conservation of momentum, and the law of conservation of energy. Here, without considering the viscous elastic behavior of the material, the polymer melt is regarded as a generalized Newtonian fluid. In the Euler coordinate system, the melt flow satisfies the viscous fluid continuity equation, the equation of motion, and the energy balance equation. However, using only these equations can not solve the complex flow behavior in the mold cavity of wax injection molding process and it is necessary to establish a model describing the melt constitutive relation and other material behaviors in the mold cavity, such as the rheological properties (viscosity model), physical properties (density of solid–liquid two phases), and thermodynamic properties (specific heat, thermal conductivity) of the wax materials established above, as well as the physical behavior of the molding process, Only by properly simplifying the above equations can the geometric characteristics be solved.

3.2.2 Numerical Simulation of Injection Process for the Ring to Ring Part

In this section, the preparation process of the wax pattern for a ring-ring characteristic part is studied using numerical simulation. Through the interpretation of the results, the wax injection channel system can be carried out, the process parameters of wax injection can be optimized, the deformation of wax die can be predicted, and the

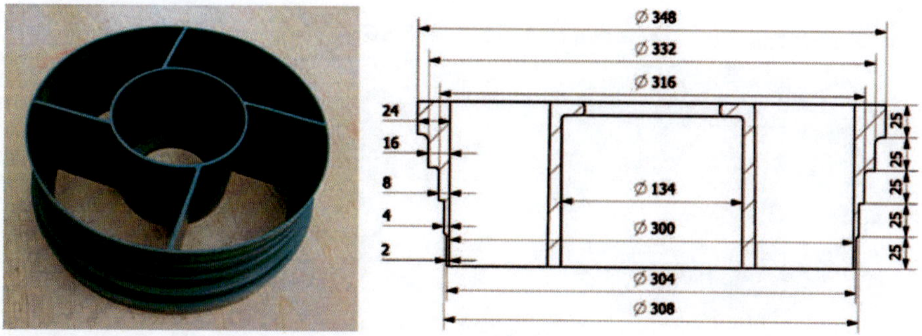

Fig. 3.9 Model and size of feature parts

shrinkage law of wax die size can be revealed. The calculated results are verified against the cavity pressure experiment.

The design of the feature is shown in Fig. 3.9.

3.2.2.1 Numerical Simulation of the Injection Process

Flow Filling System Optimization

Gate position plays a decisive role in determining the shape of melt flow front and pressure holding pressure, together with the wax mold strength and other properties of injection molding. A correct gate location can avoid some foreseeable problems. The position of the gate is based on the geometry and technical requirements of the product. The flow state, filling, shrinkage, and exhaust of the melt in the flow channel and the cavity are analyzed, in general, based on the following principles [25, 28, 38–40]:

(1) The pouring gate shall be arranged at the thicker part of the section so that the melt flows from the thick section to the thin section, ensuring that the charging mode is complete;
(2) The flow direction of the melting material filling mode should be the shortest and the flow direction change as well as the energy loss should be the minimum, in order to reduce the pressure loss;
(3) The invention is beneficial to the elimination of air in the cavity;
(4) The pouring gate is not suitable for direct vertical flushing of the melt into the cavity. Otherwise, a swirling flow is generated;
(5) The wax material is beneficial to the wax material to be filled with the cavity at a low flow rate such that the melt fracture phenomenon can be avoided and the welding marks on the wax pattern can be eliminated.

Figure 3.10 is a diagram of the filling time of wax at different gate positions. As shown in Fig. 3.10a, the gate position is at the junction of the connecting branch plate

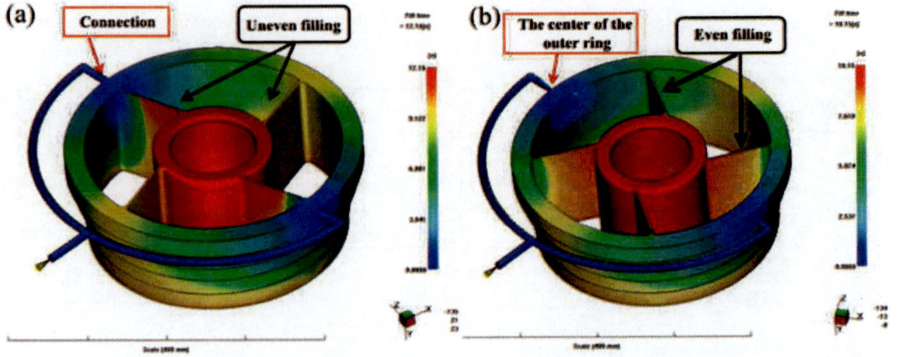

Fig. 3.10 Wax melt filling time in different channel systems. **a** Before optimization; **b** After optimization

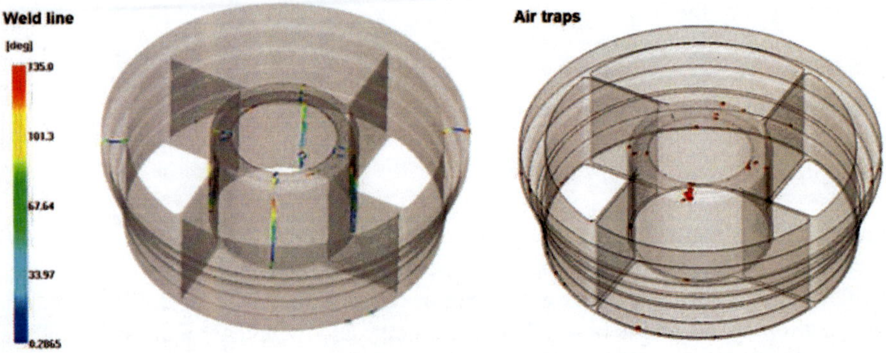

Fig. 3.11 Simulation results of the weld line and the air traps in the wax mold

and the outer ring. Since the connecting branch plate wall is thin and since the melt flows along the outer ring wall thickness, the wax melt will have a "lag effect" during the flow process of the support plate. It can be seen that the filling time of the four supports are quite different, and the uneven filling of the supports leads to uneven cooling, which will lead to the excessive warping deformation of the wax die. The optimized gate position is shown in Fig. 3.10b. The flow time of the wax melt in four branches are the same. Owing to the thin wall of the support plate, the filling time of the four branches is the same as that on the inner ring, even though the lag effect will appear on the support plate as well. Therefore, the optimized flow channel system experiences relatively uniform filling, which will not lead to excessive deformation of the wax mold size. After optimization, the filling time is short, which is 10.35 s. Thus the filling time is shortened by 2.01 s, which is more beneficial to filling.

Figure 3.11 shows the distribution of the weld line and air traps in wax die. The existence of weld line will affect the surface quality of wax die. The simulation results show that the angle of weld line is small after the optimization of the flow channel system. It can be seen from the simulation results that there is a high possibility of gas entrainment at the bottom of the outer ring and the top of the inner ring. According to the simulation results, the exhaust tank needs to be set at the bottom of the die.

Wax Mold Shrinkage Law

The shrinkage deformation is directly related to the residual stress in the injection molding process [41–44]. It is generally believed that there are two sources of residual stresses in the injection wax mold, i.e., the thermal residual stress and the flow residual stress. The former is caused by the rapid decrease of the surface temperature of the wax mold in contact with the mold, which causes the surface layer to be cured when the core of the wax mold is still in the melt state. With the gradual advance of the cooling layer from the outside to the inside, the cooling shrinkage of the internal melt is gradually limited by the external curing layer, resulting in tensile stress. The points in the article are below the glass transition temperature at different times from the higher temperature, and the shrinkage deformation experienced is not the same, resulting in a stress referred to as the thermal stress. Due to the thermal viscoelasticity of the polymer material, the product is only partially relaxed in the mold cavity after the release. The unrelaxed thermal stress is thus referred to as the thermal residual stress. This type of residual stress is due to the fact that when the flow of the wax melt is stopped, the polymer chains which are oriented in the flow direction start to develop towards the random arrangement under the action of thermal transport. Due to the rapid cooling process, the molecular orientation of the melt in the vicinity of the mold cavity wall does not have time to re-reach the equilibrium state to be frozen, resulting in the existence of the residual stress after the product is demoulded.

Figure 3.12 shows the point cloud of the wax model signature scanned by 3D laser, together with the corresponding CAD model. It can be seen that the inner and outer rings contract to the center of contraction and the contraction of the outer ring is larger than that of the inner ring.

Figure 3.13 presents the comparison between the measured and the predicted values of the flow direction of the wax mold as well as the vertical flow direction, where the flow direction is the height direction and the vertical flow direction is the thickness direction. The contraction displacement of the wax mold in the height direction is 0.4–1.2 mm; as the height increases, the contraction distance becomes greater. In the thickness direction, as the wall thickness increases, the shrinkage value of the thickness increases sharply, with a contraction displacement of 0.2–1.3 mm.

As can be inferred from Fig. 3.13, the shrinkage in the direction of flow of the wax is much smaller than that in the thickness direction because the thermal residual stress is an order of magnitude greater than the flow residual stress [42]. At the same time, for this reason, the shrinkage behavior of the wax mold shows anisotropy of the shrinkage in the flow direction of the wax and the direction of the vertical flow [41, 42].

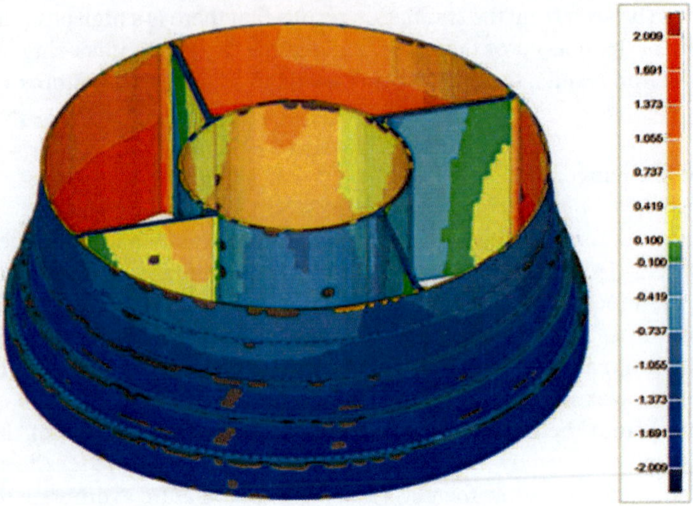

Fig. 3.12 Comparison between scanned point cloud and actual CAD model

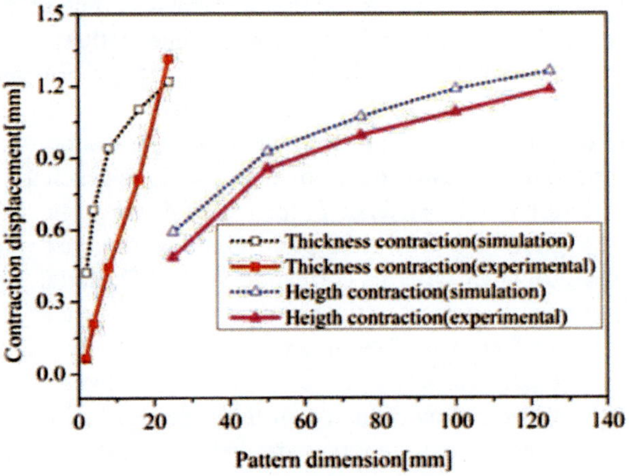

Fig. 3.13 Comparison between measured and predicted shrinkage displacements of wax mode flow direction and vertical flow direction

Figure 3.13 also shows that the shrinkage trend of wax mode predicted by numerical simulation is basically consistent with the measured value and the numerical simulation can accurately predict the shrinkage deformation trend of wax mode. In the height direction, the predicted deformation displacement value is about 0.1 mm larger than the measured value, while in the thickness direction, the predicted value is about 0.3 mm larger than the measured value. The numerical simulation shown in Fig. 3.14 can also be used to predict the deformation of wax mould. Since the analysis method is similar, it will not be repeated here.

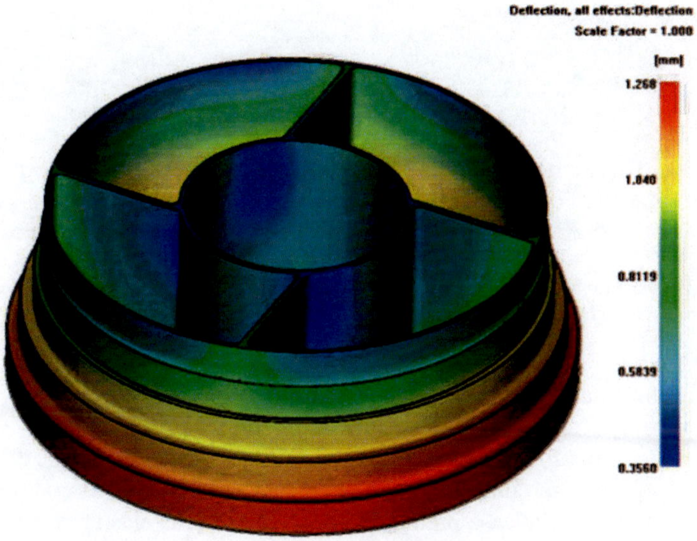

Fig. 3.14 Predicted wax die warping deformation diagram/mm

3.2.2.2 Numerical Simulation of the Storage Process

Modern engineering utilizes many types of materials, such as concrete, plastics (reinforced or unreinforced plastics), and some biological tissues, of which the underlying stress–strain relationship is related to time, a phenomenon called viscoelasticity. The polymer shows obvious viscoelastic deformation, which is essentially a deformation behavior between elasticity and viscosity. The stress in viscous elastic materials is a function of strain and time. It is described that when the strain remains unchanged, the stress will decrease with the increase of time, which is called relaxation. When the temperature is constant and deformation remain unchanged, the stress in the polymer decreases gradually with the increase of time, which is called stress relaxation. Wax usually shows linear viscoelastic properties, and there will be stress relaxation in the process of wax mold storage, which means its size will change with the increase of time.

The melting die will continue to shrink within a few hours after removal, and most of them will not stabilize until 24 h later. In order to prevent the shrinkage and deformation of the wax mold during storage, the wax mold preparation and storage should be kept at constant temperature and humidity. For wax moulds with special requirements, they should be placed in the tire mould after taking the mould, store it for a period of time, take it out after the dimension is stable, or make anti-deformation tension tendons. For some plate-like parts, the plate can be clamped to correct the deformation. Although scholars have done some research on the storage conditions of wax mould, the deformation size of wax mold in storage is not quantitatively described, and the deformation mechanism of wax mold is not described. In this

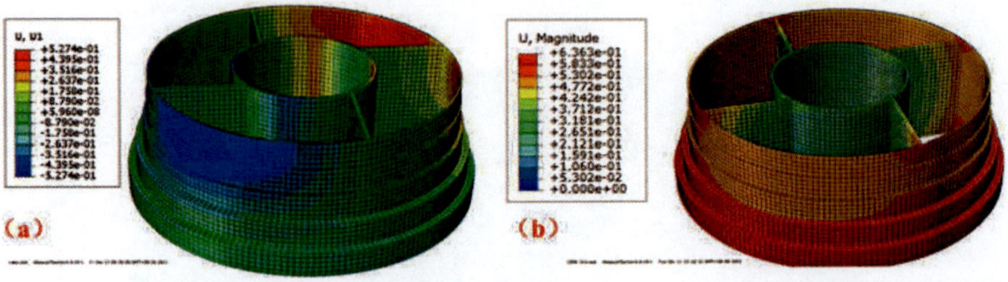

Fig. 3.15 Displacement distribution of 1000 s elastic deformation of wax mode: **a** elastic deformation, **b** viscoelastic deformation

section, the finite element software ABAQUS is used to predict the viscoelastic stress relaxation deformation size of the wax die during storage.

Figure 3.15a is the size deformation result of 1000 s wax mold elastic deformation prediction under elastic condition and Fig. 3.15b is the size deformation result of 1000 s after wax mold release predicted under the viscous elastic model. It can be seen that the maximum deformation value of viscoelastic deformation is 0.1088 mm larger than that predicted by the elastic model. The results of viscoelastic prediction show that the size shrinkage of inner ring is smaller whereas that of outer ring is larger and with the increase of thickness, the deformation of wax die increases. Under the condition of elastic model, the deformation gap between the wax die inner and outer ring and its wall thickness is not predicted. As a result, the viscoelastic model is more suitable for predicting the deformation of wax mold storage.

Figure 3.16 shows the displacement field and temperature field of wax mode after 1000 s obtained using the viscoelastic model. It can be seen that with the prolongation of the storage time from 1000 to 10000 s, the maximum deformation size of the wax mold increases from 0.6362 mm to 0.9883 mm, which increases by 55.34% due to the stress relaxation of the wax die. From the temperature field of Fig. 3.16b, it can be seen that the thick wall of wax die is not completely cooled to room temperature after 1000 s. This is because the thermal conductivity of the wax mold is relatively

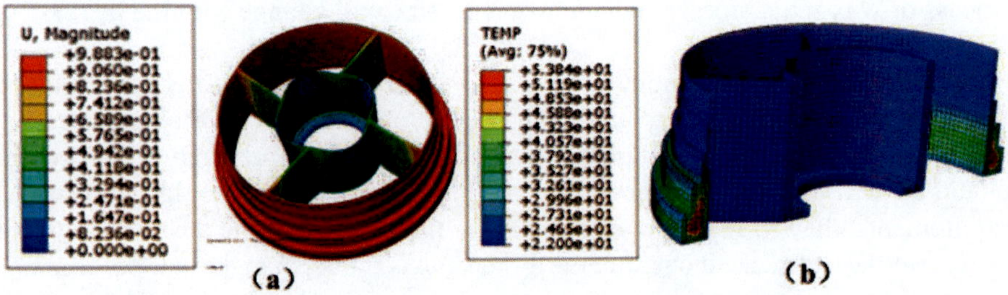

Fig. 3.16 Deformation and temperature of 1000 s after wax mold demoulding: **a** viscoelastic deformation of wax die, **b** temperature distribution

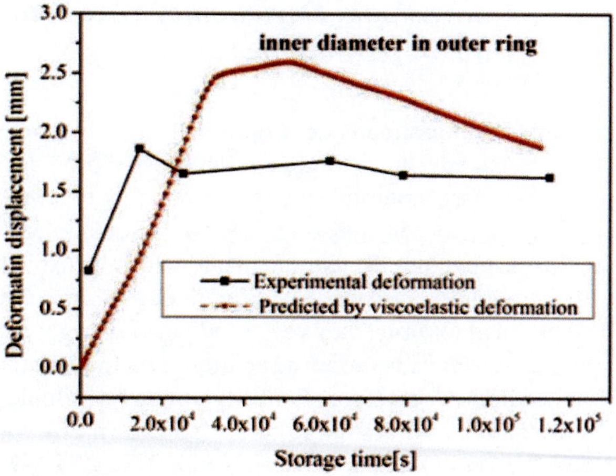

Fig. 3.17 The predicted and experimental results of the viscoelastic deformation of 100,000 s after the departing of the wax

low, which results that the wax melt in the thick wall part of the wax mold has not been completely cooled and thus continues to shrink.

Figure 3.17 is a predicted and measured deformation dimension of 100,000 s after the inner diameter of the outer ring of the wax mould is demoulded. It can be seen that the deformation dimension of the viscoelastic prediction is about twice as large as the measured result. During the actual forming process of the wax mould, the wax mould is continuously pressed into the mould cavity during the pressure maintaining of the wax mould, and the shrinkage of the wax mould is compensated. However, when the viscoelasticity is predicted, the deformation displacement is not taken into consideration during the pressure maintaining process of the wax mold, and the predicted deformation displacement therefore tends to be larger. For example, the deformation results of the Sabau [45] and other viscoelastic projections are about 2.5 times larger than the measured value. The experimental results of Chen [46] and others show that the longer the wax mold is placed, the greater the shrinkage of the wax mold, and the size of the wax mold tends to be stable after a certain period of time. The greater the wax mold thickness, the longer the wax mold shrinkage tends to be stable. It can be seen from the measured and predicted deformation displacement of the wax pattern of Fig. 3.17 before 60,000 s, the deformation size of the wax mold is increased with the increase of the storage time, and the deformation size of the wax mold after 100,000 s has a tendency to decrease, which, nevertheless, is basically stable.

3.2.3 *Dimensional Deviation of Wax Patterns with Water-Soluble Cores*

As investment casting is developing towards lightweight, complex hollow thin-wall structure castings have emerged. For these closed cavities, hollow structures, small outlets and large internal cavities, or parts that are not allowed to have draft angles, such as metal molds, they often need to be assembled from multiple pieces. Not only are the design complicated, the manufacturing cost is high, and the cycle is long, but also Demoulding is extremely difficult or even impossible. Even if some of them can be demolded forcibly, they will usually cause larger residual stresses in the structural parts, and even cause some structural parts to be damaged. As a result, the castings of these hollow structures typically employ a soluble core to form such an internal cavity. The so-called soluble core, also known as a water-soluble core, refers to as a preset wax core that can be dissolved in water. The water-soluble core is a common technique for forming a large-area thin-wall wax mold. Now, the research on the water-soluble core mold materials both domestically and abroad mainly includes the urea core, the polyethylene glycol (PEG) core, the polyvinyl alcohol (PVA) core, the starch core, the polyacrylic acid (PAA) core, the polyethylene terephthalate (PVP) core, the PEX water-soluble core, and the Aquacore core, The water-soluble core mold material is made of the Aquapor core and other adhesives.

During the injection process, the flow relationship of wax melt along the core is not the same everywhere, so the pressure distribution on the core is uneven. The uneven pressure distribution will cause the core to shift during the injection process, especially when the core is injected on the side of the core. Once the deviation occurs, the core will have a certain amount of eccentricity relative to the wax melt. If the eccentricity exceeds the allowable tolerance range, the size will be unacceptable.

3.2.3.1 Simulation Analysis

In order to determine whether or not the core deviation will occurr during the process of injection molding of large and complex thin-wall wax mold, a large ring-ring casting with the inner ring and part of outer ring is selected for simulation and analysis. For the convenience of description, it is called oblique support plate. The core migration analysis provides detailed information on the movement of the core and the interaction between the core and the wax melt during the process of flow, which allows the product designer to use this information to reduce the core deviation through, e.g., adjusting the design or processing parameters of the parts. Figure 3.18 shows the migration results of water-soluble core predicted using numerical simulation. It can be seen that the location of migration and deformation of the water-soluble core is mainly near the gate areas A and B region near the convex platform. The maximum offset of the core in area A is 0.4464 mm and the migration direction is facing gate filling direction, whereas the maximum offset value in area B is 0.2234 mm and the migration direction is backward gate filling direction. Obviously, the deviation

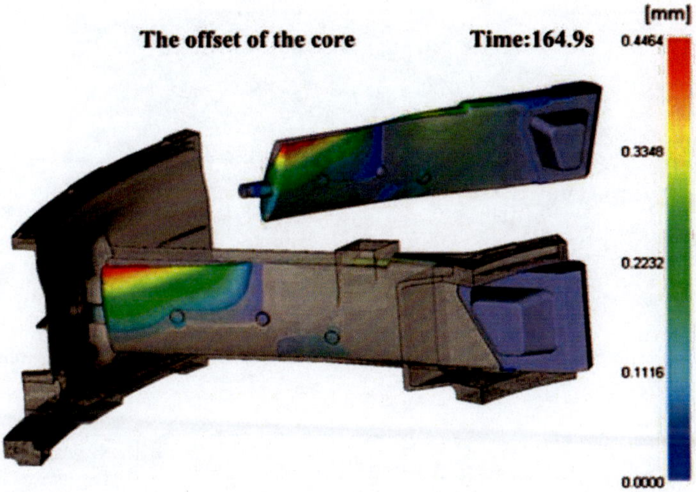

Fig. 3.18 Predicted core deflection location for the wax pattern

direction in area B is just the opposite of that in area A. In addition, the wall thickness of wax mold in area B is hollow thin-wall structure, which is not easy to be repaired with repair wax, resulting the hollow wall thickness of final castings being unpredictable and uncontrollable.

3.2.3.2 Experimental Comparison

In order to verify the prediction on core deviation from numerical simulation, the injection molding experiment of wax mold is carried out under the same injection process parameters, and the wall thickness of A, B area are measured by high precision ultrasonic thickness gauge. The data shows that there are the same regular size over the difference in the B region of 12 oblique support plate wax molds. The direction of migration in A and B regions are also consistent with the numerical simulation results. Therefore, the numerical simulation are overall consistent with the experimental results and the core deviation occurs in the water-soluble core, resulting in a change in the wall thickness of the B region.

Figure 3.19 shows the profile analysis of A and B regions. The maximum wall thickness of A area is 2.36 mm and the minimum wall thickness of B area is 1.87 mm. The wall thickness of B area exceeds the dimensional tolerance range of castings and the core offset area predicted from numerical simulation is consistent with the direction and experiment. Figure 3.20 shows the minimum and maximum wall thickness of the hollow support plate A predicted from numerical simulations and tested using experiments. Compared with the experimental results, the measured value in region A is 3.66% smaller than the predicted value and the maximum offset size in region B is 3.1% smaller than the predicted value. The predicted value is slightly larger than

Fig. 3.19 The experimental wax pattern and the real thickness in the deflection regions

Fig. 3.20 Comparisons of the experimental and simulated maximum and minimum thickness of thin-walled

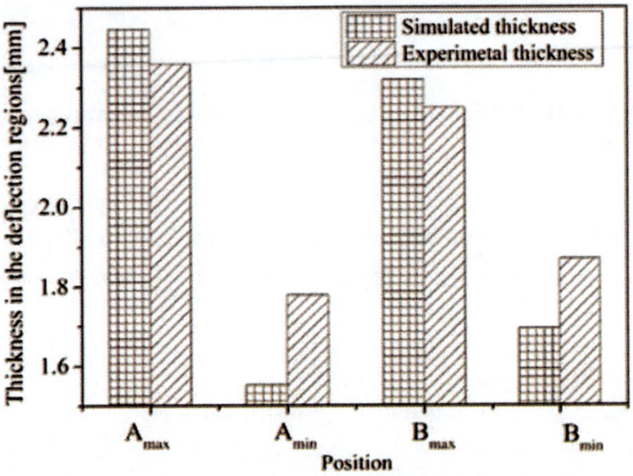

the measured value [20]. From the results of Giacomin research, the predicted value of the core deviation may be too large to consider the viscous elastic deformation of polymer melt during cooling.

3.2.3.3 Mechanism Analysis

The deviation of the core is mainly caused by the following three factors: (1) the size error caused by the injection process of the water-soluble core itself and the deformation of the mold, (2) the deformation caused by the insufficient strength at the fixed end of the water-soluble core wax mold under high injection pressure, and (3) the deformation caused by the pressure difference between the two sides of the core, which is caused by the unreasonable gate position or the change of the thickness of the parts. Due to the manufacturing error between the actual water-soluble core and the theoretical model, it is inevitable to have an uneven distribution of wax wall thickness, which then affects the wall thickness forming accuracy of precision casting

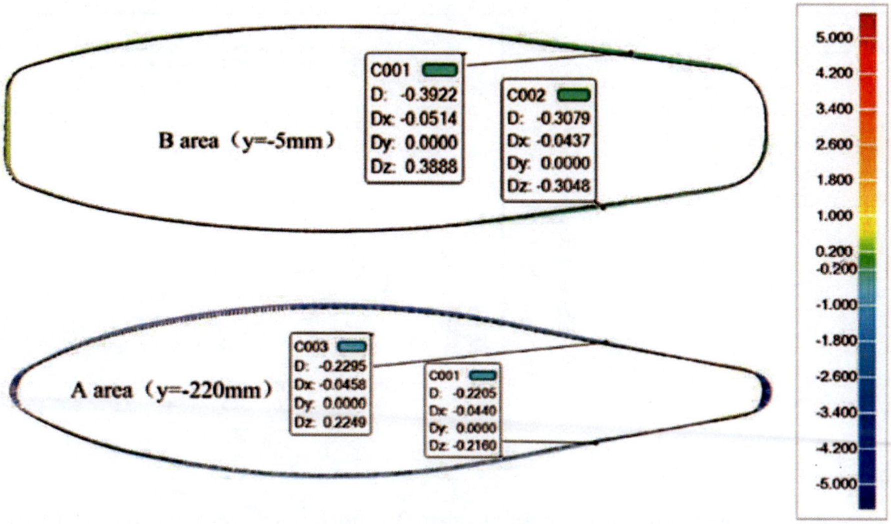

Fig. 3.21 Cross-section contour comparison between the scanned part and design model

castings. Therefore, it is necessary to match the profile data of the actual water-soluble core with the theoretical model, in order to eliminate the manufacturing error and assembly error of the water-soluble core. As shown in Fig. 3.21, for the comparison of the 2D size of the scanned actual water-soluble core with the theoretical model, the overall shrinkage and uniform shrinkage occur in the A and B regions where the water-soluble core is offset and deformed. Therefore, the size error of water-soluble core preparation is excluded, which is the cause of the super-poor hollow thin-wall size of the support plate. After the experiment, the surface of the fixed end of the water-soluble core tested is in fine quality, with no fracture nor deformation due to insufficient strength.

Obviously, the deviation of the thickness dimension of the thin wall is due to the large pressure difference on both sides of the core. The hollow branch plate is a deep cavity structure with a length of 300 mm and a width of 30 mm and the wall thickness on both sides are 2–3.5 mm. In the molding process, the core is easily offset by uneven pressure, resulting in a product that cannot be molded or does not meet the design requirements in terms of the wall thickness.

Figure 3.22 shows the relationship between the core offset and the time of the nodes in the A and B regions. It can be seen that because the viscosity of the wax material is low, the injection pressure in the filling process is small and the structure on both sides of the water-soluble core is relatively symmetrical. Note that during the process of injection filling of the wax material, the core offset is about 0.04 mm. In addition, the maximum offset of the A region occurs at the moment of switching between filling and pressure-holding, at which the pressure holding pressure increases instantly, resulting in the pressure on both sides of the A region near the gate losing balance and hence a large core deviation. As the pressure holding

Fig. 3.22 Core deflection displacement as a function of time in regions A and B

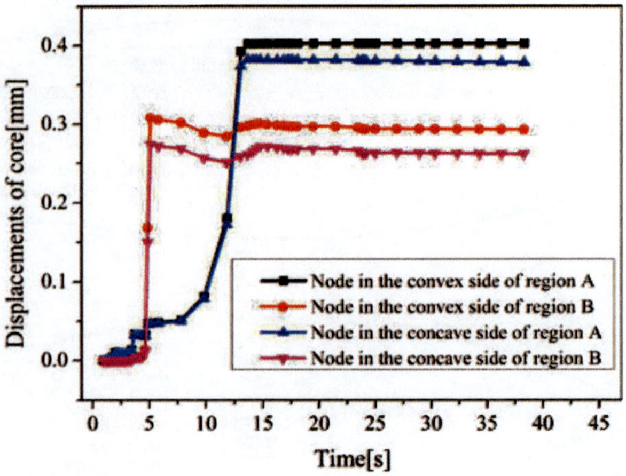

continues, the migration size remains virtually unchanged. Finally, owing to the protruding area near the support plate, there are large unmelted wax melts in the convex platform area after filling. Once the pressure holding begins, the pressure imbalance in the A region is transferred to the B region. With the holding pressure continuously applied, the wax melt in the convex platform area solidifies and the offset of the core in area B increases gradually. However, after the thin-wall wax mold has solidified in 14.01 s, the migration in the B region remains nearly unchanged.

These results show that during the wax injection molding process, the core deflection of the water-soluble core mainly occurs during the pressure holding process, rather than the filling process. It can thus be inferred that as the holding pressure increases, the core offset will also increase.

3.2.3.4 Control and Optimization

In order to overcome the phenomenon of serious eccentricity and difficulty in demolding in the process of production and trial molding. The following measures were adopted in the mold design and molding process [30]: (1) The core is offset, the root cause is insufficient stiffness, so the strength of the core must be improved; (2) Since the hollow thin-walled structure cannot be changed, then Without affecting the product molding, add positioning pins in the thin-walled parts of the small cores; add process positioning pins to the mold and leave process holes on the product, so that when the melt enters the cavity, the core can rely on, When the core is subjected to lateral force, the left and right swing deformation and bending are reduced; (3) Increase the size of the fixed part of the core and design a reasonable feeding position. (4) Optimize the wax injection molding process.

The results show that increasing the strength of the core is not effective in reducing the core offset and enhancing the strength of the water-soluble core may affect its

Fig. 3.23 The core pattern with an added fixed pin

water solubility. Expanding the size of the fixed end of the core will result in a large process hole that cannot be filled. As shown in Fig. 3.23, six symmetrical positioning pins (2 mm in diameter and 2 mm in height) are added to the two sides of the water-soluble core. From the experimental results, the process positioning pin can significantly improve the core offset.

The core offset is caused by the pressure difference during the holding process. For this reason, reducing the holding pressure reduces the core offset size. Figure 3.24 shows the two-stage low holding pressure process curve. After the filling is completed, the holding pressure is instantly reduced to 80% of the injection pressure, which ensures the dimensional stability and reduces the core offset. Finally, after the process optimization and the addition of the locating pin, the core offset size is controllable. From the optimization results in Fig. 3.25, it can be seen that the final large wax mold hollow thin-wall size is maintained between 1.93 and 2.07 mm, within the dimensional tolerances.

Fig. 3.24 The optimization packing pressure profile

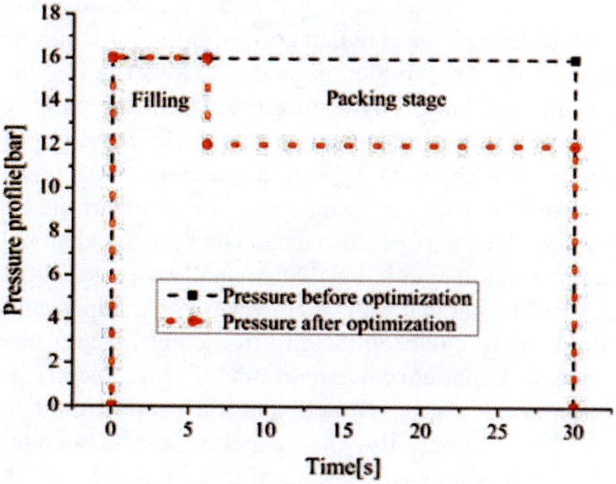

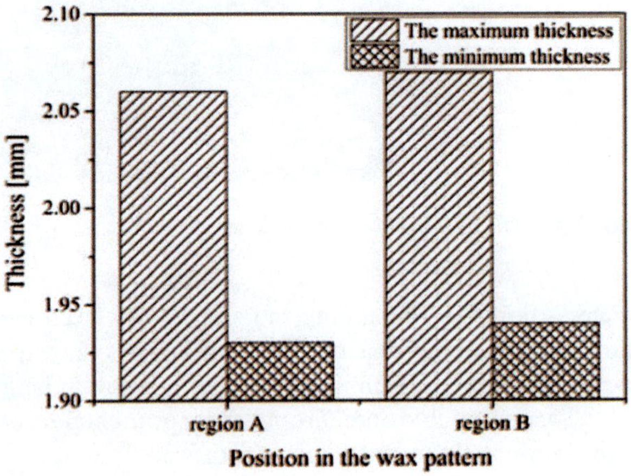

Fig. 3.25 Optimized thickness in areas A and B

3.3 Shrinkage Compensation Rate of Investment Mold Surface

Since the investment casting process is complicated, the final casting is obtained through the processes of molding, shell making, pouring, post-processing, etc. Therefore, controlling the dimensional accuracy of the investment casting is a systematic project involving all aspects of the casting process. Figure 3.26 shows a simplified diagram of the important factors affecting dimensional accuracy during investment casting [47]. It can be seen that the factors affecting the dimensional accuracy of investment castings are categorized into four major aspects: casting structure shape, size, mold, and production process. However, the dimensional accuracy of the final casting is achieved by setting the size of the mold to compensate for dimensional changes during the casting process. The principle of the mold face compensation design is to give an appropriate amount of reverse deformation at the deformation site to offset the shrinkage deformation of the casting during solidification and cooling. It is reasonable to expect that the dimensional error caused by the shrinkage compensation rate is small for small castings, but it is quite large and cannot be ignored for large castings. However, the compensation shrinkage rate design of the mold cavity has been lacking in theoretical guidance and still stays in the design stage of "experience + experiment", which mainly includes empirical data method, formula calculation method, and trial production method.

Many scholars use the "experience data method" to guide the design of the shrinkage compensation rate of casting molds. Table 3.3 shows the empirical data of shrinkage accumulated in the production of carbon steel and alloy structural steel [48]. This empirical design method is also feasible for the same casting, but the castings of different alloys and size shapes obviously cannot meet the design requirements. Therefore, the total shrinkage of the casting is preferably based on the specific

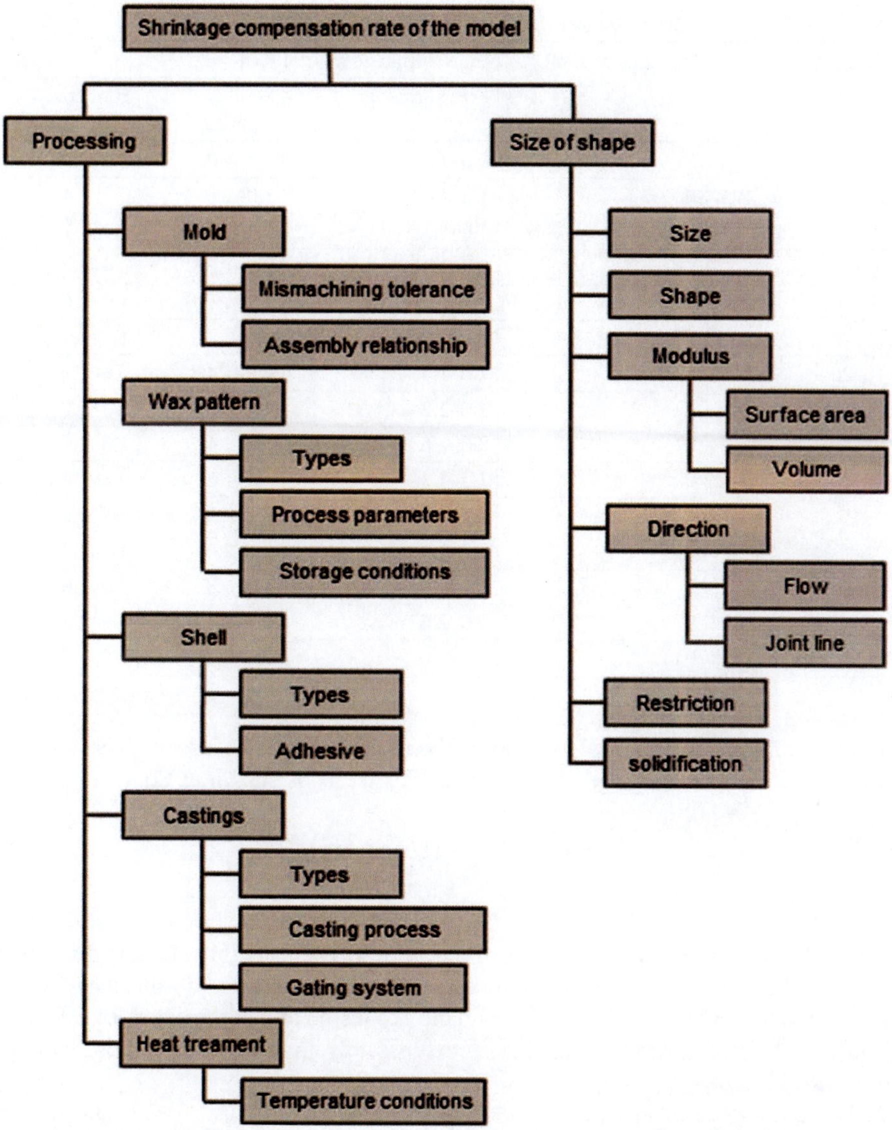

Fig. 3.26 Schematic diagram of important factors affecting shrinkage during investment casting

alloy grade and casting temperature, and the recommended empirical values are appropriately modified.

Jiang et al. [48, 49] believed that only by strictly controlling the size of the mold cavity was it possible to obtain a casting with dimensional accuracy. This requires a correct assignment of the shrinkage of each size of the casting when designing the

Table 3.3 Empirical data on total shrinkage of carbon steel and alloy structural steel

Thickness/mm	Catalogs of shell	Line shrinkage in total K/%		
		Free shrinkage	Semi-free shrinkage	Hindered shrinkage
1–3	I	0.6–1.2	0.4–1.0	0.2–0.8
	II	1.2–1.8	1.0–1.6	0.8–1.4
	III	1.6–2.2	1.4–2.0	1.1–1.7
3–10	I	0.8–1.4	0.6–1.2	0.4–1.0
	II	1.4–2.0	1.2–1.8	1.0–1.6
	III	1.8–2.2	1.6–2.2	1.3–1.9
10–20	I	1.0–1.6	0.8–1.4	0.6–1.2
	II	1.6–2.2	1.4–2.0	1.2–1.8
	III	2.0–2.6	1.8–2.4	1.5–2.1
20–30	I	1.2–1.8	1.0–1.6	0.8–1.4
	II	1.8–2.4	1.6–2.2	1.4–2.0
	III	2.2–2.8	2.0–2.6	1.7–2.3
>30	I	1.4–2.0	1.2–1.8	1.0–1.6
	II	2.0–2.6	1.8–2.4	1.6–2.2
	III	2.4–3.0	2.2–2.6	1.9–2.5

mold. Therefore, the size calculation of the mold cavity is necessary. The formula for calculating the shrinkage rate proposed by Lu [50] is shown in Eq. (3.12):

$$L = (1 + \Delta l) \pm (1/2 \sim 3/4)\Delta M \tag{3.12}$$

where L is the compression size, l is the casting blank size, $\boldsymbol{\Delta l}$ is the total shrinkage, and ΔM is the dimensional tolerance of the casting. There are three factors that affect the total shrinkage of the casting in investment casting, namely, the shrinkage of the wax mold, the shrinkage of the alloy, and the expansion and deformation of the shell. Consider the effects of these three aspects separately and then superimpose them as indicated in Eq. (3.13):

$$K = K_1 + K_2 + K_3 \tag{3.13}$$

where $\boldsymbol{K}$ (%) is the total shrinkage, $\boldsymbol{K_1}$ (%) is the shrinkage of the alloy, $\boldsymbol{K_2}$ (%) is the shrinkage of the mold, and K_3 (%) is the expansion and deformation rate of the shell (generally, $\boldsymbol{K_3}$ is a negative value). Although there have been a lot of tests and a lot of data for K_1, K_2, and K_3, it has been shown that these calculations are not accurate, which is mainly due to the following reasons. First of all, these three factors are all related to the structure of the casting and thus play restriction on each other. The structure of the casting has a great influence on the total shrinkage rate and is difficult to evaluate. Whether the shrinkage of the casting is hindered or not is related to the shape of the casting. The different parts of the complex casting have different resistance, semi-resistance, and unimpeded states. Therefore, it is necessary to analyze and assign different shrinkage rates to obtain high-precision castings. The casting structure is influenced by both the wall thickness and shape of the casting. The influence of the wall thickness of the casting is regular, and the thicker the wall, the larger the shrinkage rate. Therefore, among the three factors of total shrinkage, the thicker the wall, the more the shrinkage of the alloy and the shrinkage of the wax. In very rare cases, such as two thin-walled frames, the expansion and deformation of the shell predominate, often making the total shrinkage K negative. Second, each factor is affected by the specific process operation, so the value of total shrinkage varies with specific conditions and cannot be encapsulated in one formula. In summary, it can be seen that due to the complicated casting process and numerous influencing factors, it is difficult to select the total shrinkage rate. Therefore, the mold shrinkage compensation rate obtained by the "formula calculation method" also contains a large error, which eventually leads to an oversize of the casting.

Therefore, it is only possible to test specific parts under different process conditions from the actual situation. The product's shrinkage compensation and gating system are adjusted through trial production. Thus, it is usually necessary to adjust the shrinkage rate by repeatedly reworking the mold. Although the "trial production method" can finally obtain a suitable dimensional shrinkage compensation rate, its long production cycle and high cost seriously restrict the development of investment casting.

For important parts, always test first to find the appropriate total shrinkage under field conditions, then design and manufacture the mold, and finally put into real production. However, Zhang et al. [51] of Northwestern Polytechnical University proposed a simple and efficient reverse casting design method for precision casting molds, that is, reverse adjustment of characteristic parameters. The method integrates CAE displacement field simulation and CAD 3D modeling technology to avoid the problem of surface splicing caused by completely relying on grid information to solve the mold cavity. Its application can guarantee the dimensional accuracy and shape accuracy of precision casting, which realizes the computer-aided design and casting. Nevertheless, it only considers the solidification shrinkage deformation of the alloy during investment casting and the numerical simulation of the displacement field of the whole process is not carried out. Since it is the calculation of the displacement field based on the casting that leads to the final compensation, there is a certain error in the shrinkage rate.

The precise determination and setting cycle of precision casting mold design has always been the bottleneck problem that restricts the manufacture of high-temperature castings for aviation engines. When the traditional "experience + experiment" trial-and-error legal molds require several repairs, the CAD/CAM software and digital modeling technology are used to establish a mold surface optimization design system based on the size error displacement field of the investment casting process, which can provide a solid theoretical foundation and technical support for the rapid near-net-shape investment casting.

References

1. A.S. Sabau, S. Viswanathan, Material properties for predicting wax pattern dimensions in investment casting. Mater. Sci. Eng. A **362**, 125–134 (2003)
2. T.T. Zhou, C.G. Han, Z.T. Chai et al., Constitutive equation of investment casting materials and its application. Casting **50**, 78–80 (2001)
3. Y.Z. Xu, *Polymer Structural Rheology* (Sichuan Education Press, Chengdu, 1988).
4. Q.M. Qiu, Q.C. Wu, G.Z. Zhou, A numerical calculation method for quasi-relaxation time spectrum of materials from static stress relaxation experiment. Acta Polymerica Sinica **5**, 535–538 (1994)
5. R.J. Hao, G.B. Che, C.Y. Shen et al., Calculation of relaxation time spectrum of polymer melt. J. Zhengzhou Univ. Technol. **22**(4), 90–92 (2001)
6. Z.Y. Zhang, B.N. Liang, Factors affecting casting dimension accuracy in die making process. Mech. Res. Appl. **15**, 15–16 (2002)
7. X.P. He, Wax mold precision in investment casting. Spec. Cast. Nonferrous Alloys **24**(1), 156–159 (1999)
8. B.R. Wu, L.L.Y. Xiao, Wax mold manufacturing in precision casting. Casting **12**, 36–38 (1999)
9. R.M. Liu, X.H. Zhang, Z.F. Guo et al., Effects of EVA on properties of die materials for precision casting blades. Mater. Eng. **11**, 17–18 (1994)
10. S.J. Hao, Wax for silica sol precision casting and application of jx001-3 mold. Casting **58**, 295–298 (2009)
11. Haweifeide (2003) Mechanical properties of melt casting materials and the role of fillers, in *Proceedings of the 3rd International Conference On Nonferrous Alloys and Special Casting*, Shanghai
12. Z.T. Chai, T.T. Zhou, C.G. Han, Analysis of factors affecting rheological properties of die materials. Cast. Technol. **2**, 26–27 (2001)
13. J.C.Gebelin, M.R. Jolly, A.M.Cendrowicz et al., Simulation of die filling for the wax injection process: part I. Models for material behavior. Metall Mater Trans B**35**, 755–759 (2004)
14. A.Kelkar, Wax dimensional control with design of experiments, in *Proceedings of the 39th Annual Technical Meeting*. InvestmentCasting Institute, 4, 1991
15. M. Horacek, Accuracy of casting manufactured by the lost wax process. Foundry Trade J. **10**, 12–15 (1997)
16. M. Horacek, S. Lubos, Influence of injection parameters to the dimensionalstability of wax patterns, in *Proceedings of the Ninth World Conference on Investment Casting, San Francisco*, California, USA, 1996.
17. W. Bonilla, S.H. Masood, P. Iovenitti, An investigation of wax patterns for accuracy improvement in investment casting parts. Int. J. Adv. Manuf. Technol. **18**, 348–356 (2001)
18. A. Bakharev, Z. Fan, S. Han et al., Prediction of core shift effects using mold filling simulation, in *ANTEC Conference Proceedings*, vol. LXII (Society of Plastics Engineers, Brookfield, NJ, 2004) pp. 621–625.
19. D.G. Ahn, D.W. Kim, Y.U. Yoon, Optimal injection molding conditions considering the core shift for a plastic battery case with thin and deep wall. J. Mech. Sci. Technol. **24**, 145–148 (2010)
20. A.J. Giacomin, A.J. Hade, L.M. Johnson et al., Core deflection in injection molding. J. Nonnewton. Fluid Mech. **166**, 908–914 (2011)
21. Y.C. Chen, H.C. Liao, S.C. Tseng et al., Core deflection in plastics injection molding: direct measurement, flow visualization and 3D simulation. Polym. Plast. Technol. Eng. **50**, 863–872 (2011)
22. Y.Y.Chou, W.H.Yang, A.J.Giacominet al..Validating 3D numerical simulation of core deflection, in*Proceedings of the 24th Meeting of the Polymer Processing Society*, Salerno, Italy, June 15–19, 2008, pp. 1–5
23. V.F. Okhuysen, K. Padmanabhan, R.C. Voigt, Tooling allowance practices in the investment casting industry, in *Proceedings of the 46th Annual Technical Meeting*, Investment Casting Institute, 1998

24. P.K.D.V. Yarlagadda, T.S. Hock, Statistical analysis on accuracy of wax patterns used in investment casting process. J. Mater. Process. Technol. **138**, 75–81 (2003)
25. V. Leo, C.H. Cuvelliez, The effect of the packing parameters, gate geometry, and mold elasticity on the final dimensions of a molded part. Polym. Eng. Sci. **36**(15), 1961–1971 (1996)
26. G.H. Hu, Z.S. Cui, Effect of packing parameters and gate size on shrinkage of aspheric lens parts. J. Shanghai Jiaotong Univ. **15**(1), 84–87 (2010)
27. M. Zhai, Y.X. Gu, C.Y. Shen, Optimization design of the number and position of injection mold gate. Acta Chemica Sinica **54**(8), 1141–1145 (2003)
28. W. Fu, S.J. Fan, H. Zhang, Selection of gate position and structural form in injection mold design. Eng. Plast. Appl. **35**(10), 60–63 (2007)
29. W.G. Wang, B.S. Tian, Y.C. Tian, *Design Techniques and Examples of Plastic Injection Mould* (Chemical Industry Press, Beijing, 2004)
30. R.Z. Qiang, Several measures to overcome core deformation. Die Ind. **11**, 32–35 (1992)
31. J. Zhang, H.W. Ye, K.W. Li, Numerical simulation of mold filling process for wax pattern of the impeller in investment casting. Appl. Mech. Mater. **80–81**, 965–968 (2011)
32. Y. Shi, Formulation and process parameter optimization of temperature mold in precision casting. Master's thesis. Hefei, Hefei University of Technology
33. H.Y. Zhao, T.F. Tong, Study on linear variation of mold shell in investment casting. Spec. Cast. Nonferrous Alloys **1**, 17–20 (1990)
34. R. Sadegh, R.M. Reza, A. Javad, Design and manufacture of a wax injection tool for investment casting using rapid tooling. Tsinghua Sci. Technol. **14**, 108–115 (2009)
35. M. Ito, T. Yamagishi, Y. Oshida, Effect of selected physical properties of waxes on investment s and casting shrinkage. J. Prosthet. Dent. **75**, 211–216 (1996)
36. Z.T. Chai, T.T. Zhou, C.G. Han, Rheological properties of die materials and its application in wax injection process formulation. Thermal Process. Technol. **6**, 23–24 (2001)
37. C.F. Liu, W.K. Bao, J.L. Xu, Shear modulus approximation function and exact numerical integral expression. Shanghai Aerosp. **5**, 26–28 (2003)
38. Z.X. Zhong, *Plastic Injection Molding Technology* (Guangdong Science And Technology Press, Guangzhou, 1995)
39. W.L. Chen, *Design and Manufacture of Practical Plastic Injection Mould* (Machinery Press, Beijing, 2000)
40. M.C. Song, X.N. Jing, D.Y. Zhao, Study on shrinkage characteristics of injection molding of fan products. Eng. Plast. Appl. **2**, 31–33 (2006)
41. X. Chen, Y.C. Lam, D.Q. Li, Analysis of thermal residual stress in plastic injection molding. J. Mater. Process. Technol. **101**, 275–280 (2000)
42. W.F. Zoetelief, L.F.A. Douven, A.J.I. Housz, Residual thermal stresses in injection molded products. Ploym. Eng. Sci. **36**(14), 1886–1896 (1996)
43. K.K. Kabanemi, H. Vaillancourt, H. Wang et al., Residual stresses, shrinkage, and warpage of complex injection moulded products: numerical simulation and experimental validation. Ploym. Eng. Sci. **38**(1), 21–37 (1998)
44. W.B. Young, Residual stress induced by solidification of thermoviscoelastic melts in the postfilling stage. J. Mater. Process. Technol. **145**, 317–324 (2004)
45. N. Cannell, A.S. Sabau, *Predicting Pattern Tooling and Casting Dimensions for Investment Casting, Phase II* (Tennessee, Oak Ridge National Laboratory, 2005).
46. Y.H. Chen, J.D. Duan, Dimensional stability analysis and improvement of investment casting, in *10th Annual Meeting of China Foundry Association*, Beijing, 2012
47. H. Wan, H. Yu, Z.F. Xu et al., Study on transfer law of surface roughness and dimension accuracy of gypsum mold, mold shell and casting. Therm. Process. Technol. **39**(17), 71–73 (2010)
48. B.J. Jiang, *Practical Investment Casting Technology* (Liaoning Science And Technology Press, Shenyang, 2008).
49. L.M. Cao, X. Tang, Y. Zhang et al., Advanced high temperature alloy near-net form investment casting technology. Spec. Cast. Nonferrous Alloys **26**(3), 238–243

50. J.G. Lu, X.F. Zhang, X.T. Lu, Research on some processes in investment casting. Modern Cast Iron **6**, 67–69 (2006)
51. D. Zhang, W.H. Zhang, M. Wan et al., A reverse design method based on displacement field simulation and feature parameter extraction for precision casting die. Acta Aerologica Sinica **27**(3), 509–514 (2006)

Chapter 4
Preparation Process of Ceramic Shells

Fei Li and Fei Wang

Investment casting is one of the most critical technologies that produce can complex components used in aircraft engines [1]. The preparation of the ceramic shell is one of the essential processes in investment casting. The dimensional accuracy, surface roughness, and even internal quality of casts are all closely related to the ceramic shells [2]. In recent years, the structure of casting parts for aircraft engines has become more and more complex, and the surface quality requirements have become much higher, which demands higher standards for the performance of ceramic shells. How to further enrich and optimize the shell material system formulates a scientific and reasonable shell-making process. Besides, how to prepare a ceramic shell with excellent performance has become one of the critical points in the development of modern investment casting technology.

4.1 Overview of Ceramic Shell

4.1.1 Composition, Structure, and Performance Requirements of Ceramic Shells

The term investment casting derives from the characteristic use of mobile ceramic slurry, or 'investments,' to form a shell with an extremely smooth surface [3, 4]. These are replicated from precise patterns and transmitted in turn to the casting. Investment casting allows dimensionally accurate components to be produced and is a more cost-effective alternative than forging or machining since waste materials are kept to a minimum. Production of the investment casting ceramic shell is a crucial part of the whole process. The necessary steps in the production of an investment cast component using a ceramic shell are shown in Fig. 4.1. First, multi-component slurries are prepared, which are composed of a fine mesh refractory filler system and a colloidal binder system. A pattern wax is then dipped into the slurry, sprinkled with

B. Sun et al., *Precision Forming Technology of Large Superalloy Castings for Aircraft Engines*, https://doi.org/10.1007/978-981-33-6220-8_4

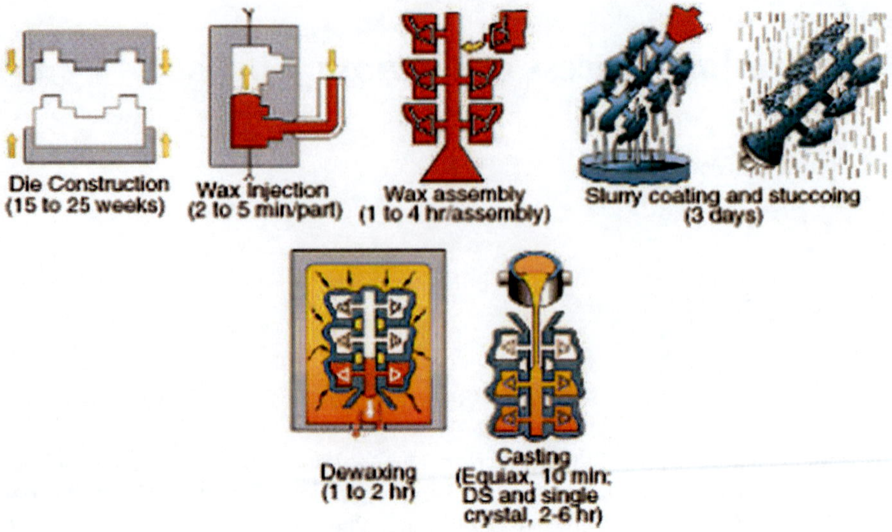

Fig. 4.1 Basic principles of the investment casting process

coarse refractory stucco, and dried. The purpose of stuccoing is to minimize drying stresses in the coatings by presenting some stress concentration centers that distribute and hence reduce the magnitude of the local drying stress. The second purpose of stuccoing is to produce a rough surface, thus facilitating a mechanical bond between the primary coating and the back-up or secondary investment. When the primary coat has set (air-dried until the binder gels), the assembly is systematically dipped into a secondary slurry and stuccoed until the required thickness of the shell is built up. The particle size of the stucco is increased as more coats are added to maintain maximum shell permeability and to provide bulk to the shell [5–7].

Thus, an investment casting shell consists of individual layers of fine refractory materials and granular refractory materials held together by a binder that has been set to a rigid gel. Flexibility exists in changing the composition of each layer. Different methods can be used to remove the wax pattern, usually steam autoclave, leaving a hollow shell. Shells are fired and filled with molten metal that solidifies inside the shell. After casting, the ceramic shell is removed through mechanical or chemical methods to obtain the parts.

It is necessary to prepare the ceramic shell with excellent properties to obtain castings of high quality. According to the working conditions of the ceramic shell, high-quality shells should meet the following basic performance requirements, including strength, deformation resistance, permeability, thermal expansion, thermal conductivity, thermal shock stability, and thermochemical stability, etc. This list is by no means exhaustive. The range of alloys cast, size, complexity of castings, and specific requirements of the cast components lead to an array of shell properties and materials, all used for a specific purpose within the casting industry. The underlying factor in all cases is the extensive range of exacting requirements to produce a successful casting.

(1) **Strength**

Strength is the most critical and fundamental property of a shell. In the different process stages of investment casting, three different strength play an essential role, namely, green strength, high-temperature strength, and residual strength. The green strength refers to the strength of the shell after drying. If the green strength of the shell is too low, cracks can easily happen during dewaxing. High-temperature strength refers to the strength of the shell when sintering or pouring. Residual strength refers to the strength of the shell after pouring, which affects the ease of casting cleaning. If the residual strength is too high, it is not easy to clean up and even cause the casting deformation or breakage during the process of cleaning. With careful consideration of the three strengths, the shell should have high green strength, sufficient high-temperature strength, and a residual strength as low as possible.

(2) **Anti-deformation**

Anti-deformation is another critical performance of the shell. The high-temperature deformation of the shell is one of the main reasons for the deformation of the casting, which mainly includes the thermal expansion and deformation of the shell associated with two aspects, i.e., during the sintering process and under the pressure of the metal liquid when pouring.

(3) **Permeability**

The air permeability of the shell refers to the ability of gas to pass through the shell wall. If the permeability of the shell is low, the efficiency of steam dewaxing decreases significantly, and the gas in the shell cannot be discharged smoothly during pouring. At high temperatures, these gases expand to form a higher air cushion pressure, which hinders the filling of liquid metal and causes defects such as porosity or insufficient pouring, especially in thin-walled castings. The permeability mainly depends on the compactness of the shell structure and the type and content of the binder. The properties and viscosity of refractories are the main factors affecting permeability.

(4) **Thermal expansibility**

The shell's thermal expansibility refers to the expansion or contraction of the shell with the increase of temperature. When the shell is heated, factors like the thermal expansion and crystal transformation often increase its size. The shrinkage is caused by the dehydration of the shell, the thermal decomposition of the material, the sintering of the materials, the liquid phase's production, and the silica gel's condensation, resulting in the densification of the shell. Thermal expansion not only directly affects the dimensional accuracy of castings but also affects the quench and heat resistance and high-temperature deformation resistance of the shell. When the refractories in the shell are heated up, some of them expand uniformly, while others expand non-uniformly. The use of refractories with small linear expansion coefficients and uniform expansion meets the requirements of high-quality shell and castings.

(5) **Other performance**

Other properties include thermal conductivity, thermal shock stability, and thermochemical stability of the shell. Thermal conductivity refers to the thermal conductivity of the shell, which affects the cooling rate of the castings, thus affecting the grain size and mechanical properties of the castings. Thermal shock stability is the ability of the shell to resist rapid temperature change without damage. The shell with poor thermal shock stability tends to crack when pouring. Thermochemical stability refers to the ability of interfacial thermochemical reactions between the shell and molten metal. It has a significant influence on surface roughness, chemical bonding, pitting, and shell peeling.

4.1.2 Raw Materials

4.1.2.1 Binder

The binder is the primary raw material for shell building in investment casting. It directly affects the quality, production cycle, and cost of the shell and castings. The choice and use of binder can be determined by casting size, shell baking temperature, pouring temperature, alloy type, and many other factors. The primary binders used to prepare investment casting ceramic shells are ethyl silicate, silica sol, sodium silicate, ethyl silicate silica sol composite binder, phosphate, refractory cement, plaster, and so on. The most widely used binders for investment casting are ethyl silicate and silica sol.

During World War II, due to the development of the national defense and aviation industry, the United States first used ethyl silicate as a shell binder to produce complex aeronautical castings such as jet turbine engine blades with a smooth surface and high dimensional accuracy. The production cycle of ethyl silicate shell is short, its wet strength and high-temperature strength are high, and its high-temperature deformation resistance is strong. Therefore, as a high-quality binder, ethyl silicate has been widely used in the investment casting industry for some time.

An ethyl silicate binder contains organic volatiles, which can easily pollute the workshop environment. Therefore, various air purification devices, such as biofiltration, have been adopted in factories using ethyl silicate as shell [8]. Hitchiner developed a system in cooperation with Environair S.I.P.A. to discharge alcohol and a small amount of ammonia gas from the air between shells into the scrubbing tower. The scrubbing liquid absorbs alcohol and ammonia gas and then enters the separator to separate alcohol and ammonia gas. Alcohol is recovered, and ammonia gas is discharged into the thermal oxidizer for oxidation [9]. The use of ethyl silicate binder has been limited in foreign countries. At the same time, the application of ethyl silicate in the investment casting industry is diminishing gradually because of its high price and high cost.

Thanks to the simple shell-making process, safety, and non-toxicity, silica sol binder was introduced into investment casting shell production in the 1960s. Besides, silica sol coating has stable performance, high-temperature strength, and good thermal shock stability. The surface quality and dimensional accuracy of the castings produced by the shell made of silica sol binder are excellent. As a result, the application of silica sol in the investment casting industry is becoming more and more extensive [10, 11]. However, silica sol also has some shortcomings, e.g., slow drying of coating, long shelling cycle, low wet strength of the shell, high residual strength, and poor shelling performance. Since the 1980s, polymer reinforced quick-drying silica sol, large particle size, high concentration fast-drying silica sol, and fiber-reinforced silica sol have been introduced, which significantly improve and optimize the performance of silica sol binder, as shown in Table 4.1.

With the increasing variety of active alloys used in investment casting as well as the increasing size of castings and the development of directional solidification technology, silicon-based binders are increasingly inadequate in the high-temperature properties of the shell, including (1) the active interfacial reaction between molten metal and shell; (2) the high creep of the shell; and (3) the transformation of silicon dioxide to shell. The thermal and mechanical stability of the shell has adverse effects. For this reason, some non-silicon-based binders have emerged, including zirconia binder [17, 18], alumina binder [19], and mullite/aluminosilicate binder [20].

4.1.2.2 Refractory Materials

Refractories account for more than 90% of the shell weight and have a significant influence on shell performance. The refractories in the shell can be divided into

Table 4.1 Several modified silica sol with excellent properties [12–16]

Name of silica sol	Characteristic	Representative products	Remarks
Polymer reinforced quick-drying silica sol	Good stability Strong adhesiveness High wet strength Low residual strength Good air permeability[22]	Ludox SK [21] LP-3301 [21]	Polymers are soluble in water, including: Polyvinyl alcohol (PVA) Hydroxymethyl cellulose, etc.
Fast-drying silica sol with large particle size and high concentration	High strength Quick-drying Good stability	Megasol [25]	The silica sol with a small particle size is subjected to high temperature and pressure Prepared by adjusting pH value [23, 24] The particle size can reach 50–100 nm
Fiber-reinforced silica sol	Strong adhesiveness Spread evenly Thickness High wet strength Good air permeability	Buntrock Industries, Inc. Wex Chemicals Co.	Nylon fibers Length: 1–1.5 mm Diameter: 19–21 μm

fine refractory powder used in the slurry preparation and coarse granular sanding materials used in stuccoing.

During the 70 years after the World War II, the industrialization of investment casting has developed rapidly. Influenced by geographical and economic factors, shell-making refractories are widely used. However, the application status of refractories has hardly changed in recent decades [21]. The main refractories used for the primary shell layer are zircon, fused corundum, and so on. The main refractories used for the back layer of the shell are fused quartz [22] and aluminum-silicon refractories (including kaolin clinker, bauxite, coal gangue, Molochite, Malay sand, etc. [23]). The physical and chemical properties of typical refractories are listed in Table 4.2.

In 2001, Minco Company, a well-known refractory manufacturer in the United States, published statistical data on the use of shell-making refractories in investment casting enterprises in the United States. The proportion of consumption of various refractories (mass fraction) is rough as follows [24]: aluminum-silicon refractories 55%, fused quartz 30%, corundum 9%, and zircon 6%. In the American investment casting industry, fused quartz has emerged as a new type of shell refractory, of which the usage is second only to that of aluminum-silicon refractories that have far exceeded other refractories such as zircon.

Fused quartz is made by rapidly cooling quartz (cristobalite or phosphorus quartz) to above 1710 °C ("melting point") [25]. Compared with other refractories, fused quartz has the following advantages.

(1) Low density

The density of fused quartz is only 2.2 g/cm^3, which is much lower than those of other refractories. Therefore, the quality of fused quartz shell with the same thickness is much smaller, which is conducive to reducing the labor intensity of workers and the operating burden of manipulators [26].

(2) Low thermal expansion

A low thermal expansion rate is a tremendous advantage of fused quartz. It is beneficial to reduce the thermal stress caused by the temperature difference between inside and outside the shell during heating, and can thus help prevent the cracking and deformation of the shell during dewaxing and sintering. Besides, a low thermal expansion rate is also beneficial to improve the dimensional accuracy of castings.

(3) Good collapsibility

About 70% of fused quartz transforms into cristobalite (crystallization) at 1100 °C. When it is cooled to 180–270 °C, cristobalite transforms from high-temperature alpha phase to low-temperature beta phase, with the volume experiencing a sudden change. The schematic diagram of crystal transformation is shown in Fig. 4.2. At this time, a large number of cracks occur in the shell, and the residual strength is significantly reduced. Improving the shelling property of castings [27, 28].

Table 4.2 Physicochemical properties of typical shell-making refractories

Name of refractory	Chemical formula	Chemical property	Melting point/°C	Density/(g cm^{-3})	Mohs hardness	Coefficient of linear expansion/× 10^{-6} °C to 120–1200 °C	Thermal conductivity/W m^{-1} K^{-1}	
							400 °C	1200 °C
Fused quartz	SiO_2	Acidic	1713	2.2	7	0.51–0.63	1.591	–
Fused corundum	Al_2O_3	Amphoteric	2050	4.0	9	8.6	12.561	5.276
Mullite	$3Al_2O_3·2SiO_2$	Amphoteric	1810	3.16	6–7	5.4	1.214	1.549
Sillimanite	$Al_2O_3·SiO_2$	Weak acidity	1545	3.25	6–7	5	–	–
Zircon	$ZrO_2·SiO_2$	Weak acidity	1960	4.5	7–8	4.6	–	2.094
Magnesium oxide	MgO	Alkalinity	2800	3.57	6	13.5	5.443	2.931
Calcium oxide	CaO	Alkalinity	2600	3.32	–	13	–	7.118

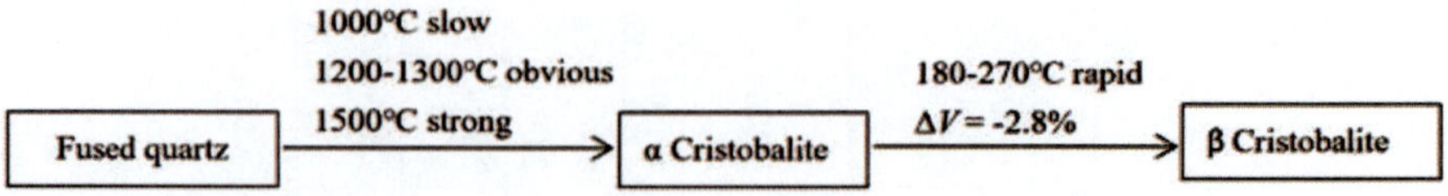

Fig. 4.2 A schematic diagram for the transformation of fused quartz to cristobalite

4.1.3 Research Status of Ceramic Shells

The preparation of a high-performance ceramic shell is a critical step in the whole investment casting process since it significantly influences the surface quality and dimensional accuracy of castings. Research on ceramic shells has always been the focus of investment casting practitioners. Many scholars and foundry engineers have carried out numerous studies on ceramic shells, which focus mainly on the following.

4.1.3.1 Relationship Between the Ceramic Shell and Surface Quality of Castings

One of the goals of investment casting is to obtain castings with good surface quality. The surface quality of castings is the result of the interaction of metal and shell. The factors that affect the surface quality of castings include (1) the surface quality of shell and (2) the interface reaction between the shell and liquid metal.

The surface quality of the shell is mainly related to the selection of raw materials and shell making process. Zengyan et al. [29] studied the effect of the binder and refractory on the surface finish of the shell. It was found that the binder type, the size distribution of powder in the slurry, and the drying process of the shell all had a significant influence on the surface quality of the shell. Rongchang et al. [30] found that if the hydrolysis and drying control of ethyl silicate shell were not good enough, the primary layer of the shell would form white SiO_2 particles after calcination at high temperature, which would prevent the contact between a refiner and liquid metal and thus affect the refining effect. Minghan et al. [31] used vacuum coating technology to coat the primary slurry, to minimize the gas content involved in the slurry, and showed that the surface finish and yield of the castings were significantly improved.

4.1.3.2 High-temperature Mechanical Behavior of Ceramic Shells

The high-temperature mechanical property of the ceramic shell is one of its most essential properties. The high-temperature strength and deformation of ceramic shells have a significant influence on the dimensional accuracy of castings. The high-temperature mechanical behavior of the ceramic shell depends mainly on its microstructural characteristics and two main factors, namely, the quantity and viscosity of the glass matrix (glass effect) and the degree and mode of contact or bonding between crystals (crystallization effect) [32]. The low content of the glass phase, high viscosity, high intercrystalline bonding, and the formation of a continuous staggered network structure is conducive to improving the high-temperature mechanical properties of the shell [33].

On the one hand, the glass phase in the shell comes from the glass phase formed during the preparation of refractories, which is cooled to room temperature after high-temperature preparation of refractories and evenly dispersed in the refractories. On the other hand, it comes from Na_2O in the binder and other refractory components in the shell in the process of shell sintering or pouring. The higher the content of Na_2O and other impurities, the worse are the high-temperature mechanical properties of the shell. For the binary shell system (SiO_2-NaO_2), the liquid glass phase has been formed below 800 °C, which harms the strength and deformation of the shell. For the ternary shell system (SiO_2-Al_2O_3-NaO_2), the feldspar or nepheline reinforcing phase forms in the shell with calcination, which can improve the mechanical properties of the shell at medium and low temperatures below 1200 °C. However, as increasing the shell temperature, the iso phase viscosity of feldspar decreases significantly, and the mechanical properties of the shell at high temperatures decrease accordingly [34, 35]. It can be seen that the number and viscosity of the glass phase in the shell must be controlled to obtain excellent mechanical properties of the shell at high temperatures.

According to the characteristics of SiO_2 binder that is widely used in ceramic shells, if the crystalline phase of refractory can react with the binder SiO_2 to form another high temperature resistant crystalline phase, the number of glass phase at the grain boundary can then be reduced, and the distribution of glass phase can be improved. Also, the so-called "direct bonding" microstructure can be formed, and the intergranular glass can thus be reduced. The viscous flowability of the glazed phase significantly improves the strength and creep resistance of the shell [36].

The amorphous SiO_2 with about 10% of the SiO_2 binder in the shell is the main source of glass phase formation. Therefore, the huge phase interface between SiO_2 binder and refractory particles becomes the weakest point of the whole shell. If the glass phase can be transformed into a crystalline phase and if the main crystalline phase can be directly combined with the high temperature stable heterogeneous phase, then the creep resistance of the shell can be effectively improved. The most commonly used method for this purpose is to add a certain amount of alumina with particle size less than 10 μm into the slurry so that the shell can form a mullite phase at high temperature, to change the intergranular bonding state [37]. A suitable amount of mineralizer can be added to promote mullite to form a certain glass phase, thus reducing the mullite phase's formation temperature facilitating the formation of mullite in the glass phase. The reaction temperature can be reduced to below 1300 °C. The mineralizers included Si-Ca [38], Al-Si-Ca [39, 40], Al-Si-Mg [41], ASM-Cr [42], B_2O_3-CaO [43], CaO [44], etc. The high-temperature strength of the shell increased by about 20 times [45] as compared with that before adding no mineralizer.

With the development of industrial gas turbines, the demand for large directionally solidified blades and other castings is continuously increasing. Simply adding mineralizers cannot meet the high-temperature strength and creep resistance requirements of large-scale shells. To solve the problem of insufficient strength and dimensional stability issue of a large-size ceramic shell at high temperature, technicians have used the method of adding fibers to strengthen the ceramic shell, wherein mullite fibers, alumina fibers, and carbon fibers are commonly used. The representative fibers are GE Company, HOWMET Company, Mitsubishi Company, Japan, etc. [46–49].

The restriction (hindrance) of the shell on the cooling shrinkage of the castings also significantly influences the castings' size [50]. The mechanical interaction between the alloy and shell during solidification may cause thermal tearing, deformation, or residual stress [51]. Excessive high-temperature strength of the shell can easily lead to thermal tearing during the solidification of castings. Therefore, the higher high-temperature strength of the shell is always desired. As long as the high-temperature strength of the shell is sufficient to support the casting forming, the high-temperature strength of the shell can be appropriately reduced by adding polymer and other additives to the shell, such that the thermal tearing of the casting can be restrained [52].

4.1.3.3 Performance Optimization of Ceramic Shell

The research on the performance optimization of ceramic shells mainly focuses on improving the permeability and collapsibility of the shell. Yuan et al. [53, 54] studied the effect of WEXPERM nylon fiber on shell thickness and air permeability. The nylon fiber was 1 mm in length and 20 μm in diameter. The addition amount was 20 g per liter of silica sol. The results showed that the coating thickness of the shell at the plane and bend corners increased by 13% and 40%, respectively, with the addition of nylon fibers. The melting point of nylon fibers is 240 °C. After steam dewaxing, the nylon fibers remain in the shell and continue to play a reinforcing role to ensure that the strength of the shell is virtually the same before and after steam dewaxing and to prevent the shell from cracking, as shown in Fig. 4.3. In the process of calcination, as long as the temperature exceeds 650 °C, the fibers are completely burned out with so many tiny holes being formed in the shell, which significantly improves the shell's permeability.

Xin et al. [55] studied the effect of impurities and calcination temperature on the residual strength of silica sol coal gangue shell. It was found that with the increase of

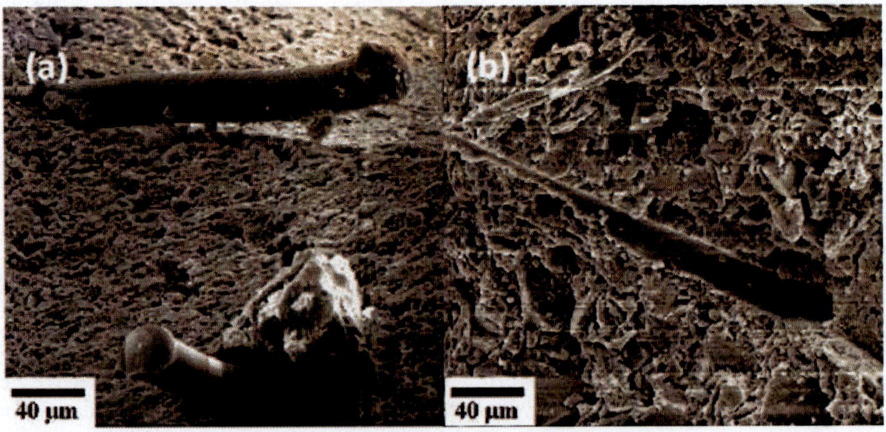

Fig. 4.3 Fracture microstructure of shell: **a** one end of the fiber is pulled out of the shell; **b** the void left by the pull-out of the fibers

calcination temperature of the shell, the glass phase content in the shell increased, and secondary mullite formation resulted in the increase of residual strength of the shell. Yuan et al. [56] changed the sanding material of the third layer of the shell from fused corundum to monoclinic zirconia. It was found that the residual strength of the shell was reduced without affecting the high-temperature strength and permeability of the shell at room temperature. Junhao [57] added fused quartz into the shell transition layer to reduce the residual strength of the shell and to improve the shell shelling performance by utilizing the characteristics of high-temperature crystallization and low-temperature phase transformation of fused quartz. Qifu and Tianfu [58] made foaming coatings and porous shells by adding a foaming agent and foam stabilizer in the back slurry. Based on ensuring enough wet strength and high-temperature strength of the shell, the residual strength was reduced, and the collapsibility of the shell was improved. Jinzhi et al. [59] used the coated sand as the sand-drenching material for the third and fourth layers of the shell, which improved the collapsibility of the shell. Besides, Guizhi et al. [60] proposed a testing method for evaluating the shelling property of investment casting mold shell. Lei et al. [61] developed a testing device for sand concession based on PC.

4.1.3.4 Improvement of Shell Making Technology and Control of Shell Defects

With the rapid development of the investment casting industry, there are more and more types of binders for ceramic shell making, and the shell making technology is also developing. The traditional silica sol shell has a low drying rate, low efficiency, and low wet strength. With the development of fast-drying silica sol [62, 63], this problem has been significantly solved. Water-soluble polymers such as latex have been commonly added to the quick-drying silica sol, and latex can also be added to the slurry of ethyl silicate hydrolysate, which can improve the wet strength and air permeability of the shell and reduce the residual strength of the shell [64].

Ethyl silicate silica sol composite binder [65] combines the advantages of both ethyl silicate hydrolysate and silica sol binder, offering better stability than that of ethyl silicate hydrolysate. As a result, it simplifies the shell making process and improves the high-temperature strength of the shell. Alternating shell-making with two binders of ethyl silicate and silica sol [66, 67] can significantly improve shell-making efficiency and shorten the shell-making cycle. The drying rate of the shell is greatly influenced by ambient temperature, humidity, and airflow [68, 69]. The determination method of shell drying has also been proposed [70].

Reducing the number of coating layers shortens the shell-making cycle and improve the shell-making efficiency. However, the decrease of the coating layer results in a decrease of coating thickness and load resistance of the shell. Increasing the strength of the shell is undoubtedly one way to reduce the number of coating layers of the shell, and the use of die materials with small thermal expansion during dewaxing can also reduce the requirement for the strength of the shell. Decheng et al. [71] used polystyrene as the mold material to prepare shell with a thickness of only 3–4 mm, which can meet the needs of low carbon steel, high alloy steel, and other metal castings. This process is called Replicast CS.

Defect control of ceramic shell has attracted much attention to precision casting enterprises. Much research has been carried out on the causes and restraining measures of defects, such as the uneven wall thicknesses [72, 73], stratification [74, 75], cracks [76, 77], and holes [78].

4.1.3.5 Green Manufacturing of Shell and Shell Waste Utilization

With the development of science and technology, more and more attention has been paid to environmental protection. Volatile organic compounds (VOC) such as ethanol and ammonia are emitted from ethyl silicate shells during shell-making, which is harmful to the environment. As early as the end of the twentieth century, western developed countries have promulgated decrees to limit the use of alcohol-based binders. However, there is no explicit regulation in this regard in China. With the continuous development and upgrading of water-based silica sol binders, various modified silica sol products with excellent properties have been introduced, and more and more alternatives of ethyl silicate are available.

How to realize the sustainable development of investment casting has become the focus of attention [79]. The treatment and recycling of waste ceramic mold shells are crucial, and the establishment of its guidelines and measures is an urgent matter for the development of precision casting [80]. As early as the 1970s and 1980s, the United States and Japan, and other industrially developed countries have begun to pay attention to the emission and reuse of waste from investment casting plants, especially the waste shell. In 1976, Orbis Fine Casting Plant, California, USA, purchased the recycled shell from Sweco Company, which produces crushing and classification equipment. In 1987, the ICI-led research project on shell recycling in the United States officially started, and there were reports that the waste shell was properly treated and processed into one-time refractory products (such as gate cups, molding runners, etc.) [81]. According to the literature [82, 83], waste shell materials can also be used as raw materials for building materials. Valenza et al. [84] used waste shell crushed materials with 3–32 μm particle size in the preparation of refractory tiles and achieved good results.

At present, according to rough estimates, the annual output of investment castings in China has reached more than 50,000 tons, and the annual output of waste shells is nearly 1 million tons. Hengyi et al. [85] of Ningbo University made a thorough and comprehensive experimental study on the chemical composition, phase composition, strength, permeability, and thermal deformation resistance of the waste shell. The research shows that as long as the treatment method is proper, the waste shell can be recycled entirely. In 2004, Jinding and Xinyu Precision Casting Co., Ltd. of Taizhou, Jiangsu Province, used the crude sand made from the recycled waste silica sol shell as the sand drenching material after the fourth layer and the powder as the slurry for the sealing layer and achieved good results.

4.1.3.6 Deformation of Ceramic Shell

Jiang et al. [86–89] systematically studied the size change and influencing factors of a simple plate-shaped shell in various technological links of shell-making and designed a stepped pyramid-shaped wax pattern. The size relationship among the wax pattern, 950 calcined shells, and 520 tool steel casting cast at 1580 was studied. The results showed that the shrinkage of shell size after calcination was 2.06%, which was more than 50% of the change of casting size. It was thus concluded as the most critical factor affecting the size of castings. However, the shell used in this experiment is silicic acid ethyl ester hydrolysate slurry formed by silicon sol-gel casting, which is very different from the traditional shell-making process. The size of the shell is measured after sintering and cooled to room temperature, which cannot wholly reflect the influence of the shell on the casting size.

Morrell et al. [90] studied the elastic creep and compressive creep of the ceramic shell using different material systems. The results showed that there were different fragile temperature ranges for different shell systems, and the fragile temperature ranges corresponded to the fragile temperature ranges of shell microstructures. For the silicon-based shell, the fragile zone was below 1000 °C. When the temperature was higher than this zone, the microstructures of the shell were consolidated and exhibited minimal distortion under load. For alumina-based and aluminum-silicon-based shell, the fragile zone was slightly above 1000 °C. Therefore, it is necessary to guard against the change of alumina-based and aluminum-silicon-based shell before pouring.

4.1.3.7 Simulation of Ceramic Shell

At present, the research on the simulation of ceramic shell mainly focuses on the stress analysis of the shell fracture, but there are few studies on the shell deformation and the subsequent impact on the accuracy of castings. Chen et al. [91], aiming at the problem that the shell of turbine blade casting is easy to break, proposed a finite element analysis method of shell based on an elastic continuum model, where deformation and stress analysis was carried out. Based on the analysis result, polyimide was added to the fragile area to increase the shell's strength, the correctness of which was confirmed by experiments.

Everhart et al. [92] targeted at the problem that hard plastic shell is easy to break in casting and established a shell deformation model by combining finite element analysis and thermodynamic analysis, obtaining a set of optimized process parameters through considering the shrinkage caused by cooling. Everhart et al. [93] used a continuum medium to simulate porous shell materials. Formulas and simulations were used to predict cracks at corners and edges of the shell, and it was concluded that the porosity of the shell affects the stress distribution. Radiation heat transfer is the primary heat transfer mode between the shell and the furnace body when the shell is sintered [94]. Zashkova [95] aimed at the fracture of the shell caused by an ultrafast heating rate, as well as the resulting physical and chemical changes of the shell during the calcination process. They studied the transient temperature of the shell in different heating time by a sequential heat transfer method, which allowed for the selection of reasonable heating time and heating rate and hence saved time and cost.

4.1.3.8 New Technology and Equipment for Shell Making

Investment casting is a traditional and modern casting process, which has been used for thousands of years. However, the integration of this old method with some new technologies is very active. The development of advanced techniques and equipment has a significant impact and promotion on investment casting ceramic shell.

Rapid prototyping (RP) is a high-tech developed in the 1990s. Mold making is an essential aspect of rapid prototyping technology in investment casting [96]. At present, the popular rapid prototyping methods include stereolithography (SLA), selective laser sintering (SLS), melt deposition (FDM), layered solid manufacturing (LOM), and three-dimensional printing (3D-P). The shell made from the above rapid prototyping patterns needs to be removed by the flash burning dewaxing method, which brings about specific requirements for the thermal shock stability of the shell.

Rapid freezing prototyping is a new rapid prototyping method. Deposited water can be selected layer by layer and solidified rapidly with the aid of the CAD model to directly prepare ice mold with a specific 3D shape [97]. Ice mold has plentiful requirements for shell making materials and technology. The ice slurry used for ice mold needs to be maintained below the freezing point. The ambient temperature of the ice mold shell should also be controlled below zero. Besides, the catalyst should be added to the slurry to shorten the gelation time of slurry on the ice mold surface. Besides, to inhibit the dissolution of the ice mold surface, the interface agent [98, 99] is also needed.

Flash burning dewaxing is a dewaxing method developed in the 1990s [100]. Flash-burning dewaxing combines dewaxing with sintering, which is probably the best choice so far for removing the materials that cannot be removed by steam dewaxing such as plastic molds (including SLA, LOM, and other rapid prototyping methods) [101]. Flash burning dewaxing can also remove wax from the shell and has a wax recovery rate of up to 80–95% [102]. Compared with steam dewaxing and sintering, it can save about 20% energy consumption [103].

Manipulator has developed into the hydraulic transmission, with 5–6 degrees of freedom. Its movement is more convenient, and automation is further improved. For example, VA Technology Ltd. has developed a new type of shell-making manipulator [104, 105]. The fully automatic shell-making machine (including drying process) introduced by MK Technologies Cyclone in Germany is reported to be able to produce shells in a few hours [106]. It is specially developed for the rapid production of investment castings. To meet the requirement of Rolls Royce Company to transform ethyl silicate shell-making into the silica sol shell-making process, Drytech Processing Ltd [107] designed and manufactured an automatic shell-drying system in 1995. The drying room is closed around, occupying small space and space, which is conducive to adjusting and controlling process parameters such as temperature, humidity, and wind speed, and is also conducive to energy saving.

Wind tunnel shell-making machines can establish the best aerodynamic wind tunnel to accelerate the air to 4 m/s and use a cyclone to dry the shell so that the shell can be dried in 30 min, thus shortening the shelling time of eight-layer shell to 4–7 h.

The strength of the shell dried by the wind tunnel shell maker is about 50% higher than that of the ordinary shell. It can be used to produce all kinds of metal castings, including magnesium and titanium alloy castings.

4.1.4 Difficulties in the Fabrication of Large Ceramic Shells

In China, the development of aero-engines and even the whole aero-industry is bumpy, and the aero-materials and manufacturing technology lag far behind the international top level. The integral precision casting technology of large aero-engine components has become a bottleneck restricting the development of aero-engines. As a critical linkage in the process of precision casting, it is urgent to have a breakthrough in the preparation technology of large ceramic shells. However, there are still many difficulties in the preparation of large ceramic shells, as discussed in the following:

(1) The physical and chemical properties of the shell have much higher requirements. The complex structural characteristics of large, thin-walled castings ask for higher requirements for the strength, permeability, deformation resistance, and yielding of the ceramic shell. The larger the casting size is, the greater is the wax pressure on the shell during steam dewaxing, and the greater is the hydrostatic pressure on the shell during pouring, which all require the shell to have sufficiently high strength at both room temperature and high temperature. However, as the structure of castings becomes more and more complex, the wall thickness of castings becomes thinner and thinner, and the requirement of shell collapsibility becomes higher and higher. If these requirements were not met, it is easy to cause thermal tearing of castings; consequently, the shell strength needs to be controlled by adjusting the shell thickness. At the same time, the increase of shell thickness leads to a decrease in air permeability, which harms dewaxing efficiency and metal liquid filling. Therefore, it is of great significance to optimize the shell formulation appropriately to achieve an optimal balance between the strength and permeability of the shell.
(2) The shell-making prefers a higher level of automation. The shell of large castings is often of heavyweight, which can weigh more than 500 kg. It is impossible to finish the soaking operation by hand. It is thus necessary to use a more automated manipulator to cooperate with the large-scale soaking barrel and sand drenching machine. These requirements make manipulators more and more expensive. Besides, the design of the manipulator's mortar and the sand spraying procedure also significantly influences the quality of the shell.
(3) It is challenging to prepare the shell with some unique structures, such as the narrow inner cavity of castings. Coating, drying, use of core-filling materials, or ceramic cores are the vital forming technologies to the shell at the narrow and long inner cavity.

(4) The control of the shell-making process is stricter. Large castings require higher internal quality and dimensional accuracy. Therefore, the control of the technological parameters of the shell-making process is more stringent, including shell drying parameters (drying temperature, drying humidity, and drying time), dewaxing parameters (dewaxing temperature and dewaxing time), and sintering parameters (sintering process curve).

Investment casting has been developed rapidly in recent decades. Scholars have made more in-depth and systematic researches on investment casting shells. However, there are still some shortcomings in the research of large-scale shells.

(1) The long production cycle has always been the disadvantage of investment casting technology. The main reason is that the time spent on die processing, die assembly welding, and shell contamination is long. However, with the rapid development of rapid prototyping technology, rapid prototyping pattern preparation technology has become more and more mature, which not only saves the opening time but also dramatically shortens the time required for pattern pressing and welding. At present, the shell-making method used in most foundries and laboratories is still the traditional multi-layer shell-making method with mortar and sand spraying. The traditional method of soaking mortar and drenching sand needs to be coated and dried layer by layer on the surface of the molten mold, and the shell-making cycle is long. In particular, for the shell of large castings, the thickness of the shell needs to be coated much thicker, and sometimes nearly 20 layers need to be coated. Because of this, it takes nearly ten days to prepare the shell alone. Although some researchers have proposed new shell making methods such as silica sol gel-forming, these methods are inefficient and have not been widely recognized for the time being. The long period of shell preparation has become a bottleneck that limits the production efficiency of investment casting. The rapid shell-making process based on a fast-drying binder and automatic shell-making drying room can improve the shell-making efficiency to a certain extent. However, this process can only be applied to small-sized shells. For large-scale shells with a large size and complex structure, the application of rapid drying technology is significantly limited because of the higher requirements for drying uniformity and wet strength of different areas of the shell. Therefore, it is of great significance for investment casting researchers to explore a new rapid shell-making technology suitable for large-scale shells.
(2) The performance of the shell is determined by the binder and refractory used in shell making. In recent decades, significant progress has been made in the field of shell-making binders. The types and brands of binders are becoming more and more abundant, which can be used to prepare ceramic mold shells meeting different casting requirements. However, the types of shell refractories are mainly confined to corundum, zirconium silicate, aluminosilicate, and fused quartz. The types of refractories used in shell making have not been expanded, which restricts the space for further optimization and improvement of ceramic shell properties. The shell system used in most foundries is a single refractory system. For example, corundum is used as a refractory powder in the

primary slurry, and mullite is used as a refractory powder in the back slurry. The advantage of this method is the simplification of the shell formulation and reduction in the workload in many aspects such as raw material procurement and slurry preparation. However, different refractories have different properties at high and low temperatures, with different effects on the binders as well. Often, a variety of refractories can be used together to make the shell obtain more balanced and excellent properties at both high and low temperatures. For example, adding a small amount of alumina to the fused silica shell can produce a secondary mullite phase at high temperature, which can further improve the high-temperature strength of the shell without affecting the permeability and yielding of the shell. Also, some additives, such as some mineralizers, ceramic fibers, have greatly improved the shell's properties. Further research in this area is still in need.

(3) With the development of investment casting, the requirement for dimensional accuracy of castings is getting higher and higher. However, most of the research on casting deformation is focused on the deformation of investment casting and metal solidification and cooling, and the attention on shell deformation is lacking, which is related to the relatively uniform and stable deformation of the shell. The larger the size is, and the more complex the structure of the castings is, the worse becomes the uniformity of the deformation of the ceramic shell, and the larger is the deformation. With the development of research, more and more attention has been paid to the deformation of shell, especially for large and complex castings. The shell's influence on the size of castings mainly manifests in (1) the deformation of the shell under the action of uneven temperature field during sintering, and (2) the deformation of the shell under the action of hydrostatic pressure during pouring. For large castings with complex shapes, the deformation caused by the hydrostatic pressure of metal can be neglected. The main factor affecting the deformation of the shell is the deformation of the shell during sintering. The deformation of a simple flat shell during sintering is characterized by a simple linear relationship, which can be measured directly by experiments and calculated using the thermal expansion coefficient of the shell. However, the size change of the shell with a complex inner cavity during baking is affected by the temperature field and deformation constraints. The size change law is thus very different from that of the simple flat shell, which cannot be directly measured from experiments and can only be studied using computer simulations. At present, the simulation analysis of the shell mainly focuses on the fracture of the shell, while the simulation of the high-temperature deformation of the shell is rarely carried out. Most foundries assume that all parts of the shell expand evenly after calcination at high temperature, but only refer to the coefficient of thermal expansion of the shell when determining the expansion rate of the shell, or rely on experience and repeated modification through experiments.

4.2 Preparation and Characterization of Large Ceramic Shells

Compared with medium and small castings, the characteristics of large castings are not only large in size but also very complicated in structure. Because of the requirement of weight reduction, a large area of the thin-walled structure is designed under the condition of meeting the mechanical properties. Taking the turbine rear frame (TRF) of nickel-base superalloy for aero-engines as an example, the contour size of this kind of large castings is more than 1 m, and there are several support plates, convex platforms, lifting lugs, and other structures, with the wall thickness being only about 2 mm. These structural characteristics of large and complex thin-walled castings demand higher requirements for the strength, permeability, and deformation resistance of ceramic shell.

4.2.1 Raw Material Composition of Large Ceramic Shells

In this section, the preparation process of two types of large ceramic shells is introduced, which are named as MS shell and WS shell, respectively. These two types of ceramic shells have the same primary layer but different back layers. MS shell is composed of a primary layer and MS back layer, whereas the WS shell is composed of a primary layer and WS back layer. For the primary shell layer, a relatively traditional and economical material combination was selected, using alkaline silica sol as the binder, high-temperature alumina powder, and zircon sand as the refractory powder, and sand-spraying material for the primary slurry. The wet strength enhancer latex and grain refiner cobalt aluminate were also added to the primary slurry to make the primary shell layer have high strength in both room temperature and high temperature, which not only ensures the thermal stability of the shell but also makes the shell surface have the function of grain refinement. This kind of primary layer can form a flat, compact, solid, and smooth surface of the shell and does not interact with the liquid metal to ensure the surface quality of the castings.

For large superalloy castings, the requirements of surface quality are very high. However, the key and difficult points in the entire investment casting process are the dimensional accuracies and deformation control of the castings, which raises higher requirements for the performance of the back layer of the ceramic shell. Two types of back-layer materials, namely, the MS back layer and the WS back layer, are designed to this end. The back layer of MS is composed of ethyl silicate and A3125 acid silica sol, which combines both water-based silica sol and ethyl alcohol silicate hydrolysate. The slurry prepared this way has good fluidity, the fast-drying speed of the shell, and a high strength of wet billet. The quick-drying silica sol is used as a binder, which is of high quality, environmental protection, high strength, and good collapsibility.

Regarding the choice of refractories for the back layer of the shell, fused quartz is undoubtedly an excellent choice. However, it has its shortcomings. Fused quartz powder has polygonal distribution, low bulk density, low slurry powder, and low strength. Therefore, fused quartz is seldom used alone. It is usually used in conjunction with other materials to enhance its advantages and avoid its disadvantages, achieving the ideal application effect. MS back layer uses fused quartz and fused mullite as a refractory powder for slurry and fused mullite as drenching material. Mullite has high refractoriness, small linear expansion coefficient, good thermal shock stability, and low price and is suitable for the use combined with fused quartz; the WS back layer uses fused quartz as the main refractory powder for slurry. A certain amount of needle petroleum coke was added to the slurry. Needle petroleum coke is beneficial to enhancing the adhesion of slurry, optimizing the coating thickness of the shell, improving the strength and permeability of green shell, and improving shelling. The back of WS still uses fused quartz as sanding material.

The main parameters of the binder for shell making are listed in Table 4.3, in which the primary layer and the filling materials are used as follows. DVSTU006 silica sol is used for the back shell of WS and 1130 C silica sol. These two silica sol are alkaline binders. The A3125 used in the back layer of MS is a composite binder. A3125 composite binder is an acid binder made of ethyl silicate, acidic silica sol with 30% silica content, and some organic solvents. DVSTU006 silica sol contains a certain amount of liquid water-soluble polymer, which is of great significance to improving the slurry coating property and the strength of the shell body. The types and sizes of refractories for shell making are listed in Table 4.4.

Table 4.3 Binder for shell making

Binder	SiO_2 content (wt%)	Particle size (nm)	Polymer content (wt%)	Density (g/cm^3)	pH	Suppliers
1130C Silica sol	29–31	8–10	–	1.21–1.22	9.7–10.5	Nalco
A3125 Composite binder	19–21	15–30	–	1.00–1.01	1.1–1.6	–
DVSTU006 Silica sol	28–30	11–16	7.7	1.19–1.20	9.8–10.2	Nalco

Table 4.4 Refractories for shell making

Refractory material	Particle size/mesh
High-temperature alumina	−325
Aluminum cobalt oxide	−200
Zircon sand	70/200
Fused silica	−325, 40/150, 120, 30/50, 50/100
Electric melting mullite	40/150, 10/70

4.2.2 *Preparation Process of Large Ceramic Shells*

Mold shell preparation is an essential part of the whole investment casting process. The dimensional accuracy, surface roughness, and even internal quality of castings are closely related to the ceramic shell. However, the process of shell preparation is complicated, the production cycle is long, and many process parameters are involved. All these bring about many difficulties for shell quality control.

4.2.2.1 Slurry Preparation

The slurry made of binder and refractory powder is the basis of the shell structure. It can be divided into the primary layer and back layer. A fast mixer can prepare the slurry. As shown in Fig. 4.4, the slurry can be mixed evenly. The capacity of the slurry mixer is 200 L, and the speed can be adjusted, with the maximum speed being able to reach 1000 r/min. Also, the control system and alarm device can adjust the temperature of slurry in the pulp mixer through an external ice water machine. When the material is added, the slurry viscosity is measured with the Cain cup every hour, and the variation of the slurry flow viscosity is recorded. It is found that the slurry flow viscosity decreases gradually until the difference between the two consecutive flow viscosity measurements is less than 1 s, as shown in Fig. 4.5. At this point, the slurry viscosity tends to be stable, indicating that the slurry flow viscosity is gradually reduced. The slurry usually needs to be stirred continuously in the mixer for at least 8 h before mixing evenly to mix all the materials evenly in a dipping barrel.

Fig. 4.4 Preparing slurry in rapid pulping machine

4.2.2.2 Dipping and Stuccoing

Dipping and stuccoing slurry are the core steps of shell preparation. Before dipping slurry and stuccoing, the viscosity and temperature of slurry should be adjusted, and the caking and ash in spreading sand materials should be screened out. Table 4.5 lists the specific shell-making process parameters. For large-scale melt molds, 500 kg and 1000 kg manipulators are, respectively, used for the dipping slurry and stuccoing operation of the primary and the back layers. The processes of dipping slurry and spreading sand stuccoing by manipulators are shown in Fig. 4.6. During shell making, the primary layer is coated and hung first. After drying the primary shell layer, the coating and hanging of the back layer are repeated until the required number of layers is reached. The last back layer is a closed layer, which is only dipped with back slurry and not stuccoed. The process parameters are given in Table 4.6.

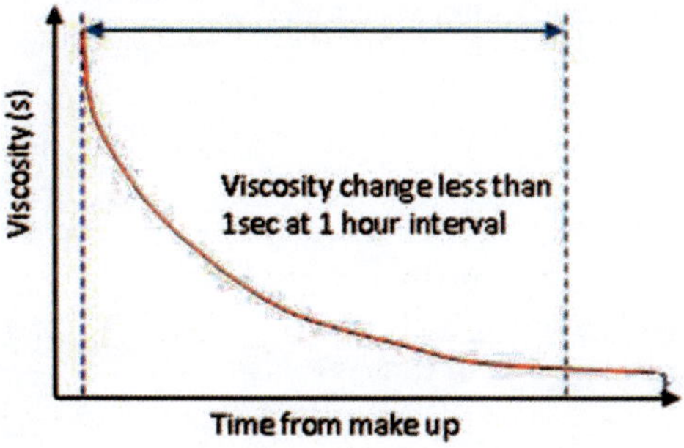

Fig. 4.5 Viscosity stability curve during slurry preparation

Table 4.5 Process parameters of dipping and stuccoing

Shell system	MS shell			WS shell		
	Primary layer	Back layer	Sealing layer	Primary layer	Back layer	Sealing layer
Type of slurry	Primary slurry	MS back slurry	MS back slurry	Primary slurry	WS back slurry	WS back slurry
Sand	70/200 zircon sand	10/70 Mullite	n/a	70/200 zircon sand	30/50 fused silica	n/a
Slurry temperature/°C	19–25	7–10	7–10	19–25	19–25	19–25
Wind speed/(m/s)	0.5	3.0	3.0	0.5	3.0	3.0
Drying time/h	24	10	24	24	8	24
Ambient temperature/°C	19–25	19–25	19–25	19–25	19–25	19–25
Relative humidity/%	50–70	50–70	50–70	50–70	50–70	50–70

Fig. 4.6 Dipping and stuccoing by using robot manipulator

Table 4.6 Types and layers of ceramic shells coated with different samples

Sample species	MS shell	WS shell
Strength/air permeability/deformation	6	6
Thermal expansion coefficient	5	5
Features/single slant plate castings	10	–
Rear casing	18	–

Fig. 4.7 Steam dewaxing kettle for shell dewaxing

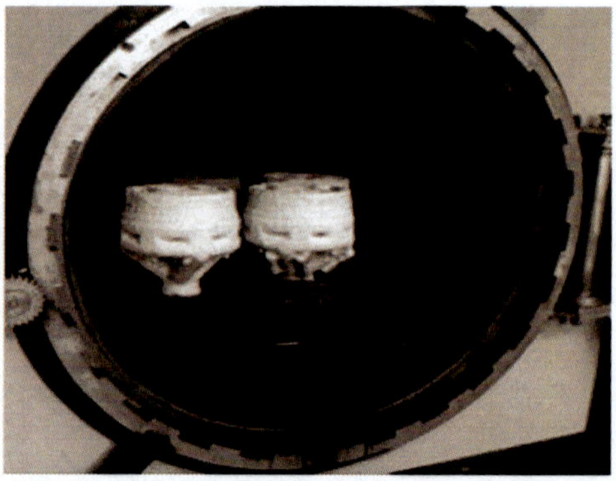

4.2.2.3 Steam Dewaxing

The process of wax pattern melting from the ceramic shell is called dewaxing. Large ceramic shells are dewaxed using steam dewaxing, which is the most widely used method at present. The sealed shell needs to be dried for 24 h before dewaxing to provide the shell with greater green strength so that it would not crack during dewaxing. Shell dewaxing is carried out in a steam dewaxing kettle, as shown in Fig. 4.7. The process parameters of shell steam dewaxing include dewaxing temperature and dewaxing time. The dewaxing temperature and pressure are a set of related parameters. The temperature and pressure in the dewaxing kettle are all provided

(a) Before sintering (b) After sintering

Fig. 4.8 Sintering and casting of TRF castings. **a** Before sintering. **b** After sintering

by water vapor. The dewaxing temperature and pressure correspond to the temperature of water vapor and the saturated vapor pressure at the given temperature. The steam dewaxing temperature and dewaxing pressure are set as 170 °C and 0.75 MPa, respectively, for precision casting shells of casing castings. However, for different sizes of shells, different dewaxing times are adopted, generally 1000–2000s. After dewaxing is finished, the pressure in the dewaxing kettle is unloaded, and the shell can be taken out.

4.2.2.4 Sintering and Pouring

Shell sintering is one of the critical processes in investment casting. Many shell samples need to be treated at a high temperature. Figure 4.8 shows the photos of large ceramic shells before and after sintering.

4.2.3 Performance Testing Method for Large Ceramic Shells

4.2.3.1 Performance Test of Ceramic Slurry

(1) **Slurry viscosity**

The flow cup viscosity of the slurry is measured using the Zahn Cup, as shown in Fig. 4.9a. Before the measurement, the flow cup is cleaned and dried, and the slurry's temperature is maintained in a proper range. The flow cup is immersed in the slurry, and the flow cup is lifted quickly. Once the upper edge of the flow cup leaves the slurry's surface, the time is started when the flow is interrupted under the small hole of the flow cup. The time used is recorded as the flow cup of the slurry viscosity. Zahn Cup has a volume of 44 mL, including five flow cups with different aperture sizes. The aperture of the Cain cup increases gradually from No. 1 to No. 5. In this chapter, Zahn Cup No. 4 is used to measure the viscosity of the primary slurry flow

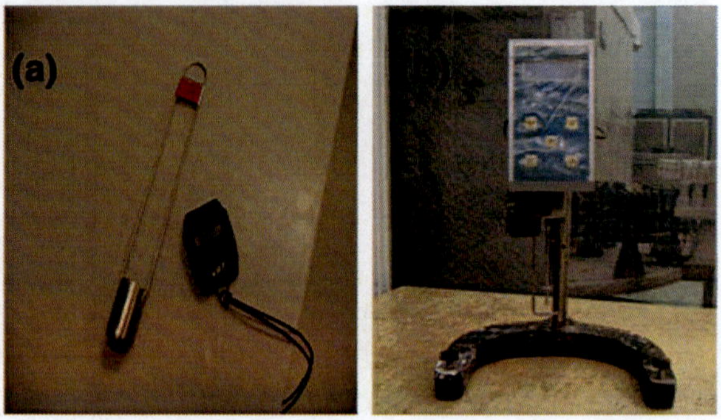

Fig. 4.9 Zahn cup (**a**) and rotary viscometer (**b**) for testing slurry viscosity

cup, and its aperture is 4.35 mm; Zahn Cup No. 5 is used to measure the viscosity of two kinds of back slurry flow cup, its aperture is 5.41 mm.

The flow viscosity of the slurry can be converted into kinematic viscosity through the equation. For the 4 # and 5 # Zahn Cups used in this chapter, the kinematic viscosity ν (mm^2/s) of the slurry can be calculated using Eqs. (4.1) and (4.2):

$$\nu = 1.48(t - 5) \tag{4.1}$$

$$\nu = 23t \tag{4.2}$$

The dynamic viscosity of the slurry of the shell and core filling material are measured using NDJ-5S and Model NXS-11A rotary viscometer, as shown in Fig. 4.9b.

(2) **Slurry specific gravity**

The process is described as following. (1) Prepare a clean measuring cylinder with a capacity of 100 mL and a weight of W_T g. (2) Pour the slurry into the measuring cylinder at the 100 mL scale and then weigh the total weight as W_G g. (3) Wash and dry the measuring cylinder and then pour the same volume of water into the measuring cylinder, weighing the weight as W_C g. The specific gravity of the slurry S_P is then calculated as:

$$S_p = (W_G - W_T)/(W_C - W_T) \tag{4.3}$$

(3) **Slurry solid content and silicon content**

One glass dish to be dried is weighed as W_T g, about 50g slurry or 10 ml slurry centrifuge (centrifuge speed > 3500 r/min, centrifuge time 30 min) is weighed into the glass dish, and the total weight W_S g of slurry or centrifuge and glass dish is weighed. The sample is put into an oven at 120 °C, and after 1 h, the sample is taken out. After recording the sample's weight when it cools to room temperature, the sample is put into an oven at 120 °C for heating again. After 30 min, take out the sample, cool and weigh, record the weight of the sample, and repeat the operation process until the weight of the sample remains unchanged. The weight is the dry weight of W_D g and then the slurry solid content or silica content percentage TSC is:

$$TSC = (W_D - W_T)/(W_S - W_T) \tag{4.4}$$

(4) **The coating weight of the slurry**

Prepare a stainless steel plate with a size of 150 mm × 150 mm × 1.5 mm, calculate the surface area C cm^2 of the steel plate, clean the stainless steel plate, dry and weigh A g, then immerse the stainless steel plate in the slurry for 5 s, quickly take the steel plate out of the slurry, as shown in Fig. 4.10, and then start recording time at the same time. After 120 s dripping time, weigh the total weight B g of the stainless-steel plate and the slurry adhered to it, and the coating amount R of the slurry is given as

$$R = (B - A)/C \tag{4.5}$$

For different types of slurries, owing to their different densities ρ, the coating amount R of the slurry on the coating sheet per unit area cannot fully reflect the adhesion amount of the slurry on the stainless-steel plate. In this case, the coating thickness D of the slurry is more suitable to represent the coating property of the slurry:

$$D = R/\rho \tag{4.6}$$

Fig. 4.10 Smear Weight Test of Slurry

(5) The wet film thickness of the primary slurry

Prepare a clean glass sheet, immerse the glass sheet in the primary layer slurry for at least 4 in (1 in = 0.0254 m) deep, and then take out the glass sheet until the slurry on the surface of the glass sheet stops flowing. After drying for 5 min, measure the coating thickness of the primary layer slurry with a wet film thickness gauge, as shown in Fig. 4.11.

(6) Slurry bubble test

Prepare one clean slide, stick one side of the slide with adhesive tape, and then dip the slide into the primary slurry and pour sand, remove the adhesive tape adhered on one side, and observe the number of bubbles on the side of the slide stained with slurry and pour sand under the light source, as shown in Fig. 4.12a. The number of bubbles should be no more than one bubble per square inch; otherwise, an appropriate amount of defoamer should be added. Take another 10 ml of the slurry centrifuge (centrifuge speed > 3500 r/min, time 30 min), put it into the centrifuge tube, shake the centrifuge tube, stop shaking after 10 s, and record the time required for bubbles in the slurry to disappear completely, as shown in Fig. 4.12b, it is generally appropriate to use 30 s or less; otherwise, add an appropriate amount of defoamer.

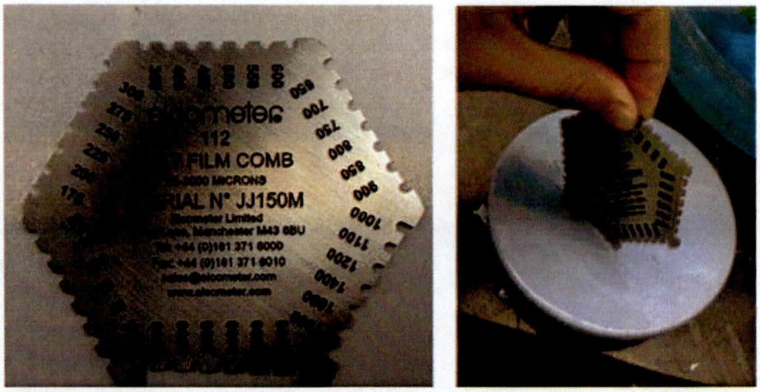

Fig. 4.11 Measurement of the wet film thickness of the surface slurry

Fig. 4.12 Test of slurry bubble quantity (**a**) and bubble rupture speed (**b**)

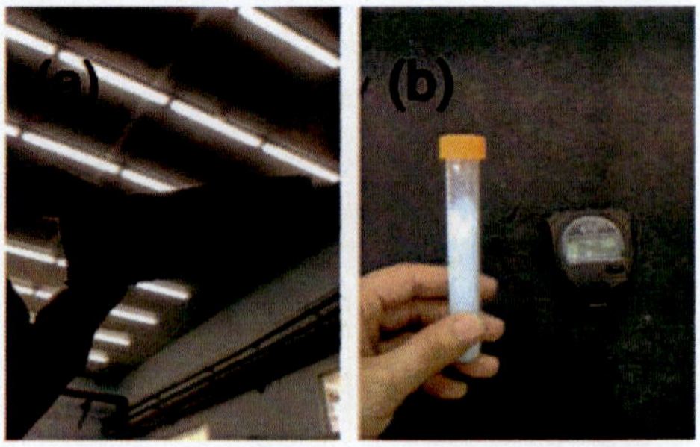

4.2.3.2 Microstructure Observation and Performance Test of Ceramic Shell

(1) **Phase analysis**

The phase of the ceramic shell was qualitatively analyzed by X-Ray Diffractometer (XRD). The diffracted samples were ground and pulverized, screened with a 100 mesh sieve, and the screened sample was taken. The test parameters are Cu target, voltage 40 kV, current 60 mA, scanning step length 0.02°, scanning speed 5°/min, diffraction angle 2θ ranging from 10° to 70°.

(2) **Microstructure analysis**

The microstructure and fracture morphology of the material is analyzed by a scanning electron microscope (SEM) equipped with an energy dispersive spectrometer (EDS).

(3) **Bending strength test**

The room temperature strength and mechanical properties of the shell and core-filled materials were tested on a universal material testing machine, and the high-temperature mechanical properties of the shell materials were tested on a high temperature bending testing machine. The sample size was 150 mm × 20 mm × 20 mm, and the loading rate was 1 mm/min. The bending strength of the material can be calculated using Eq. (4.7),

$$\sigma_{max} = 3P_{max}L/2WH^2 \tag{4.7}$$

where σ_{max} is the bending strength (MPa) of the shell, P_{max} is the maximum load (N), L is the span (mm) of the sample, W is the width (mm) of the sample, and H is the thickness (mm) of the sample. The test results are then obtained as the average of 3 or more sets of data.

AFL_B, i.e., corrected fracture load, is defined as the maximum load required to fracture a 10 mm wide and 70 mm span shell sample and is a method for standardizing the bearing capacity of the shell. It can be calculated using (4.8):

$$AFL_B = f_B\sigma_{max}H^2 \tag{4.8}$$

where f_B is a constant equal to 0.1.

(4) **Wedge strength test**

A wedge-shaped wax pattern with an internal angle of 12°, a side length of 20 mm, and a width of 10 mm is coated with a shell sample. After six layers of coating are applied, the wedge-shaped strength pattern shell sample can be obtained after sanding. The cross-sectional schematics of the wedge-shaped strength wax pattern and the shell sample are shown in Fig. 4.13.

When measuring the wedge strength of the shell, the wedge-shaped strength sample is placed on the wedge-shaped base with an internal angle greater than 12°.

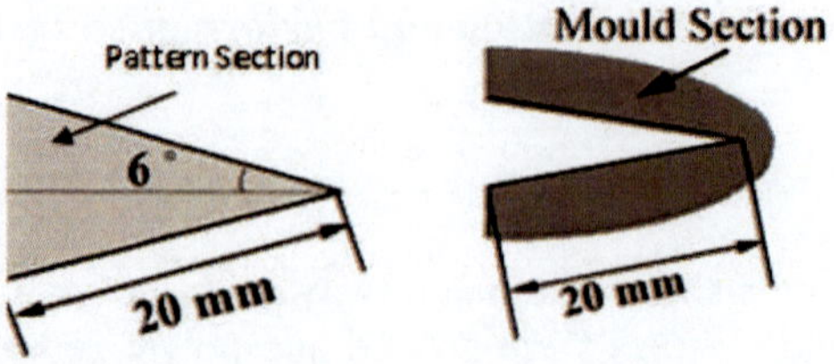

Fig. 4.13 Cross-sectional schematic diagram of wedge-shaped strength wax pattern and shell sample

The wedge-shaped strength sample applies a specific load to break the sample, and the inner surface of the wedge-shaped shell is subjected to tensile stress when loaded. The outer surface is subjected to compressive stress, as shown in Fig. 4.14. The wedge strength of the shell can be calculated by Eq. (4.9) according to the maximum load that makes the shell sample fracture:

$$\sigma_{edge} = 12.2\frac{Fd\sin\theta\cos\theta}{WT^2} \tag{4.9}$$

where σ_{edge} is the wedge strength (MPa) of the shell, F is the maximum load (N) for the shell fracture, d is the side length (mm) of the sample, T is the thickness (mm) at the tip of the sample, and W is the width (mm) of the sample.

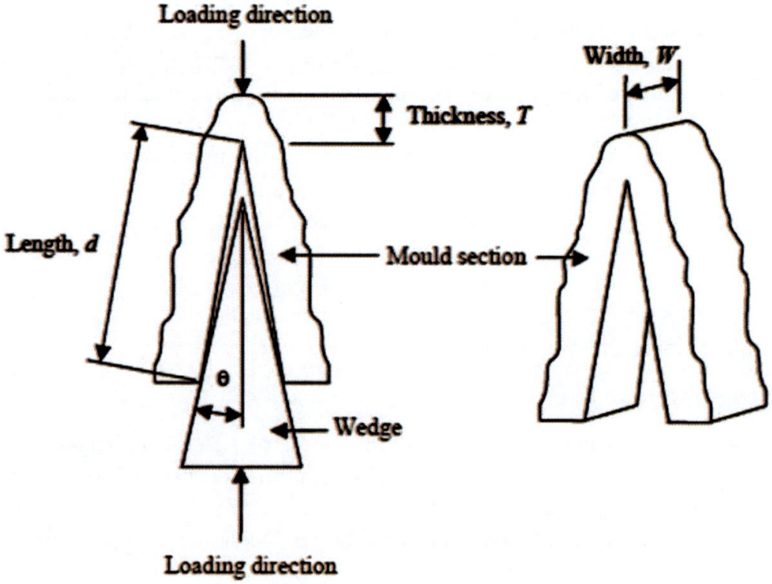

Fig. 4.14 Schematic diagram of wedge-shaped shell strength measurement

(5) Deformation measurement

The shell is coated on a wax pattern with a size of 40 mm × 10 mm × 10 mm, and the primary layer is coated with a layer. After drying for 24 h, the back layer is coated with seven layers, which are dried for 8 h. The last layer is the sealing layer, which only sticks to the pulp and does not scatter sand. After the coating is finished, the shell is steam dewaxed, cut, and polished to a size of 20 mm × 10 mm × 5 mm strips.

The flat-shell sample strip is placed on the corundum fulcrum in the roaster, and the span between the fulcrums is 150 mm. After heated to a specified temperature with the calciner, the sample is applied by placing a zirconium silicate load sheet in the middle of the sample. After holding the temperature for 1 h, the load piece is removed. Then, the sample is cooled with the furnace, as shown in Fig. 4.15. After cooled down to room temperature, the deflection, thickness, and width of the sample are measured. When the molded shell sample is loaded, the corresponding stress and elastic strain can be calculated using Eqs. (4.10) and (4.11), respectively:

$$\sigma = \frac{3F_{max}L}{2WH^2} \tag{4.10}$$

$$\varepsilon = \frac{6H\delta}{L^2} \tag{4.11}$$

where F is the loading amount (N), L is the span (mm), W is the sample width (mm), H is the sample thickness (mm), δ is the deflection (mm), and ε is the strain (%).

(6) Permeability test

Insert one end of a quartz tube, which has an inner diameter of 6 mm, an outer diameter of 10 mm, and a length of 250 mm, into a ping pong with a diameter of 40 mm. In the ball, the insertion depth is controlled within 5–20 mm, and the bonding table of the table tennis ball and the quartz tube is closed and fixed by the bonding wax. Then prepare the shell on the table tennis, and coat the layer. The back layer of the sixth layer and the back layer of the last layer is closed. Also, the quartz tube and the table tennis ball's connection end with about 30 mm length should be coated on the shell. After the shell is coated and dried, the butanone solution is poured into the table tennis from the open end of the quartz tube. The butanone solution can dissolve the table tennis ball and continuously change the butanone solution until the methyl ketone solution does not change color, which indicates that table tennis has been

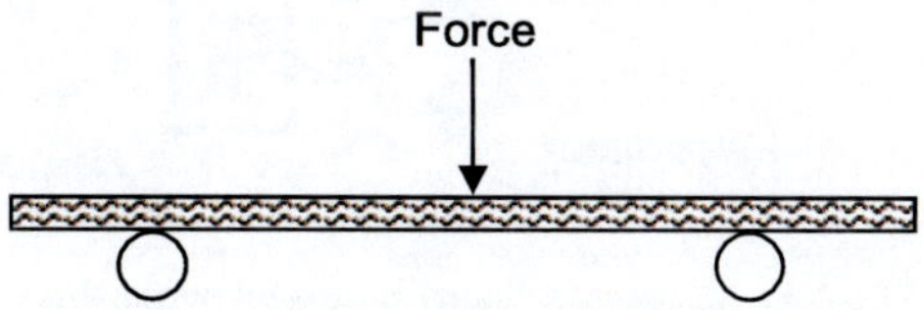

Fig. 4.15 Schematic diagram of bending deformation measurement under shell load

Fig. 4.16 Gas permeability sample of shell

completely dissolved. After drying for 2 h, the gas permeability can be measured. A sample for the gas permeability test of the shell is shown in Fig. 4.16.

The schematic diagram of the shell gas permeability's testing device is shown in Fig. 4.17. By using compressed air as a gas source, the shell sample's air pressure is kept constant at 68.95 kPa (10 psi), and the temperature rise rate of the shell sample is 3 °C/min. The test temperatures are 25, 20, 40, 60, 80, and 100 °C. Then the permeability of the shell μ (m^2) can be derived from Eqs. (4.12) and (4.13).

$$\mu = \frac{\eta V l}{\alpha p} \tag{4.12}$$

$$\eta = \eta_0 \frac{T_0 + B}{T + B} \left(\frac{T}{T_0}\right)^{\frac{3}{2}} \tag{4.13}$$

where η is the viscosity of air at the test temperature (N_s/m^2), V is the flow rate of air (m^3/s), l is the thickness of the shell sample (m), a is the inner surface area of the molded shell (m^2), p is the gas pressure passing through the sample of the mold shell (N/m^2), η_0 is the viscosity of air at room temperature (N_s/m^2), T is the elevated test

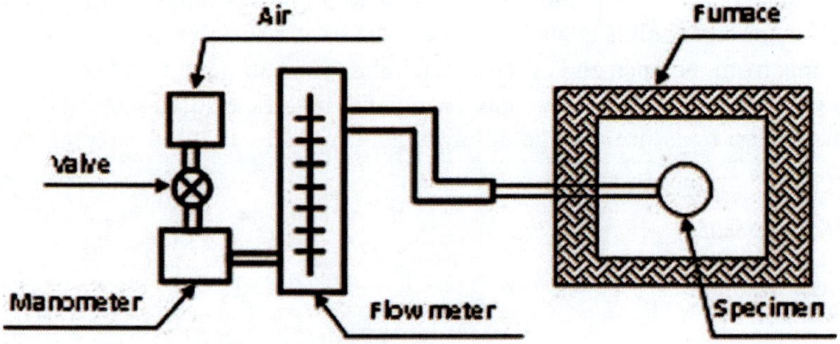

Fig. 4.17 Schematic diagram of shell permeability testing device

temperature (K), T_0 is room temperature (K), B is a constant whose number is about 10.4 K.

Each group of samples consists of six samples. Calculate the gas permeability values of the six samples separately and then take the average. Samples that exceed the average of 15% are rejected.

(7) **Density and porosity measurement**

The density of the shell was measured using the Archimedean method. Three samples are selected from each group, and the serial number is indicated. In the electronics balance M_a according to its weight. And then the specimen in a beaker of water to boil until there is no air bubbles escape, remove with a damp cloth to wipe the water on the surface of the specimen, M_b according to its weight. Place the beaker with water on the electronic balance and zero. Note that the water is not too full of exceeding the maximum range of the balance. Next, attach the sample to a thin wire and immerse it in a beaker filled with water. Note that the sample should be completely immersed in water but not in contact with the beaker wall and, at the bottom of the beaker, weigh the weight in water M_c. If the density of water is ρ_w, the sample can be calculated according to the following equation:

$$\rho = \frac{M_\alpha}{M_e} \cdot \rho_w \tag{4.14}$$

$$\theta = \frac{M_b - M_a}{M_a} \tag{4.15}$$

where ρ is the bulk density (g/cm^3), and θ is the porosity (%).

(8) **Measurement of thermal expansion coefficient**

The coefficient of thermal expansion of the shell in the range from room temperature to 1400 °C was measured using a "zero friction" dilatometer. The sample size was 4 mm × 4 mm × 20 mm, and the heating rate was 5 °C/min at standard atmospheric pressure.

4.3 Characterization of Large Ceramic Shells

4.3.1 Characterization of Ceramic Slurry

4.3.1.1 Physical Properties of Ceramic Slurry

The ceramic slurry is a vital part of the shell. The behavior of the slurry directly affects the performance of the shell. The shell can be divided into a primary layer and a back layer, the slurry is also divided into a primary layer slurry and a back slurry. The MS and WS shells described in this chapter have the same primary

Table 4.7 Properties of ceramic slurry

Test item	Primary	MS secondary	WS secondary
Temperature (°C)	18–25	7–10	18–25
Viscosity time (s)	26–28	21–25	12–15
pH value	9.6–10.5	1.5–3.0	9.6–10.5
Density (g/cm^3)	2.4	1.4	1.5
Solid content (%)	>78	>76	>72
SiO_2 content (%)	18–23	18–23	26–32
Plate retention (μm)	89	146	174
Wet film thickness (μm)	75	–	–
Bubble test (in.$^{-2}$)	<1	<1	<1
Foam test (s)	28	18	26

slurry but different back layer slurries. The physical properties of the three slurries are shown in Table 4.7. Test items include slurry temperature, viscosity, pH, density, solids content, silica content, amount, wet film thickness, bubble content, and bubble rupture rate.

The functionality of the ceramic slurry is mainly determined by the formulation of the slurry, which means the type of binder and refractory materials of the slurry, e.g., solid content, silica content, etc. However, for a particular formulation of the slurry, the focus is on its process and rheology. Among the properties listed in Table 4.7, temperature, pH, and viscosity are the three most basic performance indicators of the slurry. Temperature and pH mainly affect the stability of the slurry, and viscosity mainly affects the work of the slurry.

The temperature of the slurry is easier to control. The shell making workshops of large precision casting plants are generally equipped with constant temperature and humidity systems. A temperature control device can also be installed in the barrel to control the temperature of the slurry more accurately.

The pH value of the slurry has many influencing factors, and the electrolyte in the non-deionized water, refractory material, and the investment cleaning agent has a significant influence on the pH value of the slurry. Besides, low renewability, high silicon content, over-mixing, bacterial content of the slurry, and excessive standards also have a more significant impact on the pH of the slurry. Therefore, when the pH in the slurry deviates too much, acid adjustment is required. The acidic slurry is generally adjusted with dilute hydrochloric acid, and alkaline slurry can be used with ammonium hydroxide or tri-ethanol.

The viscosity of slurry can be divided into flow viscosity, kinematic viscosity, and dynamic viscosity. The viscosity of slurry listed in Table 4.7 is flow viscosity, which reflects the fluidity of the slurry. The so-called flow viscosity corresponds to the scenario where the slurry flows out completely from a specific volume of flow cup at the bottom of a certain aperture in the nozzle.

The primary layer slurry is used to form a smooth and compact surface of the ceramic shell surface. In general, the size of the refractory powder is relatively small, and the solid content of the slurry is relatively high. Therefore, compared with the back slurry, the primary layer slurry generally has higher viscosity of flow cup. The flow viscosity of MS back slurry is higher than that of the primary layer slurry. However, since MS slurry is alcohol-based, its fluidity and wettability are better than those of the water-based slurry, and the higher viscosity does not harm the coating of the shell. For the water-based WS slurries with the same back layer, the viscosity of the flow cup is required to be lower than that of the primary layer slurries and MS slurries to obtain a better coating effect.

The flow viscosity of the slurry is mainly affected by the particle size of the refractory material and its volume content in the binder. In the case of a fixed grade of refractory, the size distribution and shape of refractory are fixed, and the viscosity of the slurry flow cup is mainly affected by the volume content of refractory in the binder. The larger the volume of refractory in the slurry, the greater the slurry flow cup's viscosity and the thicker the slurry, as shown in Fig. 4.18a.

It can be seen that the flow viscosity of the primary slurry and the two back layer slurries increase as the volume fraction of the refractory material increases. The flow viscosity can be converted into kinematic viscosity, which can be used for kinematic viscosity on No. 4 and No. 5. The flow viscosity between the flow cups is converted, as shown in Fig. 4.18b. Based on Fig. 4.18a, b, it can be seen that under the same flow viscosity, the back slurry has the most considerable solid volume content, whereas the WS back slurry has the second largest. The layer slurry has the smallest volume content of solids; this is determined by the particle size and grain shape of the refractory material in the three slurries. The refractory material used in the primary slurry is 325 mesh alumina, and the particle size is much smaller than that in the two back layer slurries, resulting in the lowest solid volume content. The refractory particle size used in the two back layer slurries do not differ much, but the WS slurry is a full fused silica slurry. In addition to fused silica, the MS slurry contains a certain amount of mullite. Therefore, compared with WS slurry, MS slurry has a higher solid volume content.

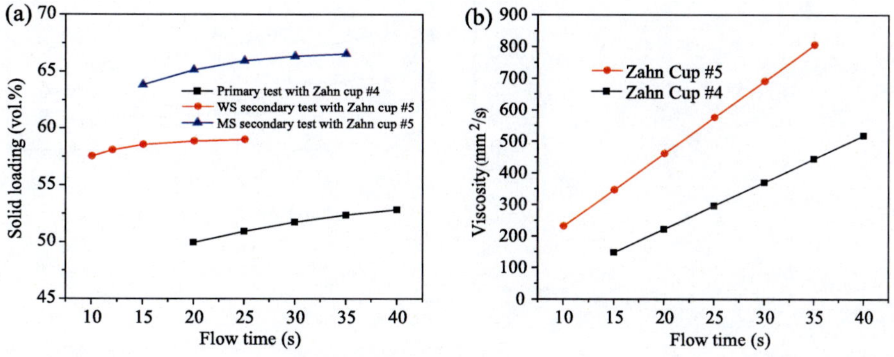

Fig. 4.18 Relationship between solid content and viscosity of the slurry. **a** The relationship between slurry solid content and flow time; **b** relationship between kinematic viscosity and low viscosity

4.3.1.2 Rheological Properties of Ceramic Slurry

Flow viscosity is one type of technological viscosity commonly used in the daily maintenance of slurry, which can reflect the flow energy of slurry but cannot characterize the fluid type of slurry. To this end, the kinetic viscosity of the three slurries is measured using a rotating viscometer, and the result is shown in Fig. 4.19. The kinetic viscosity curve can reflect the rheological properties of the slurry. The silica sol and ethyl silicate hydrolysate slurry are generally pseudo-plastic fluids with low yield value [108]. The yield value of slurry refers to certain minimum shear stress needed to drive the fluid to flow. The strength of the internal structure of slurry is mainly caused by a network structure with low strength formed by the mutual attraction between dispersed phase particles of the whole system.

It can be seen from Fig. 4.19 that when the rotor speed is constant, the viscosity of the slurry is higher, so is the dynamic viscosity. The viscosity of the three ceramic slurries decreases with the increase of the rotor speed, which is characterized by shear thinning. It is a typical pseudoplastic fluid, which is most suitable for the use as an investment casting slurry. When the slurry is stored in the pulp tank daily, most of the slurry in the pulp tank is in a low shear rate state. In this case, the slurry viscosity is high, and the slurry can be suppressed to the maximum extent. The precipitation of the refractory powder keeps the slurry uniform and stable. However, the shear rate at the contact point between the rotating wax mold/shell and the slurry is larger, and the viscosity of the slurry at the interface is lower, which is beneficial to the slurry coating.

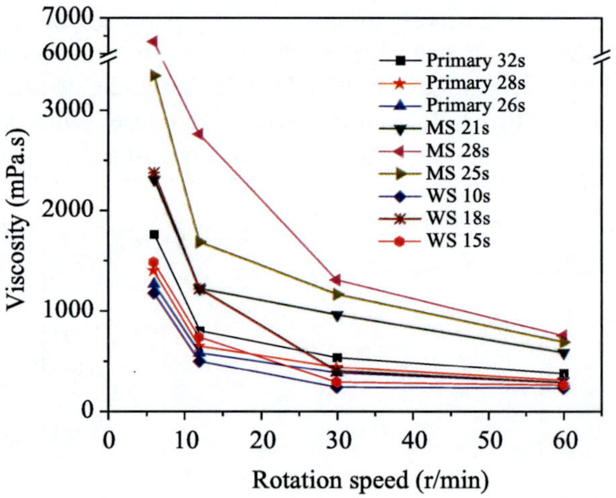

Fig. 4.19 Kinetic viscosity curve of ceramic slurry under different viscosities of flow cup

4.3.2 Microstructure Analysis of Ceramic Shell

The ceramic shell is structurally composed of a top layer and a back layer. Since the primary layer and the back layer have different functional requirements, the material compositions of the primary shell layer and the back layer generally can vary significantly, and different binders and refractory powders are typically used. Therefore, the composition of ceramic shells is complex, and some complex shell systems can contain nearly ten elements.

The preparation process of ceramic shells is complex, which requires a layer-by-layer coating. For a certain layer of the shell, different coating areas are limited by the technical characteristics of the slurry coating and sand spraying, e.g., uneven slurry coating thickness and uneven sand spraying, which occur inevitably. The structural difference between different layers is significant, not only between the primary shell layer and the back layer but also between the back layer. It is hard to achieve a uniform structure in the shell due to different coating time under the slurry process and rheological property in the fluctuation.

The cross-section structure of a typical ceramic shell is shown in Fig. 4.20. The structure of the shell is generally divided into a primary layer and a back layer. There are two points worth mentioning regarding the cross-section structure of the molded shell: the combination of the primary layer and the back layer, distribution of slurry, and sand-spraying materials in the back layer.

4.3.2.1 Combination of Shell Primary Layer and Back Layer

The bonding condition between the shell primary layer and back layer has a great influence on the casting quality. The main concern of the primary layer and back layer is the requirement of its bonding strength.

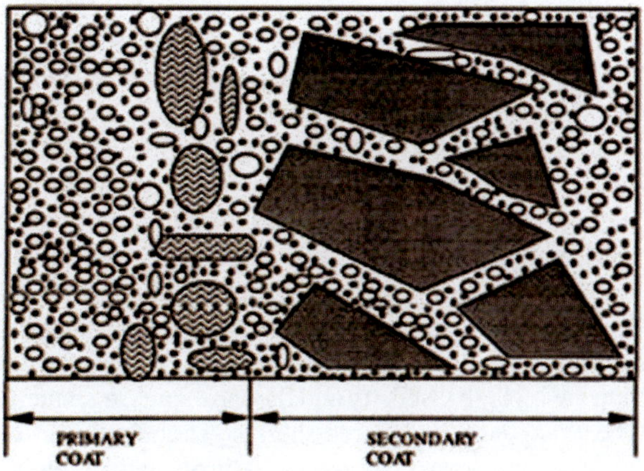

Fig. 4.20 Typical structure diagram of ceramic shell [109]

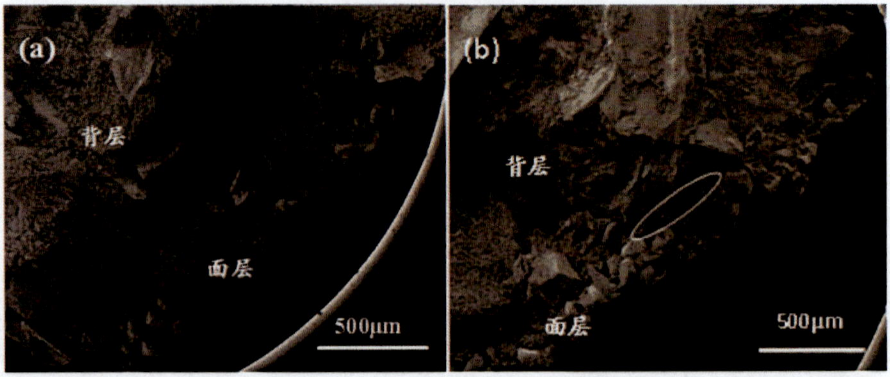

Fig. 4.21 Combination of primary layer and a back layer. **a** MS shell; **b** WS shell

The refractories used by the primary layer and the back layer are generally different, and their expansion coefficients are different. If the bonding strength between the primary layer and the back layer is not high enough, the former is likely to fall off when the shell is subjected to high temperature or metal liquid impact, resulting in leakage and inclusion defects. The particle size of the primary layer is generally small, which is not conducive to the combination of the primary layer and back layer. The cross-section microstructure of the MS-shaped shell and WS-shaped shell is shown in Fig. 4.21. It can be seen that the primary layer and back layer of the MS-shaped shell are closely bound, and no signs of poor bonding between the primary layer and back layer can be observed in SEM images, as shown in Fig. 4.21a. However, in the cross-section structure of the WS shell, traces of clearance can be faintly seen at the junction between the primary layer and back layer, as shown in Fig. 4.21b. It can be seen that the bonding condition of the MS back layer and the primary layer is better than that of the WS back layer and primary layer, which is related to the good wettability and strong permeability of alcohol-based slurry of MS shell.

4.3.2.2 The Layered Structures of the Two Shell Back Layers

In general, there is no need to worry about the bonding strength between the back layers, except for the occurrence of excessive dust content of sand-coated materials and extremely rough inter-layer drying conditions. The distribution of slurry and sand-spraying materials in the back layer directly depends on the slurry viscosity and sand-spraying grain size. When the slurry with low viscosity and good fluidity is used, and the sand spraying material with appropriate granularity is used, a thin and compact overall shell can be obtained. The layered shell can be prepared using the slurry with high viscosity and the sand spraying material with appropriate granularity. The strength of the layered shell is generally lower than that of the monolithic shell, but when cracks appear, the crack propagation can be effectively prevented through the gap in the sand layer.

The cross-section structure of the back layer shell can be divided into three types: monolithic structure, semi-layered structure, and fully Layered structure [110]. Generally speaking, the permeability of monolithic shells and layered shells is better than that of semi-layered shells, while the strength of monolithic shells is weaker than that of layered shells and semi-layered shells because their structures are more conducive to crack propagation [111]. The cross-section structures of the two back shell types are shown in Fig. 4.22. It can be seen that the MS back layer shell is a typical layered shell due to its high slurry viscosity, whereas the WS shell presents the characteristics of the overall shell. However, for WS shells, the addition of needle petroleum coke significantly improves the billet strength and air permeability of the shells. Additionally, the refractory material of fused quartz, which has extremely strong thermal shock resistance, can reduce the generation and propagation of cracks in the shell during dewaxing, sintering, and pouring.

The slurry viscosity was adjusted from 25 to 19 s, and the MS-shaped shell was shown in Fig. 4.22c. It is evident that the MS-shaped shell in Fig. 4.22d presented a semi-stratified structure since the slurry viscosity was adjusted to thin by using a granular-grained sand-coated material. Compared with the medium-sized shell in Fig. 4.22c, the MS-shaped shell in Fig. 4.22d presented an obvious stratified structure. It can be seen that the smaller viscosity of the slurry and the thicker granulation of sand are conducive to the formation of the overall shell, while the thicker viscosity of the slurry and the finer granulation of sand are conducive to the formation of layered shell.

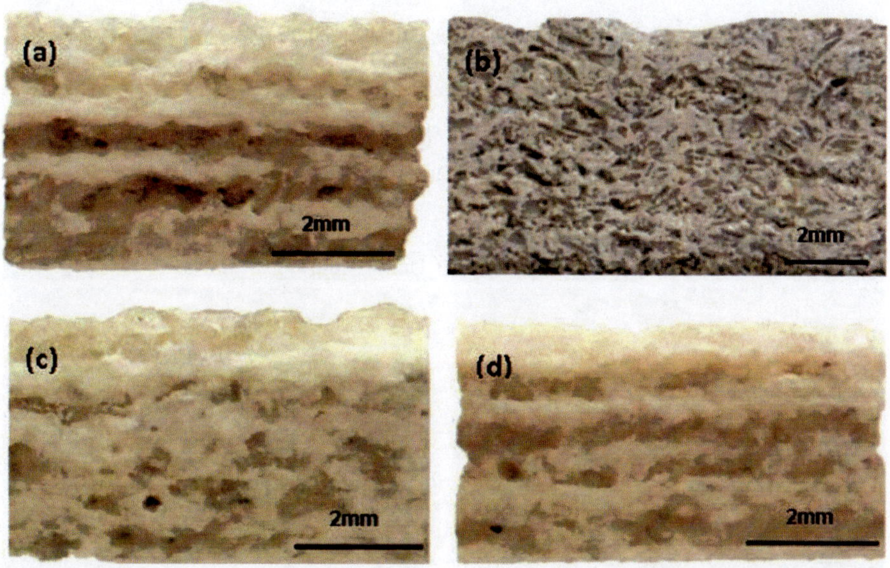

Fig. 4.22 Cross-section structure of the ceramic shell back layer. **a** MS back layer shell (slurry viscosity of 25 s, sand granule of 10/70 mesh); **b** WS back layer shell (slurry viscosity 15 s, sand granule 30/50 mesh); **c** MS back layer shell (slurry viscosity 19 s, sand granulation 10/70 mesh); **d** MS back layer shell (slurry viscosity: 19 s, sand granulation: 50/70 mesh)

4.3.2.3 Microstructure and Phase Analysis of Ceramic Shells

The primary layer of the shell directly is contacted with the metal liquid, which requires high chemical stability and smoothness of the primary layer. The microstructure of the primary shell layer is shown in Fig. 4.23. It can be seen that after the primary layer is dried, its organization is flat, dense, and uniform. After the shell is sintered at 1000 °C, the primary layer's porosity increases significantly with the loss of solvent and latex in the shell. With the increase of sintering temperature to 1200 °C, the viscosity of the nepheline phase formed in the surface shell of SiO_2-Al_2O_3-Na_2O during sintering decreases sharply. The content of the nepheline phase is very small, but it has a great influence on the sintering of the shell, which intensifies the sintering of the surface shell and makes the structure dense. When the sintering temperature of the shell continues to rise to 1400 °C, a denser sintered structure is achieved. With the increasing of sintering temperature, the grain size of the primary layer also increases.

For ceramic shells, the primary layer is usually only one layer, and the back layer contains more than one layer. The back layer determines the strength, permeability, thermal conductivity, and other indicators of the shell. Therefore, the microstructure of the back layer is also observed. The microstructure of the MS back shell is shown in Fig. 4.24. Because MS back layer shell uses alcohol-based adhesives, most of the solvent has evaporated in the drying process. As a result, compared with the primary

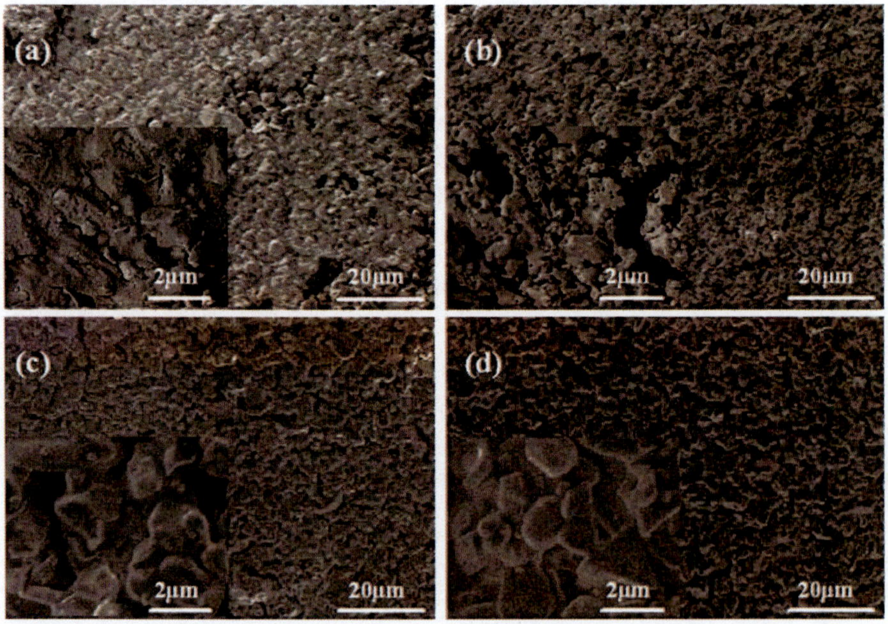

Fig. 4.23 Microstructure of primary shell layer. **a** Blank; **b** roasted at 1000 °C for 2 h; **c** roasted at 1200 °C for 2 h; **d** roasted at 1400 °C for 2 h

layer, after drying of the shell body, a higher porosity is presented; after 1000 °C sintering, shell porosity increase is not obvious, and the organization also no obvious change after sintering. With the increase of sintering temperature, the shell is sintered gradually, and the quartzite is precipitated.

The addition of needle focus has a great impact on the structure and properties of the WS shell. Therefore, it is necessary to understand the microstructure of needle focus. Figure 4.25 shows the microscopic structure of the needle coke. It can be seen from Fig. 4.25 that the length of the needle coke is about 1–2 mm, and the width is about 0.2–0.5 mm. Additionally, the needle coke surface is distributed in needle-like areas. The needle-like fibers are tightly combined, and the surface of needle-like areas is very rough, with many small bumps.

The cross-section of needle coke is composed of a needle-like area and mosaic area. The needle-like area is composed of fine needle-like materials with a diameter of 3–30 μm. The mosaic area is the main structure of the needle coke section. Compared with the needle area, the mosaic area shows less anisotropy and has a porous structure with pore diameter ranging from several hundred nanometers to tens of microns.

The cross-section of the needle coke is a composite structure consisting of a needle-shaped area and a mosaic area. The needle-shaped area is composed of small needles with a diameter of 3–30 μm. Moreover, the surface of the needle is also distributed with many small protrusions. The mosaic area is the main structure of the needle-shaped focal section, which is less anisotropic than the needle-like area and has a porous structure with pore diameters ranging from hundreds of nanometers

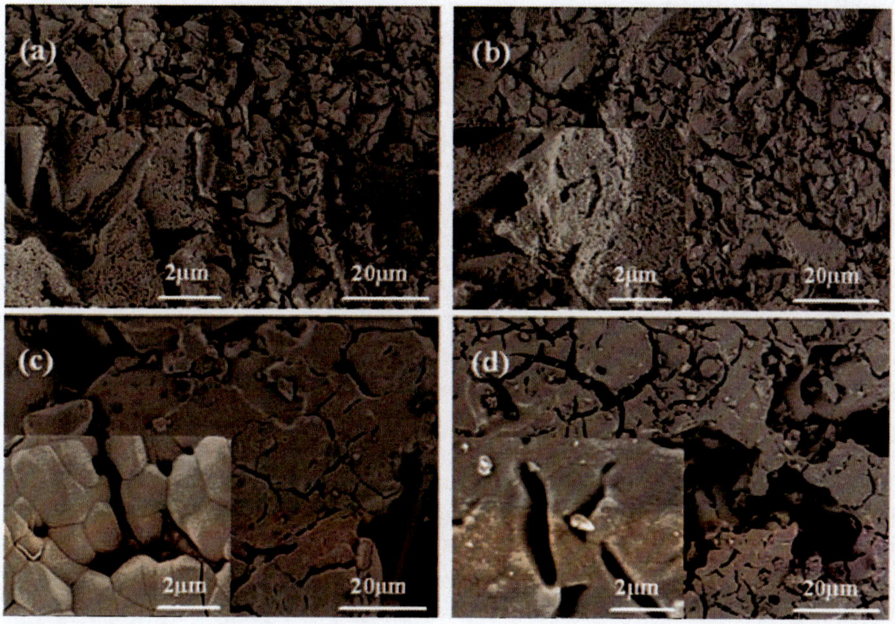

Fig. 4.24 Microstructure of MS back shell

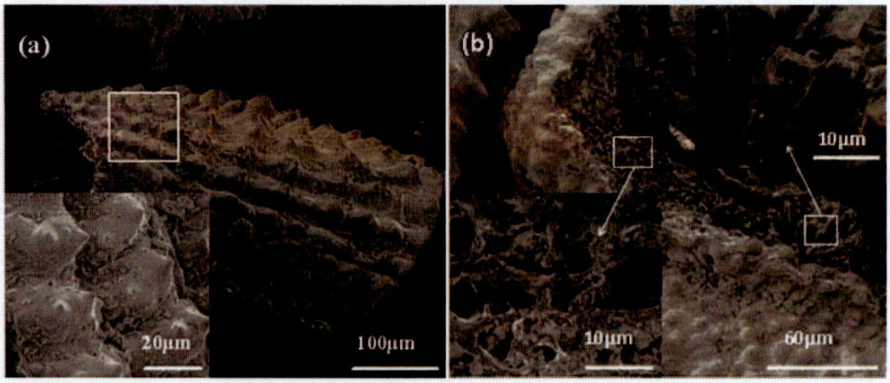

Fig. 4.25 The microscopic structure of needle coke. **a** Outside of the needle; **b** needle coke section

to tens of micrometers. This special structure of needle coke greatly increases the specific surface area. The solvent and tiny particles in the slurry can enter into the pores of the cross-section, causing the needle coke and the shell base to bond more tightly, which is of great significance to improve the strength and permeability of the shell.

The microstructure of the WS shell is shown in Fig. 4.26. It can be seen that there are many pores in the shell body, especially those with a diameter of about 100 nm, which is closely related to the porous honeycomb structure inside the needle coke. Small pores in the porous honeycomb structure in the needle coke deny the entrance of the fused silica particles in the refractory powder, so only the silica particles in the silica sol can enter. Since the solid content in the pores is too low, these pores would appear after drying.

After calcination at 1000 °C, most of the needle petroleum coke of the shell has been burned out, and the fused silica has been substantially converted into cristobalite. As the calcination temperature increases, the pores in the shell gradually merge, leading to increased pore size and a decreased pore number.

Both the MS shell back layer and the WS shell back layer contain a large amount of fused silica, which is prone to crystal transformation at high temperature and cristobalite precipitate, thus posing a significant influence on the performance of the shell. Therefore, it is necessary to study the phase analysis of the high-temperature crystallization of the two types of back shell layers.

Fused silica remains a metastable state below the melting point but is unstable from a thermodynamic point of view. At high temperatures, it gradually transforms into α cristobalite, also known as crystallization. Factors like impurity content, dispersion, temperature, and holding time all affect the crystallization rate. During crystallization, with a volume expansion of 0.9%, the volume change due to crystallization causes the ceramic shell to produce a network crack, which to some extent, reducing the strength of the shell. However, crystallization also changes the silica from a glassy state to a crystalline one, which can effectively inhibit the softened viscous

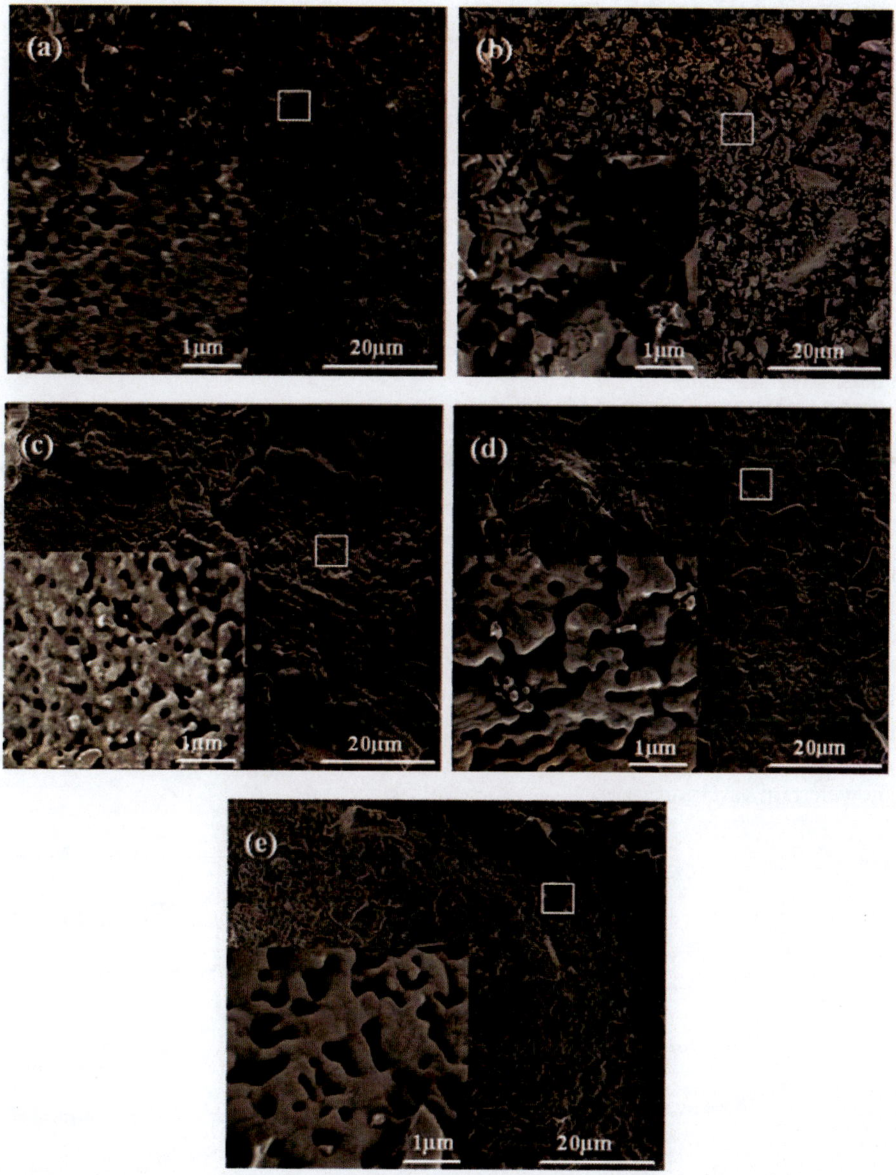

Fig. 4.26 Microstructure of the WS shell. **a** Green body; **b** calcined at 800 °C for 2 h; **c** calcined at 1000 °C for 2 h; **d** calcined at 1200 °C for 2 h; **e** calcined at 1400 °C for 2 h

flow of fused silica, reduce the high-temperature deflection value of the shell, and significantly improve the high-temperature deformation resistance of the shell during crystallization. The volume expansion can also offset some of the shrinkage caused by sintering. The fused silica is transformed into α-cristobalite at high temperatures. When the shell is cooled to 170–220 °C after the casting, the α cristobalite transforms from the high-temperature type to the low-temperature type with a volume shrinkage of 2.8%, causing cracks in the shell. This sharp drop in strength is conducive to the removal of the shell. Therefore, the crystallization of fused silica during calcination and casting is of great significance to improving the high-temperature deformation resistance and improving the peel back resistance of the shell. In general, it is desirable to melt as much as possible during the calcination and casting process to convert quartz to cristobalite.

The XRD patterns of the MS back layer shells at different calcination temperatures are shown in Fig. 4.27a, when the heating rate is 5 °C/min, and the holding time is 1 h. It can be seen from the figure that no obvious cristobalite precipitates over 1000 °C, and as the calcination temperature increases to 1250 °C, crystallization begins to become apparent. The XRD patterns of the WS shells at different calcination temperatures are shown in Fig. 4.27b. The heating rate of the shells during calcination is 5 °C/min, and the holding time is 1 h. It can be seen from the figure that XRD cannot detect the diffraction peak due to the low content of needle petroleum coke in the body, as well as a large amount of cristobalite has been precipitated after 1000 °C. As the calcination temperature increases, the crystallization is more dominant. All the diffraction peaks in the figure are cristobalite phases. Since the fused silica powder grades used in the two back-layer shells are different, the crystallization conditions show certain variation.

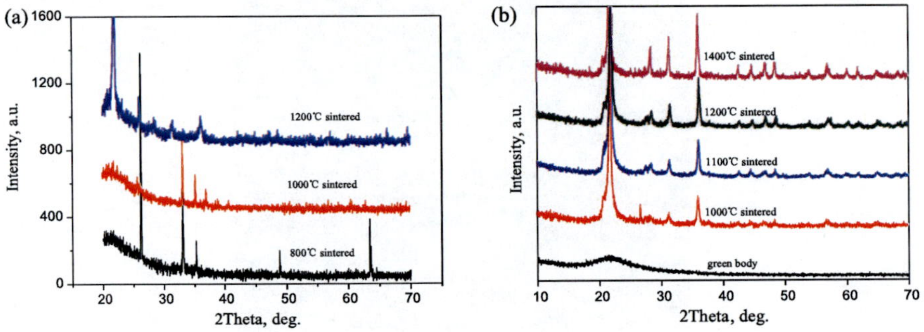

Fig. 4.27 XRD spectra of the shell at different calcination temperatures

4.3.3 *Characterization of Coating Performance of Ceramic Shells*

The coating properties of ceramic shells have to be considered in the research of ceramic shell. If the same strength shell is coated with the same number of layers, it has a thicker thickness, which undoubtedly makes the shell have a stronger load resistance. The shell is composed of slurry and sand spraying. The adhesion thickness of slurry and sand spraying affects the coating thickness of the shell.

Many important parameters need to be controlled when the wax pattern is soaked in the slurry. For instance, the time the wax pattern is immersed in the slurry, the time it takes to be removed from the slurry, and the time the slurry flows, etc. Among them, the drain time of the wax pattern from the slurry is related to the slurry layer's thickness and the adhesion ability of the slurry to the sand. For this reason, the weight of smear and the amount of sand-spraying adhesion of surface slurry, MS back slurry, and WS back slurry are measured.

Figure 4.28 shows the slurry retention and coating thickness on the stainless-steel plate at different drain times. It can be seen that as the drain time increases, the retention of the slurry on the stainless-steel plate is continuously reduced until a steady-state is reached. The retention and adhesion thickness of WS slurry are the highest, MS slurry is in the middle, and the amount of primary layer slurry is the lowest. It can be seen from the figure that when the drain time reaches 80 s or more, the retention of the slurry on the stainless-steel plate is stabilized. The thicker slurry is beneficial to the coating thickness of the shell.

For the slurry of different sanding materials, the particle size and density of the sanding material can vary. As a result, the adhesion effect of the slurry on the sanding is not comparable, and it is therefore difficult to directly compare the thickness difference between the different shells. Due to the limited area of sanding adhesion at the sharp corners of the shell, the amount is insufficient, so the corners tend to have a lower coating thickness than the plane. Consequently, wedge samples of MS

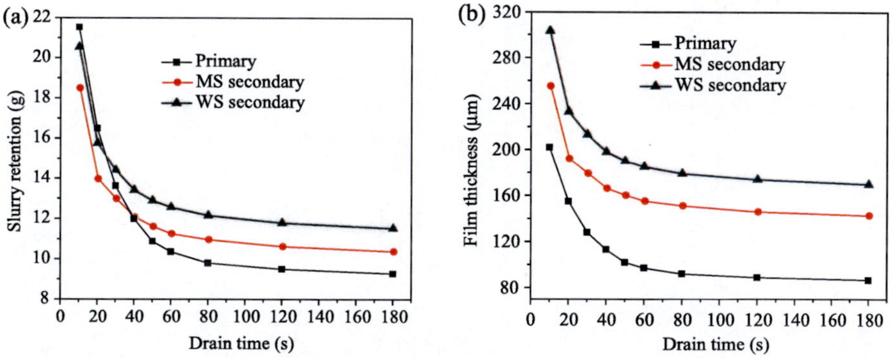

Fig. 4.28 **a** The relationship between the weight of the ceramic slurry smear, **b** the coating's thickness and the slurry's flow time

and WS shells are prepared to measure the thickness of the shell at the plane and sharp corners. The thickness of the coating is six layers, including a primary layer and a layer of sealing, as shown in Fig. 4.29.

The shell sample's thickness at the plane and the sharp corners are measured separately, and the results are shown in Fig. 4.30. It can be seen from the data that the WS shell has a higher coating thickness than the MS shell at both the plane and the sharp corner. The 6-layer WS-shell has a coating thickness of 5.90 mm at the plane, which is 38% higher than the 4.29 mm of the MS-shell. The thickness of the two shells at the sharp corners is lower than that of the corners. The thicknesses of the WS and MS shells at the sharp corners are 4.53 mm and 2.46 mm, respectively. At the flat surface, the coating's thickness is respectively reduced by 23 and 43%, while the thickness of the WS-type shell at the sharp corner is 84% higher than that of the MS shell. It can be seen that the WS shell has a higher coating thickness and better coating uniformity at the plane and sharp corners.

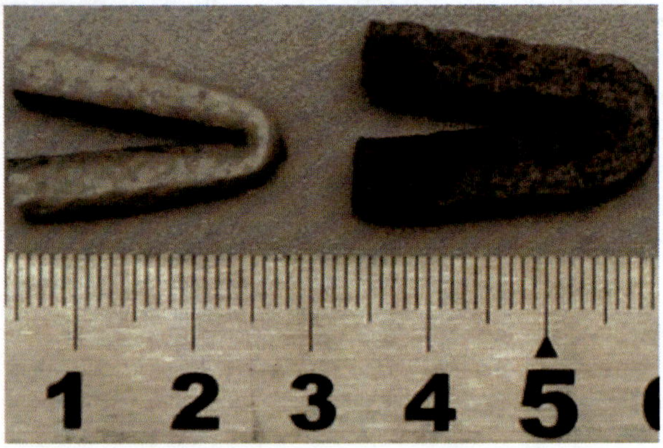

Fig. 4.29 Thickness test sample of ceramic shell

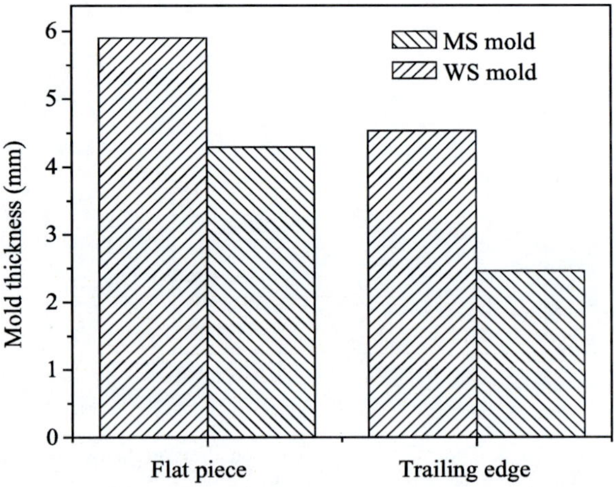

Fig. 4.30 Comparison of the thickness of the ceramic shell

4.3.4 *Strength Characterization of Ceramic Shells*

Strength is the most important and basic property of the shell. The strength of the shell should be able to withstand the following conditions: mechanical treatment stress during coating and moving; wax mold shrinkage and expansion force caused by temperature change during drying; expansion stress of wax mold during heating dewaxing; thermal shock, erosion, and static pressure during molten metal casting; shrinkage stress when the metal solidifies and cools [112]. Therefore, there are three strength indicators at different stages of investment casting, namely, green body strength, high-temperature strength, and residual strength. Shells with excellent performances should have high green body strength, sufficient high-temperature strength, and low residual strength.

4.3.4.1 Drying Mechanism of Ceramic Shell

The drying condition of the ceramic shell has a great influence on the strength of the shell. For the constant temperature and humidity in shell room, the temperature and humidity can be maintained within a suitable range, and the drying time becomes a key factor affecting the shell strength. The slurry prepared by different types of binders lead to huge differences in drying time preparing shells. Therefore, it is very important to determine the dry time of the shells. The relationship between the evaporation amount of the solvent, under different dry time, in the single layer, the MS back layer, and the WS back layer shell with the dry time of the shell was studied, as shown in Fig. 4.31. The relationship between the strength of the dry layer of the six-layer primary layer, the MS back layer, and the WS back layer shell as a function of the dry time of the shell is shown in Fig. 4.32.

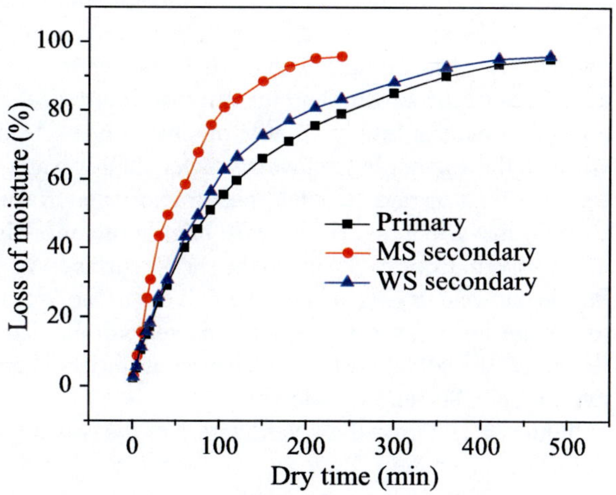

Fig. 4.31 The relationship between solvent volatilization and dry time in the shell

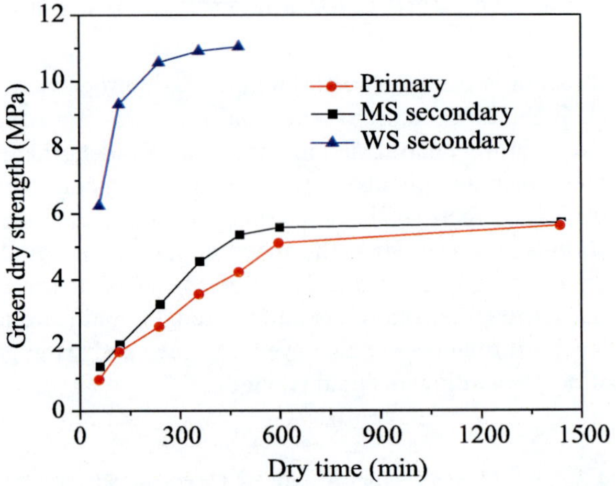

Fig. 4.32 The relationship between the green strength and the dry time of the shell

The volatilization of the solvent in the shell can be roughly divided into two stages, i.e., the constant-speed stage and the deceleration stage [112]. It can be seen from Fig. 4.31 that the first 100 minutes of drying of the primary layer and the back-layer shells is approximately a constant speed stage. The solvent volatilized at this stage is mainly a free solvent encapsulating the ceramic particles on the surface of the shell, as to dry the surface. The evaporation rate of the solvent in the shell is relatively fast and constant. During drying, the solvent migrates from the inside to the surface of the shell by capillary attraction. As the drying process continues, the refractory particles squeeze up against each other, which impedes the solvent's diffusion from inside to the surface. As the shell enters the deceleration stage, during which the evaporation rate of the solvent in the shell is continuously reduced until the solvent exchange between the shell and the external environment is balanced.

As can be seen from Fig. 4.31, the solvent volatilization rate of the MS shell is the fastest, the WS type shell in the middle, and the primary layer shell the slowest. The volatilization rate of the solvent in different shells is related to the volatilization characteristics of the solvent and the microstructure of the shell. The MS shell contains a large amount of highly volatile organic solvent, leading to generally a faster rate, whereas the solvent in the primary layer shell is water, leading to a relatively slower the volatilization rate. Besides, the refractory material used in the MS shell has a small particle size and a relatively dense microstructure, which also hinders the diffusion of the solvent from the inside to the shell's surface. As a result, the primary layer shell has the slowest drying speed. The solvent in the WS shell is also water. The refractory material is fused silica with coarser particle size and irregular particle shape; therefore, the solvent diffusion channel is wider. Therefore, the solvent evaporation rate in the WS shell is moderate.

Figure 4.31 shows the volatilization of the solvent as a function of the dry time in the single-layer shell. However, for the multi-layer shell, the diffusion of the solvent in the shell is much more complicated. In the beginning, the solvent can diffuse to

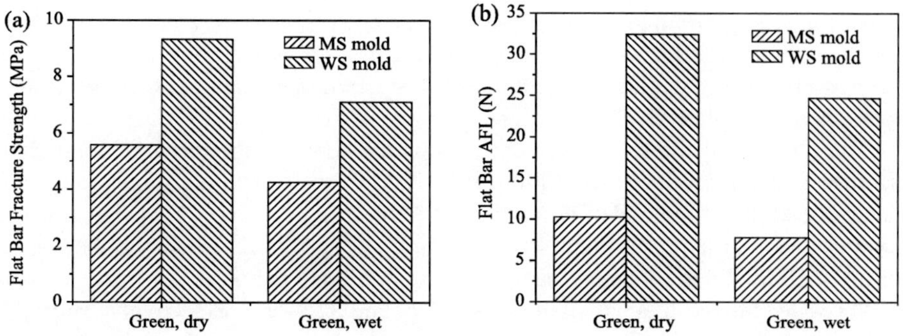

Fig. 4.33 **a** Strength and **b** fracture load of ceramic shell

both the surface and the inside of the shell. As the drying process continues, the solvent previously diffused into the interior gradually diffuses back to the surface again, and then drying is achieved. It can be seen that as the number of coating layers increases, the time required for the shell to dry completely is gradually increased. Thus, Fig. 4.32 shows the strength of the dry body coated with a six-layer shell as a function of the dry time. It can be seen from the figure that as the drying time of the shell prolongs, the strength of the dry body of the shell gradually increases. When the drying time reaches a certain value, the shell strength tends to be stabilized and gradually approaches the maximum. The six-layer surface shell and the MS back layer shell need to be dried for at least 10 h and 8 h, respectively, for the dry body strength to be stabilized. For the six-layer WS back layer shell, the stable strength of the shell can be achieved for 6 h.

4.3.4.2 Bending Strength of Ceramic Shells

The strength of the green body is divided into two types, namely green dry strength and green wet strength. The so-called green dry strength refers to the strength measured after steam dewaxing of the shell at a temperature of 20–25 °C, relative humidity of 50–70%, and drying for 24 h. The green wet strength refers to the strength measured after 15 min of steam dewaxing.

It can be seen from Fig. 4.33 that the green dry strengths of the MS shell and the WS shell are 5.59 MPa and 9.32 MPa, respectively, and the green dry strength of the WS shell is 67% higher than that of the MS shell. The green dry strength of the shell is not sufficient to measure its ability to resist cracking during steam dewaxing [113]. The dewaxing steam atmosphere (temperature and moisture) has a significant effect on the shell properties [114]. The study shows that the green wet strength of the two shells is both lower than the green dry strength, which is 4.25 MPa and 7.10 MPa, respectively, about a quarter lower; this is mainly because that the colloidal network in the shell adsorbs water vapor during steam dewaxing. When the adsorbed water vapor reaches a certain level, the strength of the shell is weakened, and the polymer-like latex in the shell is softened after exposure to high-temperature water vapor, so the combination of the high polymer and the surrounding silica gel particles is also weakened, which adversely affects the strength of the shell.

The (AFL_B) of the shell takes the various thickness of the different shells into account and truly reflects the ability of the shell to resist external forces in certain coated layer. It is a method of normalizing shell bearing capacity [115]. The fracture load of the shell is proportional to the flexural strength of the shell and the square of the shell thickness. It can be seen from Fig. 4.33 that the dry body fracture loads of the MS shell and the WS shell are 10.27 N and 32.44 N, respectively. Since the thickness of the WS shell is 38% higher than that of the MS shell, the fracture load of the WS shell is much higher than the MS shell, with an increase of 2.16 times. The wet body fracture load of the shell is also lower than the dry body fracture load, and the wet body fracture loads of the MS shell and the WS shell are 7.82 N and 24.71 N, respectively. It can be seen that under the same number of coating layers, the fracture load of the WS shell is much higher than that of the MS shell.

The high-temperature strength of the ceramic shell is shown in Fig. 4.34a, and the data of the MS shell and the WS shell at 800 °C are 7.81 MPa and 3.07 MPa, respectively. When the test temperature rises to 1000 °C, the high-temperature strength of the MS shell is abruptly increased to 10.32 MPa, and that of the WS shell is also slightly increased to 3.27 MPa. The high-temperature strength of the ceramic shell is closely related to the number and viscosity of the glass matrix in the shell (glass effect), as well as the degree and mode of contact or bonding between the crystals (crystallization effect) [32]. The low glass phase content, high viscosity, high degree of crystal-to-crystal bonding, and the formation of a continuous interlaced network structure are beneficial to improve the high-temperature strength of the shell [33]. For both the MS shell and the WS shell, the factors affecting the high-temperature strength are mainly the quantity and viscosity of the glass phase. The mullite powder used in the MS shell contains a relatively high content of Na_2O. Therefore, the MS

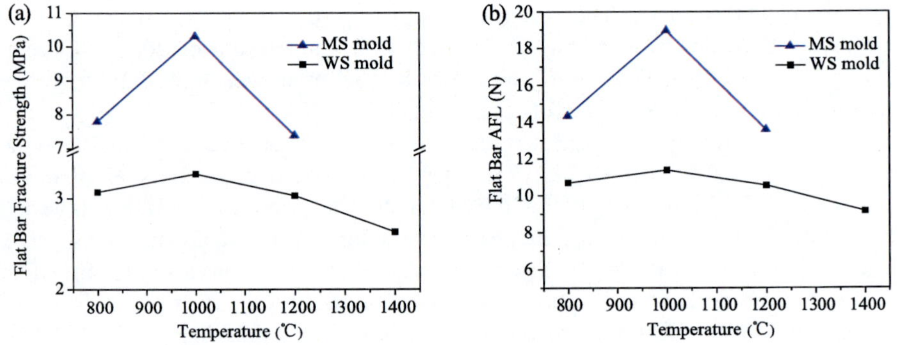

Fig. 4.34 **a** Shell high-temperature strength and **b** shell fracture load

shell can be regarded as a ternary (SiO_2-Al_2O_3-Na_2O) shell system. As the calcination proceeds, the MS shell forms a feldspar or nepheline-reinforced phase at a lower temperature, increasing the strength of the shell [34]. Therefore, the MS shell has higher high-temperature strength at 800–1000 °C. As the calcination temperature increases, the number of reinforcing phases also increases, which greatly improves the high-temperature strength of the MS shell. The content of Na_2O impurities in the WS shell is extremely small. Although SiO_2 and Na_2O form a liquid glass phase at 800 °C, the formation of the glass phase has little effect on the high-temperature strength of the shell. At 800–1000 °C, the WS shell sintering can only rely on the surface diffusion between the refractory particles, not much on the glass phase, and the burning loss of the needle petroleum coke in the shell can then adversely affect the high-temperature strength of the shell. Under the combined effect of all these factors, the high-temperature strength of the WS shell showed a small increase.

When the test temperature is raised from 1000 to 1200 °C, the high-temperature strength of both shells is lowered. For the MS shell, when the calcination temperature reaches 1050 °C, the viscosity of the feldspar phase is significantly reduced [35], resulting in a huge decline of the strength of the MS shell, from 10.32 to 7.40 MPa, while the WS shell's strength, limited by the refractoriness of the fused silica, declines from 3.27 to 3.03 MPa.

When the test temperature rises to 1400 °C, with the further decrease of the viscosity of the glass phase in the MS shell, it has undergone significant softening deformation, and the strength has been impossible to measure. For the WS shell, the strength is further reduced due to insufficient refractoriness of the fused silica, from 3.03 MPa at 1200 °C to 2.63 MPa.

For the high-temperature fracture load of the shell, the strength of the MS shell is higher, but the thickness of the WS shell is larger, which closes the gap between the high-temperature fracture loads of the two shells. The MS shells' fracture loads are 14.37 N, 18.99 N, and 13.62 N at, respectively, 800 °C, 1000 °C, and 1200 °C, while the fracture loads at WS shells' are 10.69 N, 11.38 N, 10.55 N and 9.16 N, respectively at 800 °C, 1000 °C, 1200 °C, and 1400 °C, as shown in Fig. 4.34b.

The collapsibility of the ceramic shell depends largely on its residual strength. The influencing factors of the residual strength are similar to those of the high-temperature mechanical properties of the shell. However, contrary to the high temperature, the higher the glass phase content in the shell matrix, the higher the residual strength of the shell. The contacting or bonding state between crystals in the shell, such as the precipitation of silica, generation of secondary mullite, conversion from α silica to β silica, cracks generated during sintering cooling, etc., also significantly influences the residual strength. It can be seen that the impurity content and the calcination temperature in the shell are the main factors affecting the residual strength.

The residual strength of the shell is shown in Fig. 4.35a. The residual strength of the MS and the WS shell after calcination at 1000 °C and cooling to room temperature are 3.35 MPa and 2.35 MPa, respectively. The higher glass phase content of MS shell leads to a relatively high residual strength. The residual strengths of the MS and WS shells after calcination at 1200 °C and cooling to 400 °C were 5.53 MPa and 3.01 MPa, respectively, which was much higher than the residual strength measured

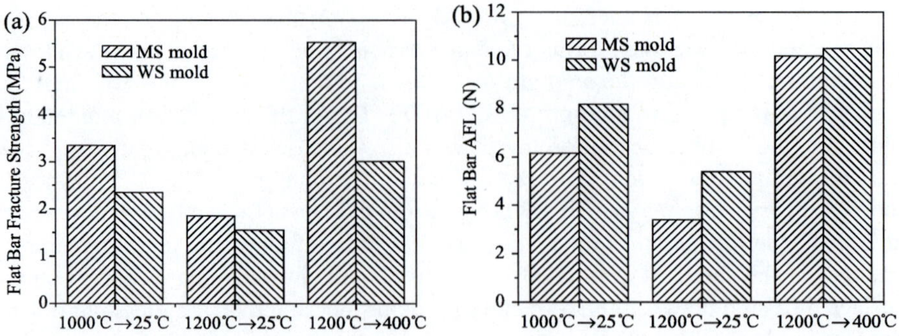

Fig. 4.35 **a** Residual strength of the shell and **b** residual fracture load of shell

after calcination at 1200 °C and then cooled, which are 1.85 and 1.55 MPa. It can be seen that since the conversion from α cristobalite to β cristobalite has not occurred when the shell is cooled to 400 °C, it still has high residual strength. When the shell is cooled to room temperature, because a large amount of α cristobalite has been formed when the shell is calcined at 1200 °C, so at the temperature of 170–220 °C, the conversion from α cristobalite to β cristobalite occurs in the shell, resulting in cracks out of crystal transformation. Therefore, the residual strength of the MS and the WS shell is lower when cooled from 1200 °C than cooled from 1000 °C.

The residual fracture load of the shell is shown in Fig. 4.35b. Although the residual strength of the WS shell is lower than that of the MS shell, the residual fracture load is slightly larger because of the thick shell. The residual fracture loads of the MS and WS shells after cooling to room temperature from 1000 °C were 6.16 N and 8.18 N, respectively. And 3.40 and 5.40 N after calcination from 1200 to 400 °C. The residual fracture loads cooled to room temperature from calcination at 1200 °C were 10.18 N and 10.48 N, respectively.

4.3.4.3 Effect of Needle Petroleum Coke on Bending Strength of WS Shell

The green strength of the WS shell and the cristobalite shell without petroleum coke is shown in Fig. 4.36. The green dry strength of the full-quartz shell is 7.59 MPa, which is 19% lower than the 9.32 MPa of the WS shell. The green wet strength of the full-quartz shell is only 4.86 MPa, which is much lower than the 7.10 MPa of the WS shell. It can be seen that the addition of needle petroleum coke is beneficial to the green body strength of the shell. Promoting the green body strength of the shell has always been the focus of the research community. With the increasing size of the shell and the use of new molds with high expansion coefficients like organic foam, which increase the possibility of cracking during shelling and dewaxing, and raises higher requirements [116]. Latex is usually added to the slurry to enhance the strength of the shell [117]. However, latex is expensive, and the shell with latex

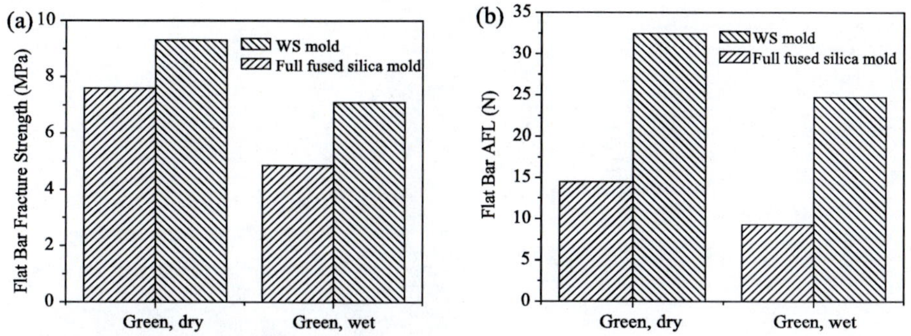

Fig. 4.36 **a** Strength and **b** fracture load of ceramic shell

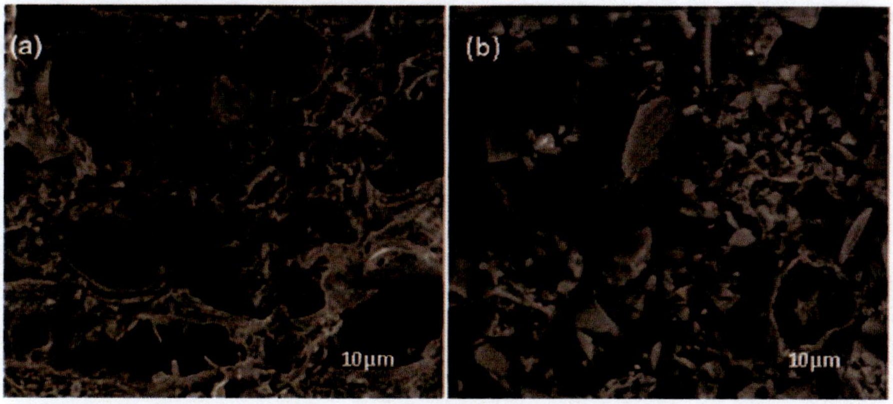

Fig. 4.37 **a** Needle petroleum coke mosaic zone in section and **b** needle petroleum coke mosaic zone in WS shell fracture

added shows a significantly weakened wet body strength under steam. Nylon fiber can inhibit the reduction [118], but the surface of the nylon fiber is too smooth, and the stress used to pull the nylon fiber out of the shell base is small, which has little practical use on promoting the body strength.

The green strength enhancement of the spheroidal petroleum coke is mainly determined by the physical properties and microstructure of the needle petroleum coke. The needle petroleum coke is a special petroleum coke produced through delayed coking of petroleum residue and coal tar. Due to its high degree of graphitization, high strength, low expansion rate, good thermal shock resistance, abundant sources, and low cost, it is the primary filler material for graphite electrodes in electric arc furnaces [119, 120]. The outer surface of the needle petroleum coke is rough and has a porous cross-sectional structure, as shown in Fig. 4.37a, where the pore diameter varies from hundreds of nanometers to tens of micrometers. This unique structure of the needle coke significantly increases the specific surface area; therefore, the silica

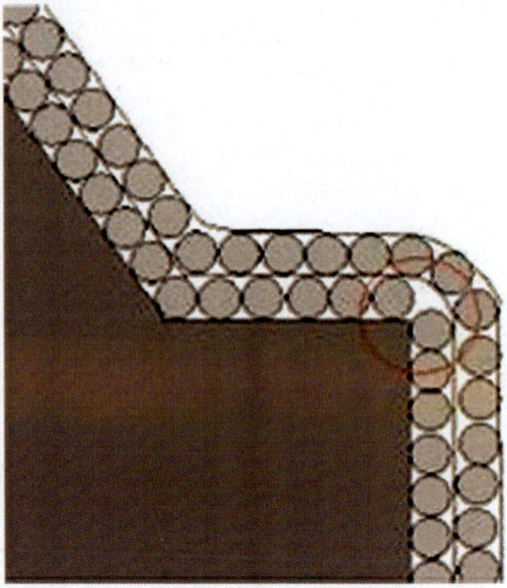

Fig. 4.38 Diagram of stucco distribution at the corner of the shell

sol particles and the fine refractory particles in the shell slurry can enter into the pores of the cross-section; so that the needle coke and the shell base has a tighter degree of bonding, as shown in Fig. 4.37b, which is vital to improving the green body strength of the shell.

4.3.4.4 Wedge Strength of the Ceramic Shell

Strength differs in different parts of the shell. Cracking and breakage of the shell during dewaxing and casting occur mainly at the cusps and corners of the shell, such as the trailing edge of the turbine blade, where the shell has low strength and is prone to stress concentration, which easily leads to the rupture [121–123]. The cusps of the shell have different structures on the same plane, which is determined by the distribution of the sand particles, as shown in Fig. 4.38. As the number of coating layers increases, the area at the cusps of the shell increases gradually. Close to the investment mold, the area at the cusps is small, and the distribution of the sand particles is sparse, which makes the cracks highly possible to be observed [124].

The wedge strength test results of the shell are shown in Fig. 4.39. The wedge strengths of the MS and WS shell body specimens are 1.25 MPa and 1.96 MPa, respectively. The sample is cooled down from 1000 °C to room temperature. The residual strengths obtained were 1.15 MPa and 1.02 MPa, respectively. After cooling from 1200 °C to room temperature, the strength was further reduced, to 0.78 and 0.63 MPa. It can be seen that the wedge strength of the shell is much lower than the flexural strength of the flat sample in the same dry or heated state. In the WS

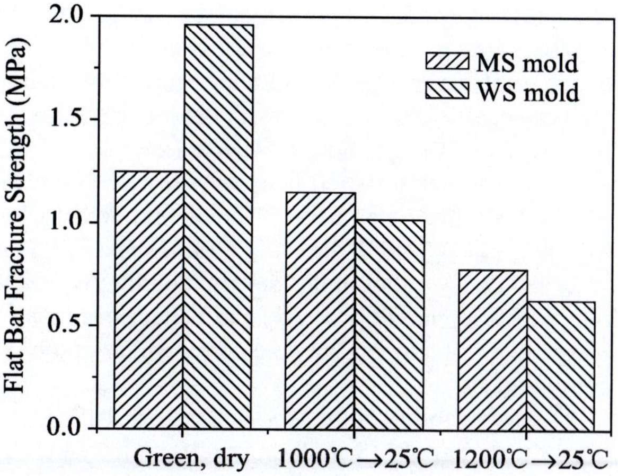

Fig. 4.39 Wedge strength of mold shell

shell, due to the addition of the needle petroleum coke, the coating property at the cusps is improved, and the thickness significantly increased, which is beneficial for improving the wedge strength.

4.3.5 Air Permeability Characteristics of Ceramic Shells

Air permeability refers to the ability of gas passing through the shell wall, which is one of the essential properties of the shell. The permeability of the shell has a significant influence on the dewaxing and pouring process. During steam dewaxing, water vapor enters the surface of the wax pattern through the void of the shell, and the wax can also penetrate the void of the shell after melting to alleviate the expansion stress of the wax material. The shell with good permeability can improve the dewaxing efficiency of the shell and reduce the possibility of cracking when the shell is dewaxed. When pouring, the permeability determines the filling capacity of the shell [125, 126]. The air in the cavity must have the outlet of discharge. Otherwise, the backpressure would delay the filling speed of metal and even cause insufficient pouring [127]. The air permeability can affect the backpressure of trapped gas in the cavity. It is estimated that at least 70% of the dimensional accuracy of castings is related to the permeability of the shell [128].

4.3.5.1 Permeability of Ceramic Shell

At a constant pressure difference and a specific test temperature, the gas viscosity and the inner surface area of the shell are fixed. In this case, the shell's permeability is proportional to the gas flow rate and the thickness of the shell. The gas flow rate is affected by the apparent porosity of the shell and the thickness of the shell. Therefore, apparent porosity and shell thickness are two key factors affecting the permeability of the shell. At different test temperatures, the change of gas viscosity also affects the permeability of the shell; with the increase of test temperature, the gas viscosity increases, and the gas flow rate decreases.

In this chapter, the shell samples' permeability from room temperature to 1000 °C is tested, and the data of the MS shell and WS shell are shown in Fig. 4.40. For the MS shell and WS shell, the sample's thickness is 4.29 mm and 5.90 mm, respectively.

It can be seen that under the test conditions, the WS shell has a larger gas flow rate, higher apparent porosity, and better permeability than the MS shell. With the increase of test temperature, the permeability of the WS shell and MS shell increases continuously.

The permeability of the WS shell is 0.15×10^{-14} m^2 at room temperature, which is higher than that of MS shell 0.09×10^{-14} m^2. With the increase of test temperature, solvents, polymers, and petroleum coke in the shell are evaporated and burn out gradually, so the apparent porosity of the shell increases continuously. However, the increase of apparent porosity of the MS shell is much less than that of the WS shell, which is mainly due to the formation of feldspar or nepheline glass phase in the ternary system (SiO_2-Al_2O_3-Na_2O) MS shell, which is not conducive to the

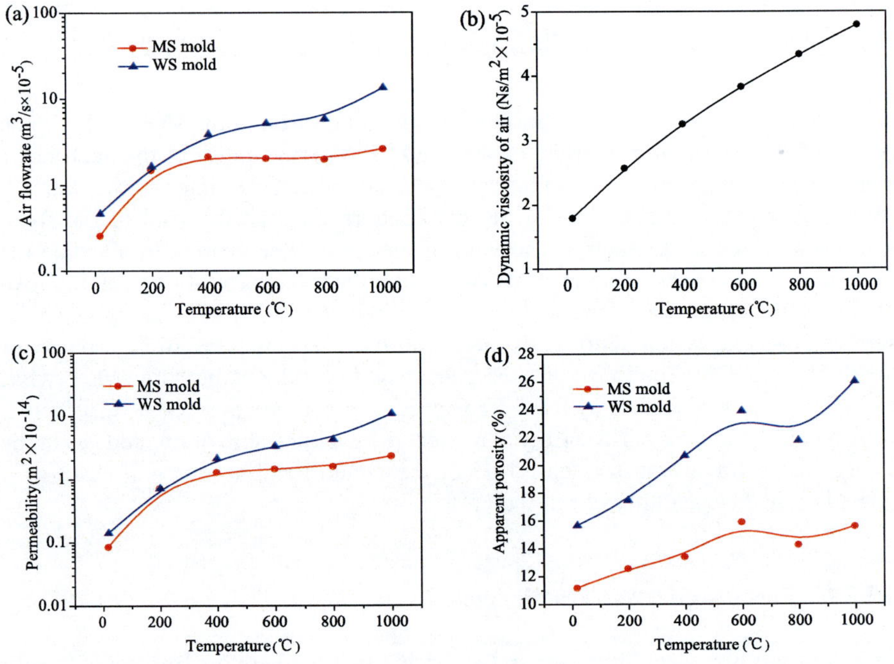

Fig. 4.40 **a** Gas flow rate at different temperatures, **b** air viscosity at different temperatures, **c** permeability of ceramic shell at different temperatures, **d** apparent porosity of shell samples calcined at different test temperatures

increase of apparent porosity of shell. The apparent porosity of the WS shell and MS shell increased to 23.96% and 15.96%, respectively, when the test temperature reached 600 °C. When the test temperature reaches 400 °C, the gas flow rate increases significantly. In the interval of 400–600 °C, the gas flow of the two shells showed a distinctly different trend. The gas flow rate of the WS type shell continues to increase, while the gas flow rate of the MS type shell shows a slight downward trend; this is mainly due to the gas viscosity increases with the increase of test temperature, as shown in Fig. 4.40b. The MS shell is more affected by the increase of gas viscosity, and thus the gas flow rate of MS shell in the range of 400–600 °C is lowered. With the increase of test temperature, the air permeability of the two types of shells is continuously improved. The air permeability of WS shells at 200 °C, 400 °C, and 600 °C is respectively 0.73×10^{-14}, 2.19×10^{-14}, and 3.43×10^{-14} m^2. The permeability of MS shell at the three temperatures increases to 0.72×10^{-14}, 1.33×10^{-14}, and 1.51×10^{-14} m^2.

When the test temperature is from 600 to 800 °C, the number of glass phases in the MS shell increases, further reducing viscosity. The binary system (SiO_2-Na_2O) type WS shell also generates the glass phase at 793 °C, introducing the glass phase into the pore type shell. These glass phases are easy to enter into the pores of molded shells and aggravate the sintering of molded shells. Therefore, when the temperature rises to 800 °C, the apparent porosity of the two types of shells presents a downward trend, and the apparent porosity of WS shells and MS shells decreases to 21.82% and 14.32%, respectively. Due to the increase of gas viscosity, the gas flow in the MS shell continues to decrease slightly, while the gas flow in the WS shell continues to increase. The permeability of WS shell and MS shell at 800 °C was 4.37×10^{-14} m^2 and 1.65×10^{-14} m^2, respectively.

When the test temperature reached 1000 °C, cracks were gradually formed along with the sintering, and the apparent porosity of the shell increased significantly. The apparent porosity of the WS shell and MS shell reached 26.06% and 15.64%, respectively, and the gas flow volume also rose to 13.53×10^{-5} m^3/s and 2.66×10^{-5} m^3/s, respectively. At this temperature, the permeability of the MS shell is 2.40×10^{-14} m^2. The permeability of the WS shell is about four times that of the MS shell, which is 11.13×10^{-14} m^2.

4.3.5.2 Effect of Needle Petroleum Coke on Gas Permeability of WS Shell

The methods that help improve the permeability of shell include the following. (1) The use of granular polygonal, loosely packed low-density refractories as shell materials [129], which can significantly improve the porosity of the shell and make the shell have better air permeability in dewaxing and sintering. (2) Adding high polymer to the shell size [2]. The loss of polymer during shell sintering leaves pores, which are helpful to improve the permeability of the shell. However, the above two ways of improvement are inadequate. The use of polygonal refractory material seriously affects the strength of the shell; the use of latex, nylon fiber, and other high polymers can improve the shell at high-temperature permeability, but cannot improve the permeability of the shell in steam dewaxing.

Permeability was tested to study the effect of needle petroleum coke on the shell. The sample thickness of the WS shell and all quartz shells were 5.90 mm and 4.37 mm, respectively. Air permeability data are shown in Fig. 4.41. It can be seen that WS shells have more massive gas flow and higher apparent porosity than all-quartz shells under the same test conditions. The addition of needle petroleum coke significantly improved the permeability of WS shell by nearly 4 times.

The permeability of the WS shell was 0.15×10^{-14} m^2 at room temperature, which was higher than 0.04×10^{-14} m^2 for all quartz shell. The blank body of the WS shell is evenly distributed with many holes with diameters of tens to hundreds of nanometers. These holes are derived from the cross-section Mosaic area of needle petroleum coke, which can form additional aeration channels in the shell, allowing the WS shell to have greater apparent porosity, gas flow, and better permeability. Studies show that not all the pores in the molded shell affect the air permeability of the elevated shell, and the closed holes in the molded shell or the tiny cracks in one layer of the molded shell do not have a significant effect on the air permeability of the elevated shell [56]. Only the holes with good connectivity are beneficial to

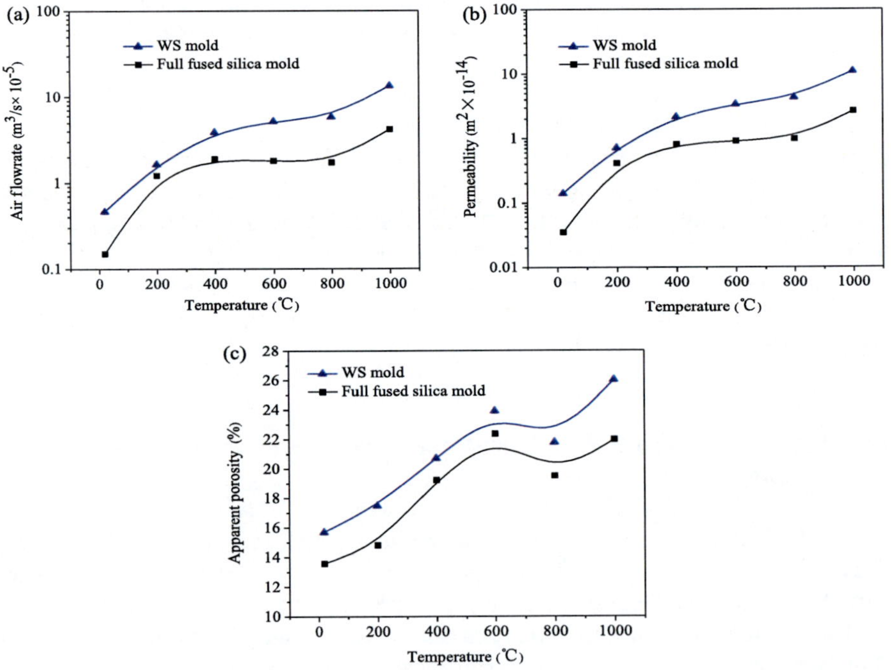

Fig. 4.41 **a** Gas flow rate of shell samples at different temperatures at 10 psi gas pressure; **b** gas permeability of ceramic shell samples at different temperatures; **c** apparent porosity of shell samples after calcination at different test temperatures

the permeability of the raised shell. High air permeability at room temperature can improve the efficiency of steam dewaxing of shell and greatly inhibit the cracking of shell during steam dewaxing.

With the increase of the test temperature, the moisture in the shell gradually evaporates, the polymer and petroleum coke gradually lose, and the apparent porosity of the shell increases continuously. Petroleum coke in the WS shell was continuously burned out with the increase of sintering temperature of the shell, thus forming well-connected holes and improving the apparent porosity of the shell. When the test temperature reaches 600 °C, the apparent porosity of the WS shell increases to 23.96%, higher than the apparent porosity of all quartz shell 22.36%. When the test temperature reached 400 °C, the gas flow rate increased significantly. In the interval of 400–600 °C, the gas flow of the two shells showed a distinct trend. The gas flow of the WS shell continued to increase, whereas the gas flow of the all-quartz shell showed a slightly downward trend; this is mainly due to the gas viscosity increases with the increase of the test temperature, as shown in Fig. 4.40b. It is apparent that the all-quartz shell is more affected by the increase of gas viscosity, thus causing a decrease in the gas flow rate of the entire quartz shell in the range of 400–600 °C. With the increase of test temperature, the two types of shells' air permeability are continuously improved. The permeability of WS shell at 200 °C, 400 °C, and 600 °C are 0.73×10^{-14} m^2, 2.19×10^{-14} m^2, and 3.43×10^{-14} m^2, respectively; and the all-crystal shell's gas permeability at these three temperatures increased to 0.41×10^{-14} m^2, 0.81×10^{-14} m^2, and 0.91×10^{-14} m^2, respectively.

When the test temperature increases to 800 °C, both types of the shell form sodium silicate glass, then the low viscosity of the glass phase flows into the pore in the shell and reduces the porosity. The WS type and quartz type shell's porosity decreases to 21.82% and 19.51%, respectively. Due to the increase of gas viscosity, the gas flow in

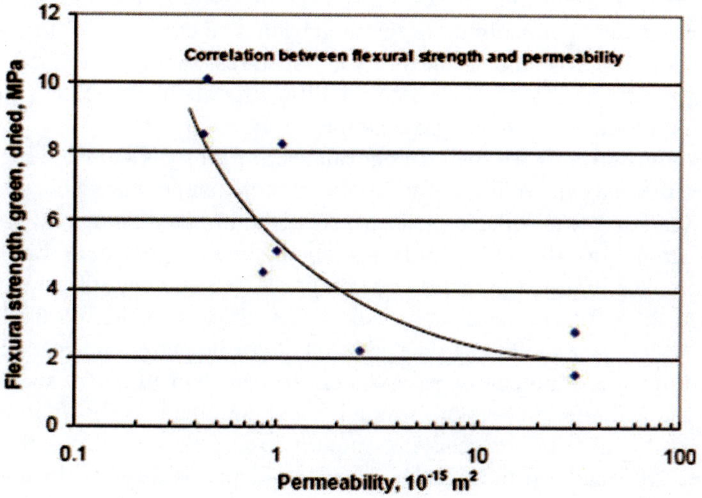

Fig. 4.42 Relationship between strength and permeability of a typical ceramic shell [90]

the all-quartz shell continues to decrease slightly, while the gas flow in the WS shell continues to increase with the further loss of needle petroleum coke in the WS shell and the further increase of well-connected holes in the WS shell. The permeability of the WS shell reached 4.37×10^{-14} m^2 at 800 °C, much higher than the whole quartz shell at this temperature permeability 0.98×10^{-14} m^2.

When the test temperature increases to 1000 °C, the cracks formed in the sintering process of the shell gradually increased, and most of the needle petroleum coke was burned away, leading to a significant increase in the apparent porosity of the shell. The apparent porosity of the WS shell increased more significantly, and the apparent porosity of the WS shell and the all-quartz shell reached to 26.06% and 21.98%, respectively. At 1000 °C, the permeability of the whole quartz shell is 2.64×10^{-14} m^2, and the gas permeability of the WS shell reaches 11.13×10^{-14} m^2. It can be seen that the addition of needle petroleum coke greatly improves the high-temperature air permeability of the shell, which is of great help to the filling integrity of the raised shell.

4.3.5.3 Overall Optimization of WS Shell Strength and Permeability by Needle Petroleum Coke

As being the two most important performance indexes of ceramic shells, the strength and permeability of the shell are often not fully taken into account, and it is often necessary to make a balance between the strength and permeability of the shell [90]. The typical relationship between shell strength and permeability is shown in Fig. 4.42. It can be seen that the shell with good permeability often has high porosity and low green body density, resulting in insufficient shell strength, while the shell with high strength generally has poor permeability.

In this chapter, the design idea of the WS shell is to break the traditional concept that the strength and permeability of the shell cannot be controlled at the same time. It is intended to improve the permeability and load resistance of the shell.

Adding needle petroleum coke into the all-quartz shell can achieve the overall optimization of the strength and permeability of the shell.

The relationship between shell strength and permeability is shown in Fig. 4.43a. It can be seen that the permeability of WS shells at room temperature, 800, and 1000 °C are better than that of all-quartz shells, especially at high temperatures. Meanwhile, the green strength of the WS shell is also significantly higher than that of the all-quartz shell. The high-temperature strength of the WS shell at 800 and 1000 °C is also comparable to that of the all-quartz shell. The thickness of the WS shell is thicker than that of all quartz shell due to the addition of needle petroleum coke. Therefore, WS shell still has a significantly better load resistance than all-quartz shell under the condition that its high-temperature strength is not dominant, as shown in Fig. 4.43b. In summary, the addition of needle petroleum coke greatly improves the high and low-temperature load resistance and high and low-temperature air permeability of the WS shells, realizing the overall optimization of shell strength and air permeability.

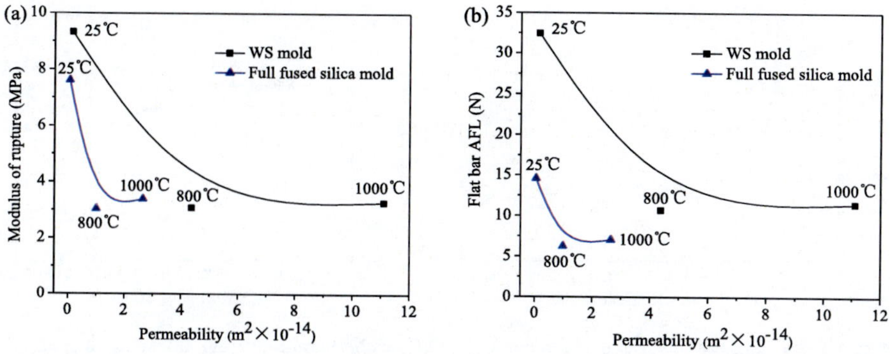

Fig. 4.43 Relation between **a** shell strength and **b** fracture load and permeability of the ceramic shell

4.3.6 Linear Deformation Characteristics of Ceramic Shells

The thermal expansion of shell refers to the expansion or contraction of the shell with the increase of temperature. Linear expansion coefficient is usually used to represent the thermal expansion of shell and refractories in investment casting. The increase of shell size is caused by the thermal expansion of shell material and isomeric crystal transformation. In contrast, the shrinkage of shell size is caused by the dewatering of the shell, thermal decomposition of material, sintering, production of the liquid phase, and condensation of silicone gel.

Thermal expansion is one of the important properties of the shell, which has a direct influence on the dimensional accuracy of casting and also affects the resistance to quench and heat and high-temperature deformation of the shell. The shell with a small linear expansion coefficient and uniform expansion is in line with the requirements of high-quality shell and casting.

4.3.6.1 Linear Expansion of the Shell

The thermal expansion coefficient of MS and WS ceramic shells was measured, and the temperature range was from room temperature to 1400 °C. WS shell sample length is 19.94 mm, and the MS shell sample length is 20.14 mm, with a heating rate of 5 °C/min under the air atmosphere. The obtained average linear expansion coefficient of shell and the average linear expansion amount are shown in Fig. 4.44. It is shown that with the test temperature goes up to 200 °C the coefficient of expansion of shell fell sharply. When the temperature is higher than 200 °C, the MS shell and thermal expansion coefficient of thermal expansion amount of basic constant before plunging trends change, and the shell of the thermal expansion coefficient and the change of the thermal expansion amount except at 800 °C or so appear a little inflection point, change in other temperature range are relatively uniform.

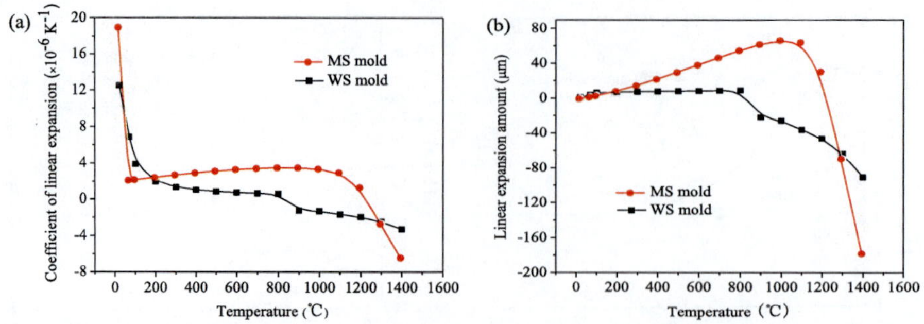

Fig. 4.44 **a** Average linear expansion coefficient of the shell and **b** average linear expansion

Also, the sintering shrinkage of ceramic shells after cooling to room temperature at different sintering temperatures and holding time is tested. The samples are coated with 8 layers, and the size of the block samples was 50 mm × 40 mm × 6 mm, as shown in Fig. 4.45. When measuring the sintering shrinkage of samples in the two directions of length and width, the dimensional accuracy of the shell billet in the two directions shall be ensured, and the dimensional error of the shell in the two directions shall not be greater than 0.04 mm.

The experimental results are shown in Table 4.8. It can be seen that the sintering shrinkage of the mold shells are all negative, which indicates that the sintering shrinkage of the mold shells show size shrinkage when the mold shells are cooled to room temperature. The sintering shrinkage of the mold shells increases with the increase of sintering temperature and the extension of insulation time, while the influence of sintering temperature on sintering shrinkage is greater than that of insulation time. When the sintering temperature of the shell is lower than 1100 °C, the sintering shrinkage is small, and the sintering shrinkage increases little with the extension of insulation time. However, when the sintering temperature rises above 1250 °C, the

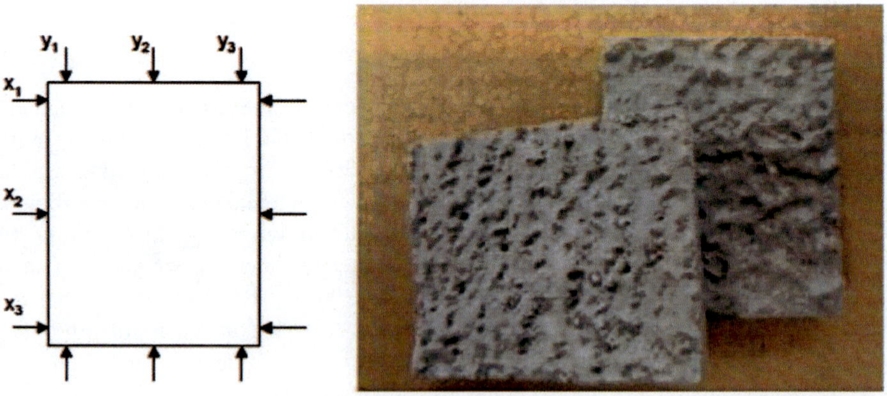

Fig. 4.45 Shell sample used to measure the shrinkage of the shell

Table 4.8 Sintering shrinkage of ceramic shells at different sintering temperatures and holding times

Sintering temperature/holding time	WS shell		MS shell	
	Size change *X*/%	Size change *Y*/%	Size change *X*/%	Size change *Y*/%
1000 °C/1 h	−0.04	−0.02	−0.07	−0.04
1000 °C/3 h	−0.16	−0.18	−0.16	−0.13
1000 °C/5 h	−0.28	−0.22	−0.18	−0.20
1100 °C/1 h	−0.52	−0.50	−0.42	−0.48
1250 °C/1 h	−1.08	−1.11	−1.12	−1.11
1400 °C/1 h	−1.32	−1.29	−1.48	−1.40

sintering shrinkage starts to increase significantly, which is related to the increase of the number of glass phases in the shell, the decrease of viscosity, and the more significant crystallization of fused quartz.

4.3.6.2 Linear Deformation Mechanism of the Shell

Linear deformation closely relates to changes in shell's physical phase and microstructure. Processes like the solvent's evaporation, the crystal transformation, the liquid glass phase formation, and the decrease of the glass phase's viscosity all have an important impact on the linear deformation of the shell.

According to the linear expansion coefficient and linear expansion amount of the shell in Fig. 4.44, the rapid decrease of the linear expansion coefficient between the MS type shell and the WS type shell from room temperature to 200 °C is caused by the volatilization of the solvent in the shell. The linear expansion coefficient of the solvent is 1–2 orders of magnitude higher than the shell. Therefore, as the solvent in the shell evaporates, the linear expansion coefficient of the shell decreases sharply. At this stage, although the linear expansion coefficient of the shell changes greatly, the linear expansion of the shell is not significant because of the low temperature of the shell.

After the solvent in the shell is substantially volatilized, the shell's linear expansion coefficient mainly depends on the viscosity and quantity of the glass phase in the shell. For the ternary system (SiO_2-Al_2O_3-Na_2O) MS type shell, when the temperature rises to 500 °C, the mullite in the shell reacts with the Na_2O impurity to form new crystalline phase nepheline. It is a rack-mounted structure of sodium aluminosilicate without additional anion, melting point of 1526 °C, which has a significant enhancement in the shell. Therefore, with the increase of temperature, the coefficient of linear expansion of the shell remains stable. When the temperature rises above 1050 °C, the nepheline phase adhesion in the MS shell decreases sharply, causing the linear expansion coefficient of the MS shell to drop sharply. When the temperature exceeds

1240 °C, the coefficient of thermal expansion of the shell is negative, and the shell exhibits shrinkage.

For the binary system (SiO_2-Na_2O) WS shell, the inflection point appears near 800 °C, and the expansion coefficient at the inflection point decreases sharply; this is mainly because Na_2O in the shell reacts with SiO_2 at 793 °C to form a liquid glass phase, thereby reducing the expansion coefficient of the shell.

Shell calcination at high temperatures produces liquid glass phases, which leads to the sintering shrinkage in the sintering temperature. At the same time, due to the crystallization of fused silica in the shell at high temperature, cristobalite is precipitated out. The calcined shell undergoes a transition between high and low-temperature crystal forms of cristobalite when cooled to between 170 and 220 °C, along which a volume shrinkage also occurs. These factors lead to a certain amount of shrinkage after calcination of the shell. These shrinkages can react to some extent to the extent of the glass phase in the shell and the high-temperature crystallization of the shell at a certain calcination temperature and holding time the contraction can, to a certain extent, reaction under a certain sintering temperature and holding time, which has an important significance to the research process of the shell.

References

1. Z. Litong, C. Lamei, L. Guoli et al., *Near Net Shape Investment Casting Precision Casting Theory and Practice* (National Defense Industry Press, Beijing, 2007)
2. S. Jones, C. Yuan, Advances in shell molding for investment casting. J. Mater. Process. Technol. **135**(2–3), 258–265 (2003)
3. N. Cannell, A.S. Sabau, *Predicting Pattern Tooling and Casting Dimensions for Investment Casting, Phase III* (Oak Ridge National Laboratory, US, 2007)
4. N. Cannell, A.S. Sabau, *Predicting Pattern Tooling and Casting Dimensions for Investment Casting, Phase II* (Oak Ridge National Laboratory, US, 2006)
5. J. Buju, *Practical Investment Casting Technology* (Liaoning Science and Technology Press, Shenyang, 2008)
6. B. Yanzhen, C. Caijin, Z. Jinlun, *Investment Casting Precision Casting Technology* (Zhejiang University Press, Hangzhou, 2012)
7. China Foundry Association, *Investment Casting Handbook* (Mechanical Industry Press, Beijing, 2000)
8. Y. Shuangjing, W. Jianli, L. Zhigang et al., International investment in casting adhesives and surface coatings additive, in *Proceedings of the 6th Annual Meeting of China Foundry Association Casting Branch*, Wuhan, 1999
9. M. Oles, R. Sharkey, A new approach to the treatment and recovery of alcohol from the exhaust air steam of ethyl silicate shell building, in *9th World Conference on Investment Casting*, San Francisco, 1996
10. J. Buju, Z. Zeheng, C. Bing et al., *Investment Casting Handbook* (Mechanical Industry Press, Beijing, 2000)
11. W. Jianli, *Application of Quick-Drying Silica Sol in Investment Casting* (Tsinghua University, Beijing, 1997)
12. A.T. Bozzo, Remasol Abdond SP-3301 and LP-3301 binders, in *48th Annual Technical Meeting*, 2000
13. C. Bing, Polymer and fiber-reinforced silica sol—a review of the progress of precision casting technology at home and abroad (4). Spec. Cast. Nonferrous Alloys **25**(4), 231–233 (2005)

14. Z. Yangzheng, Method for producing large particle size and low viscosity silica sol. CN: 86104144A, 1998
15. S. Weide, W. Xiujing, D. Yirui et al., Method for producing large particle silica with high purity, high concentration and high uniform particle distribution. CN: 1155514A, 1997
16. J. Vandermeer, A new silica binder for investment shell slurries. INCAST **3**, 18–19 (1999)
17. N. Ichinose, *Introduction to Fine Ceramics—Applications in Engineering* (Wiley, 1987)
18. Magna Industrials Limited Product Specification MIL/21. Zirmast-zirconia binder of refractory grade
19. R.C. Feagin, Method of making a ceramic shell mould. US Patent 4,216,815, 12 Aug 1980
20. A.J. Skoog, R.E. Moore, Refractory of the past for the future: mullite and its use as a bonding phase. Ceram. Bull. **67**(7) (1988)
21. R.L. Wood, D. Von Ludwig, *Investment Casting for Engineers* (Reinhold Publishing Corporation, New York, 1952)
22. J.D. Snow, D.H. Scott, Comparing fused silica and alumino-silicate investment refractories. INCAST **4**, 21–25 (2001)
23. J. Buju, Y. Shuangjing, L. Zhigang et al., Today's investment casting shell list, in *Proceedings of the 6th Annual Meeting of China Foundry Association Precision Casting Branch*, Wuhan, 1999
24. C. Bing, A new rooster of shell-making refractories—a review of the progress of precision casting technology at home and abroad (5). Spec. Cast. Nonferrous Alloys **25**(5), 294–297 (2005)
25. W.E. Worrall, *Ceramic Raw Materials—Second Revised Edition* (Pergamon Press, 1982)
26. D. Jerry, H.S. David, Comparison fused silica to alumino2silicates in water based shell systems, in *Investment Casting Institute: 48th Annual Technical Meeting*, USA, 2000
27. J. Junhao, X. Guangmin, Application of fused silica in silica sol shells. Foundry **57**(8), 788–792 (2008)
28. C. Meiyi, L. Zide, Cristobalite transformation and its significance in investment casting. Casting **4**, 9–13 (1993)
29. Z. Zengyan, S. Zhixiong, H. Jianping, Surface quality of precision cast shell. Model. Mater. **3**, 44–45 (1997)
30. G. Rongchang, L. Bangsheng, Y. Yongcun et al., Analysis of white matter on the inner surface of nickel-based superalloy and ceramic shell. Foundry Technol. **24**(6), 517–518 (2003)
31. Z. Minghan, Z. Bo, X. Yugui et al., Precision casting surface coating vacuum coating process, in *Proceedings of Shanghai International Conference on Nonferrous Alloys and Special Casting*, 1998, pp. 357–360
32. L. Xiaofu, L. Yanchun, S. Guiqiao et al., Research status of high temperature mechanical properties of ceramic shells for directional solidification. Spec. Cast. Non-Ferrous Alloys **30**(10), 913–918 (2010)
33. S. Gengchen, Z. Xiangchong, High temperature mechanical properties of Al_2O_3-SiO_2 refractories. Adv. Mater. Sci. **2**(4), 61–68 (1988)
34. Z. Hengyi, B. Yin, T. Tianfu, Study on thermal deformation characteristics of water glass shell. Spec. Cast. Non-Ferrous Alloys **1**(Supplement), 24–25 (1999)
35. T. Tianfu, C. Defu, Z. Hengyi et al., Study on the behavior of investment casting water glass shell Na_2O. Spec. Cast. Non-Ferrous Alloys **5**, 17–22 (1990)
36. Z. Litong, Y. Xinghua, Creep resistance and microstructure of fused corundum shell. Foundry **3**, 25–30 (1985)
37. D.G. Charis, Ceramic investment molds embodying a metastable mullite phase in its physical structure superalloys. US Patent, 4188450, 1980
38. H. Huiwen, *Study on Investment Casting Technology of Directionally Solidified Nickel-Based Superalloy Blades* (Harbin Institute of Technology, Harbin, 2002)
39. X. Mingren, H. Deyuan, D. Zhenrui et al., Study on high temperature shell of directional solidification investment mold. Aeronaut. Mater. **3**, 25–26 (1986)
40. X. Ke, Improvement of shell forming process of directional solidification hollow blade. Mater. Eng. **3**, 36–38 (1998)

41. X. Mingren, Z. Yong, Study on high-strength thin-walled shell of single crystal blade. Mater. Eng. **9**, 74–77 (1999)
42. Y. Dingao, C. Shufeng, Effect of adding Cr_2O_3 on the sintering and properties of mullite spinel composites. Foreign Refract. **29**(1), 48–52 (2004)
43. Z. Shidong, C. Zhongqiang et al., Study on thermal strength properties of alumina shells for directional solidification casting. Foundry **60**(4), 338–340 (2011)
44. P.M. Current, M.H. Fassler, J.S. Perron, Calcia modified ceramic shell mold system. CA Patent, 1080428A1, 1980
45. L. Lianqing, Blade directional solidification process. Aerosp. Mater. Technol. **1**, 11 (2004)
46. J.M. Lane, C. John, Reinforced ceramic investment casting shell mold and method of making such mold. US Patent, 4998581, 1991
47. G. Asish, J.K. Frederic, H.M. Philip et al., Reinforced ceramic shell mold and related processes. US Patent, 6352101, 2002
48. G. Asish, A.G. Robert, J.K. Frederic et al., Ceramic shell molds provided with reinforcement and related process. US Patent, 6431255, 2002
49. V.N. Rajeev, C. John, Reinforced ceramic shell mold and method of making same. US Patent, 6568458 B2, 2003
50. C. Bing, Dimensional stability and precision of investment castings. Spec. Cast. Nonferrous Alloys **1**, 53–58 (2003)
51. S. Norouzi, H. Farhangi, M. Nili-Ahmadabadi et al., *Proceedings of International Conference on Manufacturing Science and Technology, ICOMAST2006*, Maleka, Malaysia, 2006, pp. 454–459
52. S. Norouzi, H. Farhangi, The impact of ceramic shell strength on hot tearing during investment casting, in *International Conference on Advances in Materials and Processing Technologies*, 2011, pp. 662–667
53. C. Yuan, S. Jones, Investigation of fibre modified ceramic moulds for investment casting. J. Eur. Ceram. Soc. **23**(3), 399–407 (2003)
54. C. Yuan, S. Jones, S. Blackburn, The influence of autoclave steam on polymer and organic fibre modified ceramic shells. J. Eur. Ceram. Soc. **25**(7), 1081–1087 (2005)
55. G. Xin, L. Xiaoyang, L. Zhigang et al., Morphology and strength change of silica sol shell after high temperature action. Spec. Cast. Non-Ferrous Alloys **31**(7), 636–639 (2011)
56. C. Yuan, P.A. Withey, S. Blackburn, Effect of incorporation of zirconia layer upon physical and mechanical properties of investment casting ceramic shell. Mater. Sci. Technol. **29**(1), 30–35 (2013)
57. J. Junhao, An effective way to improve the shelling performance of silica sol shells. Spec. Cast. Non-Ferrous Alloys **30**(3), 245–251 (2010)
58. Z. Qifu, T. Tianfu, Investment casting foam coating and porous shell research. Spec. Cast. Non-Ferrous Alloys **5**, 6–9 (1994)
59. Z. Jinzhi, M. Xueming, Y. Xiaohong, Application of coated sand in investment casting of silica sol. Spec. Cast. Non-Ferrous Alloys **30**(11), 1035–1037 (2010)
60. L. Guizhi, T. Tianfu, B. Yin, Test method for shelling of precision cast shell. Spec. Cast. Non-Ferrous Alloys **3**, 24–25 (1991)
61. R. Lei, L. Ke, Z. Ying et al., Development of a sand-based repellent detection device based on PC. **57**(1), 33–35 (2008)
62. Y. Shuangjing, J. Buju, W. Jianli et al., FS-30 quick-drying silica sol shell process research. Spec. Cast. Non-Ferrous Alloys **2**, 13–16 (1998)
63. L. Zhigang, Y. Shuangjing, C. Xulong et al., Investment casting shell manufacturing and reinforced fast-drying silica sol. Foundry **10**, 28–31 (2005)
64. H. Jingtan, The addition of ethyl silicate shell to the strength enhancer of the body—latex research. Spec. Cast. Non-Ferrous Alloys **3**, 21–23 (1992)
65. L. Shouxin, S. Yuhai, L. Younian, Ethyl silicate-silica sol hybrid binder and its new precision casting process, in *China Foundry Association Precision Casting Branch Ninth Annual Meeting*, 2005, pp. 127–135

66. Y. Shuangjing, Z. Hongmei, J. Buju et al., Discussion on the composite shell of silica sol and ethyl silicate—one of the methods for shortening the shelling period of silica sol. Charact. Cast. Non-Ferrous Alloys **3**, 15–17 (1997)
67. Y. Shizhou, F. Jun, A preparation process of high performance investment casting composite shell. Foundry Technol. **31**(5), 628–630 (2010)
68. L. Zhigang, J. Buju, Y. Shuangjing et al., The effect of wind speed and humidity on the drying process of silica sol. Spec. Cast. Non-Ferrous Alloys **4**, 48–49 (2003)
69. B.K. Chakrabarti, Drying conditions and their effect on ceramic shell investment casting process. Mater. Sci. Technol. **18**(8), 935–940 (2002)
70. W. Leyi, S. Zhengxing, P. Xingshu, Drying of silica sol bond type shell and its determination. **2**, 33–39 (1981)
71. L. Decheng, S. Guirong, H. Jingfu et al., ReplicastCS ultra-thin high-strength shell preparation process. Foundry Technol. **5**, 26–28 (1993)
72. H. Dejun, The effect of local thickening of shell on the shrinkage of castings. Spec. Cast. Non-Ferrous Alloys **3**, 43 (2003)
73. Z. Yong, X. Mingren, A way to improve the uniformity of the wall thickness of investment casting shell. Mater. Eng. **10**, 39–40 (1998)
74. C. Wei, Measures to prevent delamination of investment casting shells. Locomot. Roll. Stock Eng. **3**, 43–45 (2002)
75. G. Yusheng, T. Junshan, Q. Xuan, The influence of shell stratification on the quality of investment castings and countermeasures. China Foundry Equip. Technol. **4**, 16–17 (2002)
76. R. Zhongxiang, Elimination of investment in investment casting shell. Foundry **12**, 49 (1998)
77. Z. Minhua, L. Bing, C. Jianhua, Discussion on cracking of ceramic casting shell baking process based on SL prototype. China Foundry Equip. Technol. **4**, 16–18 (2003)
78. G. Yingmin, Research on ant hole defects of precision casting shells. Foundry Technol. **12**, 1170–1171 (2005)
79. C. Bing, The sustainable development of investment casting in China. Spec. Cast. Non-Ferrous Alloys **2032**(1), 24–26
80. Yamaguchi, Basic application technology of precision casting method, in *Heat Treatment of Casting and Forging*, Vol. 7, 1992, pp. 45–46
81. R.C. Oberst, M. Oles, Benefits of using recycled ceramic shell materials in the manufacture of technical refractories, in *Investment Casting Institute: 49th Annual Technical Meeting*, USA, 2001
82. Z. Zhuxing, Recycling cycle of foreign foundry waste. Model. Mater. **2**, 41–42, 36 (2000)
83. Z. Weiqiang, *Solid Waste Treatment and Utilization* (Chemical Industry Press, Beijing, 2001)
84. F. Valenza, R. Botter, P. Cirillo et al., Sintering of waste of superalloy casting investment shells as a fine aggregate for refractory tiles. Ceram. Int. **36**(2), 459–463 (2010)
85. Z. Hengyi, The composition and re-use analysis of investment casting shell waste. Spec. Cast. Non-Ferrous Alloys **25**(1), 52–54 (2005)
86. J. Jiang, X.Y. Liu, M. Yeung, Factors affecting dimensional changes in solid ceramic moulds. Trans. AFS **109**, 975–986 (2001)
87. J. Jiang, X.Y. Liu, Observations of mould dimensional changes during gelling in the ceramic mould process, in *Proceedings of the 10th Word Conference on Investment Casting*, Paper 22, Monte Carlo, 14–17 May 2000
88. J. Jiang, X.Y. Liu, The burning-out process of ceramic moulds. J. Int. Metal Res. **17**(2), 1–7 (2004)
89. J. Jiang, X.Y. Liu, Dimensional variations of castings and moulds in the ceramic mould casting process. J. Mater. Process. Technol. **189**(1–3), 247–255 (2007)
90. R. Morrell, P.N. Quested, S. Jones et al., *Studio Project DISIC-Dimensional Stability of Ceramic Casting Moulds* (National Physical Laboratory, UK, 2006)
91. X.J. Chen, D.C. Li, H.H. Wu et al., Analysis of ceramic shell cracking in stereolithography-based rapid casting of turbine blade. Int. J. Adv. Manuf. Technol. **55**(5–8), 447–455 (2011)
92. W.A. Everhart, S.N. Lekakh, V.L. Richards et al., *Foam Pattern Aging and Its Effect on Crack Formation in Investment Casting Ceramic Shells* (Transactions of the American Foundry, Columbus, OH, 2012)

93. W. Everhart, S. Lekakh, V. Richards et al., Corner strength of investment casting shells. Int. J. Metalcast. **7**(1), 21–27 (2013)
94. C. Miao, C. Haiggeng, X. Wanda, Analytical solution of temperature field of one-dimensional, non-steady-state, variable heat flow billet in heating furnace. China Metall. **17**(3) (2007)
95. L. Zashkova, Mathematical modeling of the heat behaviour in the ceramic chamber furnaces at different temperature baking curves. Simul. Model. Pract. Theory **16**, 1640–1658 (2008)
96. C. Bing, Application of rapid prototyping technology in investment casting—review of progress in foreign precision casting technology (12). Spec. Cast. Non-Ferrous Alloys **25**(12), 732–735 (2005)
97. Q. Ming, M.C. Leu, V.L. Richards et al., Dimensional accuracy and surface roughness of rapid freeze prototyping ice patterns and investment casting metal parts. Int. J. Adv. Manuf. Technol. **24**, 485–495 (2004)
98. Q. Liu, G. Sui, M.C. Leu, Experimental study on the ice pattern fabrication for the investment casting by rapid freeze prototyping. Comput. Ind. **48**, 181–197 (2002)
99. S. Pattnaik, D.B. Karunakar, P.K. Jha, Developments in investment casting process—a review. J. Mater. Process. Technol. **212**(11), 2332–2348 (2012)
100. F. Glenn, Flashfire dewax enters new generation. INCAST **2**, 14–15 (1995)
101. Pacific Kiln, Insulations, Inc., Flashfire dewax system responds to rapid prototyping needs. Addresses traditional concerns. INCAST **2**, 13 (1994)
102. C. Bing, A new generation of flash dewaxing furnace and waste shell utilization—a review of the progress of precision casting technology at home and abroad (7). Spec. Cast. Non-Ferrous Alloys **25**(7), 433–435 (2005)
103. J. Tom, Dewax and burnout energy efficiency in the investment casting foundry. INCAST **3**, 19–21 (2002)
104. C. Bing, Automating the shell making process—a review of the progress of precision casting technology at home and abroad (6). Spec. Cast. Non-Ferrous Alloys **25**(6), 366–368 (2005)
105. VA Technology Ltd., Offers shell room automation for the new millennium. INCAST **3**, 22–23 (1999)
106. MK Technology # s Cyclone, German Company develops machine for automatic shellbuilding and drying; reduces process to hours rather than days. INCAST **1**, 23 (2005)
107. N.M. Ford, Environment regulations spur advances in water based shell systems and shell room automation. INCAST **8**, 14–15 (1995)
108. B. Yankun, Y. Liancong, K. Agen et al., Study on rheological properties and process properties of silica sol coatings for investment casting. Spec. Cast. Non-Ferrous Alloys **5**, 4–8 (1993)
109. S. Jones, *Improved Sol Based Ceramic Moulds for Use in Investment Casting* (University of Birmingham. 1993)
110. R.S. Doles, A new approach to the characterization and optimization of investment casting shell systems, in *9th World Conference on the Investment Casting*, San Francisco, California, USA, 1996, p. #8
111. B.S. Snyder, D.H. Scott, J.D. Snow, A new combination shell strength and permeability test. Invest. Cast. Inst. **51**(11), 1–26 (2003)
112. *Three-Machine Engine Technical Information Network Casting Blade Research Group. The Second Part of the Investment Casting Translation Set* (Shanghai Institute of Science and Technology Information Publishing, Shanghai, 1977)
113. T. Branscomb, The importance of green MOR for autoclave cracking, in *Proceedings of 50th Investment Casting Institute (ICI) Technical Meeting*, Chicago, 2002
114. F. Wang, F. Li, B. He et al., Microstructure and strength of needle coke modified ceramic casting molds. Ceram. Int. **40**(1), 479–486 (2014)
115. J. Vandermeer, A unique silica binder for investment shell systems, in *Proceedings of 10th World Conference on Investment Casting*, Monte Carlo, Monaco, 2000, p. 3
116. D. Kline, S. Lekakh, C. Mahimkar et al., Crack formation in ceramic shell during foam pattern firing, in *Proceedings of the Technical and Operating Conference*, Chicago, Illinois, USA, 2009

117. S. Jones, S. Leyland, The use of conductivity as a means of assessing the extent of wet back in an investment casting mold, in *Proceedings of 22nd BICTA Conference on Investment Casting*, Bath, 1995, p. 4
118. D. Duffey, A. Wexcoat, Totally new concept in water based binders, in *Proceedings of 25th BICTA Conference on Investment Casting*, Cheltenham, 2001, paper 11
119. C.W. Park, S.H. Yoon, S.M. Oh, An EVS (electrochemical voltage spectroscopy) study for the comparison of graphitization behaviors of two petroleum needlecokes. Carbon **38**(9), 1261–1269 (2000)
120. H.G. Kang, J.K. Park, B.S. Han et al., Electrochemical characteristics of needle-coke refined by molten caustic leaching as an anode material for a lithium-ionbattery. J. Power Sources **53**(1), 170–173 (2006)
121. S.P. Leyland, R. Hype, P.A. Withey, The fitness for purpose of investment casting shells, in *Proceedings of 8th International Symposium on Investment Casting (Precast95)*, Czech Republic, Brno, 1995, pp. 62–68
122. R. Hyde, S.P. Leyland, P.A. Withey et al., Evaluation of the mechanical properties of investment casting shells, in *Proceedings of the 22nd BICTA Conference*, Bath, Sept 1995, p. 7
123. W. Everhart, S. Lekakh, V. Richards, Corner strength of investment casting shells. Int. J. Metal Cast. **7**(1), 21–27 (2013)
124. D.M. Kline, *Controlling Strength and Permeability of Silica Investment Casting Molds* (Missouri University of Science and Technology, US, 2010)
125. H.J. Mueller, Particle and pore size distributions of investments. J. Oral Rehabil. **13**, 383–393 (1986)
126. F. Reza, H. Takahashi, N. Iwasaki et al., Effects of investment type and casting system on permeability and castability of CP titanium. J. Prosthet. Dent. **104**(2), 114–121 (2010)
127. M. Syverud, H. Hero, Mold filling of Ti castings using investments with different gas permeability. Dent. Mater. **11**(1), 14–18 (1995)
128. H. Matysiak, R. Haratym, M. Kbczyk, Gas flow through a multilayer ceramic mould in lost wax foundry process. Arch. Foundry Eng. **9**(2), 155–158 (2009)
129. C. Yuan, S. Jones, S. Blackburn, Development of a new ferrous aluminosilicate refractory material for investment casting of aluminum alloys. Metall. Mater. Trans. A **43**(13), 5232–5242 (2012)

Chapter 5
Filling and Solidification Process Control of Large Castings

Jiao Zhang, Yongbing Dai, Jianbo Ma, Qing Dong, Yanfeng Han, Taiwen Huang, and Guangyu Yang

Due to the overall large size, complex structure, significant difference of wall thickness, and large proportion of thin-walled structure, the filling behavior and solidification process of large castings show different characteristics from conventional precision castings. How to ensure the complete filling of the thin-walled structure of the casting and to realize the balanced control of the solidification structure at different structures is one of the technical difficulties in the precision forming of large castings.

5.1 Melt Characteristics and Their Influence on Solidification Structure and Properties

5.1.1 Effects of Compositional Fluctuation on the Thermophysical Properties of Molten K4169 Alloy

The thermophysical properties of molten alloy, such as specific heat, viscosity, surface tension, diffusion coefficient, molar volume, etc., are crucial during the casting process, which affect not only the fluidity of the melt but also the solidification behavior. Numerical simulations have become one of the most important methods of improving casting technology. To obtain accurate values of those thermophysical properties is a matter of great concern for simulating the casting filling and defect formation.

K4169 is a typical cast superalloy with excellent medium-temperature performance, machinability, and welding performance. It has been widely used in important assembling parts in terrestrial, aerospace, and seafaring engines. The main elements of the chemical composition of K4169 alloy are Ni, Cr, Fe, Mo, Al, Ti, C, and Nb. It has a wide allowable range in its nominal composition. The thermophysical properties of the melt will change when the composition is varied. The alloying elements

B. Sun et al., *Precision Forming Technology of Large Superalloy Castings for Aircraft Engines*, https://doi.org/10.1007/978-981-33-6220-8_5

may be at the upper or lower limit of the compositional range, which results in a large number of possible compositional combinations. As a result, it is possible to optimize the alloy composition from the perspective of casting process design. In order to improve the alloy casting performance through optimizing the alloy composition, it is necessary to figure out how fluctuation influences the thermophysical properties of melt. Experimental determination of the thermophysical properties is a comprehensive, complex, and time-consuming task. At present, owing to the development of computational material science and the richness of various alloy databases, it is a common practice to calculate the thermophysical properties using some thermodynamic software for phase diagram calculation, e.g. JMatPro.

Table 5.1 shows the variation trends of eight thermophysical parameters (i.e. density, molar volume, specific heat, surface tension, viscosity, thermal conductivity, diffusion coefficient, and thermal expansion coefficient) as the concentrations of several alloying elements (Ni, Cr, Nb, Mo, Al, and Ti) increase from the lower to the upper limits for molten K4169. These features were obtained on the basis of the calculation results by the software JMatPro. The temperature is in the range from 1380 to 1500 °C. A fixed concentration of C, 0.04%, is used. The remainder is the alloying element Fe. The rising (↗) and falling arrows (↘) in the table indicate the trend of the corresponding thermophysical properties when the content of an element increases from the lower to the upper limit, respectively (*Note: The contents of other elements always take the lower limits*). Table 5.2 shows the fluctuation range for the eight thermophysical parameters of molten K4169 alloy at 1400–1500 °C.

The liquidus temperature (**LIQT**) of the K4169 alloy is also greatly affected by the composition. The calculated LIQT by JMatPro varies in the range from 1342 to 1380 °C, which is consistent with the experimental values reported in the literature [1–4]. There are 64 types of combinations when the eight kinds of elements mentioned above take the upper or lower limits, respectively. The concentrations of Nb, Cr, and Ti greatly influence the LIQT. When one of them takes the upper limit, the LIQT is about 10 °C lower than that when the chosen element takes the lower one.

Table 5.1 Variation trends of thermophysical properties of molten K4169 with increasing alloying element (Ni, Cr, Nb, Mo, Al and Ti) concentrations from the lower to the upper limits

Element	Ni	Cr	Nb	Mo	Al	Ti
Mass fraction (wt.%)	50.0↗55.0	17.0↗21.0	4.4↗5.4	2.8↗3.3	0.3↗0.7	0.65↗1.15
Density	↗	↘	↗	↗	↘	↘
Molar volume	↘	↗	↗	↗	↗	↗
Average expansion coef	↗	↘	↗	↗	↗	↗
Specific heat	↘	↗	↗	↘	↗	↗
Viscosity	↘	↗	↗	↗	↘	↘
Surface tension	↘	↘	↘	↗	↘	↘
Thermal conductivity	↗	↘	↘	↘	↘	↘
Diffusion coef	↗	↘	↘	↘	↗	↗

The difference is about 33 °C between the LIQTs when all of them take the upper limits and take the lower ones (Note that in this and the following cases, all of the rest alloying elements take their lower limits when they are not considered in the comparison study). The LIQT decreases about 3 °C when the concentration of Mo (or Al) shifts from the lower to the upper limit. The LIQTs when both of Mo and Al take the lower limits and take the upper ones show a difference of about 5 °C. Increasing the contents of alloying elements such as Cr, Nb, Ti, Mo, Al, etc., can lower the **LIQT**, for the alloy being closer to the eutectic composition.

The fluidity is mainly affected by the viscosity of the melt and the **LIQT**. Although the viscosity increases with increasing Nb and Mo contents, as shown in Table 5.1, the alloying elements in K4169 including Nb and Mo (except for Cr) should take the upper limits, considering that lowering the LIQT is more favorable for improving the fluidity. According to the thermodynamic calculations, the alloy with higher Cr content tends to form body-centered α-Cr phase. The precipitation of α-Cr at grain boundaries will reduce the impact toughness of the alloy [5]. Therefore, the Cr content in the alloy should not take the upper limit, and it can be selected between 17.0 and 19.0, which is also advantageous for reducing the viscosity of the melt.

Table 5.2 Fluctuation range of thermophysical properties of molten K4169 at 1400–1500 °C

	1400 °C	1500 °C
Density (g/cm^3)	7.3670–7.5426	7.2880–7.4425
Molar volume (cm^3/mol)	7.7307–7.8214	7.8153–7.9771
Specific heat (J/(gK))	0.7120–0.7259	0.7394–0.7536
Surface tension (mN/m)	1786.86–1812.05	1742.02–1766.61
Viscosity (mPa·s)	8.0503–8.4743	6.5857–6.9242
Thermal conductivity (W/(m·K))	25.2603–27.4655	26.5448–28.8421
Diffusion coef. (10^{-9}m^2/s)	0.9637–1.0144	1.2499–1.3141
Average expansion coef. (10^{-6} K^{-1})	25.7999–27.4303	26.7551–28.3133

5.1.2 *Effect of Superheating Treatment on the Melt of Nickel-Base Superalloy*

5.1.2.1 Experimental Study of Superheating Treatment

The phase transition temperature of K4169 was measured under different superheating conditions by differential thermal analysis (DTA). The **superheating temperatures (SHT)** of 1400, 1550, 1600, and 1680 °C were selected, and the DTA curves of heating and cooling processes are shown in Fig. 5.1.

According to Fig. 5.1a, the **LIQT** of the K4169 alloy during the heating process is 1349 °C and the solidus temperature **(SOLT)** is about 1270 °C. Figure 5.1b shows that during the cooling process, when the **SHT** is 1400 °C, the γ phase nucleation temperature is 1326 °C, and the MC carbide precipitation temperature is 1274 °C. When the SHT is increased to 1550 °C, the nucleation temperature of the γ phase is reduced from 1326 to 1284 °C, and the nucleation undercooling is increased from 23 to 64 °C. When the **SHT** is further increased to 1600 °C, the nucleation temperature of the γ phase is 1301 °C and the nucleation undercooling is changed to 47 °C. When

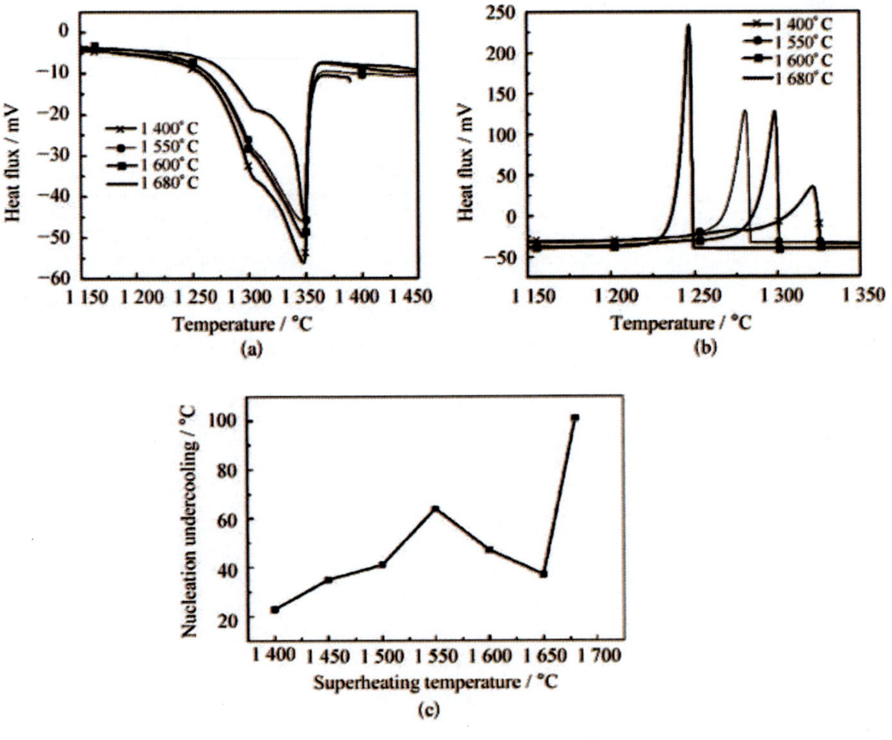

Fig. 5.1 Effect of melt superheating treatment on the solidification process of K4169 superalloy evidenced by **a** heating DTA curve, **b** cooling DTA curve, and **c** nucleation undercooling variation

the **SHT** is raised to 1680 °C, the nucleation temperature of the γ phase becomes 1249 °C and the nucleation undercooling becomes 101 °C. This phenomenon is similar to the findings by Kolotukhin et al. [6], that is, the nucleation temperature of γ phase tends to decrease firstly, followed by an increase, and finally decrease again as the **SHT** continues to increase. Therefore, with increasing the **SHT**, the nucleation undercooling of the K4169 alloy firstly rises, then declines, and finally rises again, and the non-equilibrium **SOLT** range drops at first, then is raised and finally drops once more. It is shown that the superheating treatment for the melt affects the nucleation undercooling and non-equilibrium solidification temperature intervals by changing the melt structure, which has an important influence on the solidification structure and properties of the alloy.

With the development of solidification technology and cluster physics, researchers have realized that there is a significant correlation between the solidification structure and melt structure. The melt structure is affected by not only the composition but also the thermal history. Liu et al. [7] studied the effect of melt homogenization temperature on the grain size of superalloy In 738LC and found that a relatively low homogenization treatment temperature is beneficial to grain refinement. Yin et al. [8, 9] studied the effect of melt superheating treatment on the grain size of M963 alloy and found that when the **SHT** reached 1850 °C, a very fine equiaxed grain structure was obtained. Therefore, the melt superheating treatment has an important effect on the grain size, and the effect is closely related to the **SHT** and holding time. The appropriate **SHT** and holding time can effectively refine the grains.

The preheating temperature of the mold shell in the thermal control solidification (**TCS**) process is set between the solidus and liquidus, which greatly inhibits the nucleation on the inner wall of the mold shell. Using **TCS** process, large nucleation undercooling can be obtained after superheating treatment. The solidification microstructure obtained by the combination of melt superheating and **TCS** process is presented in Fig. 5.2.

According to Fig. 5.2, with increasing **SHT** of melt, the average grain size of K4169 decreases firstly, followed by an increase, and finally decreases once again. When the **SHT** is 1680 °C, the average diameter of grain is 82 μm. At the same time, the dendritic structure changes to granular one. The shrinkage cavity is greatly reduced with the decreasing of grain size.

The melt superheating treatment at a certain temperature results in the structural transition of molten alloy, which affects the nucleation undercooling. The nucleation rate and growth rate, during the process of metal crystallization, increase with increasing the undercooling. The effect on the former is much more significant than that on the latter, which leads to the grain refinement.

5.1.2.2 Molecular Dynamics Simulation

The structure of the nickel-base superalloy melt is undoubtedly very complicated due to the existence of various alloying elements and their intricate interactions. There

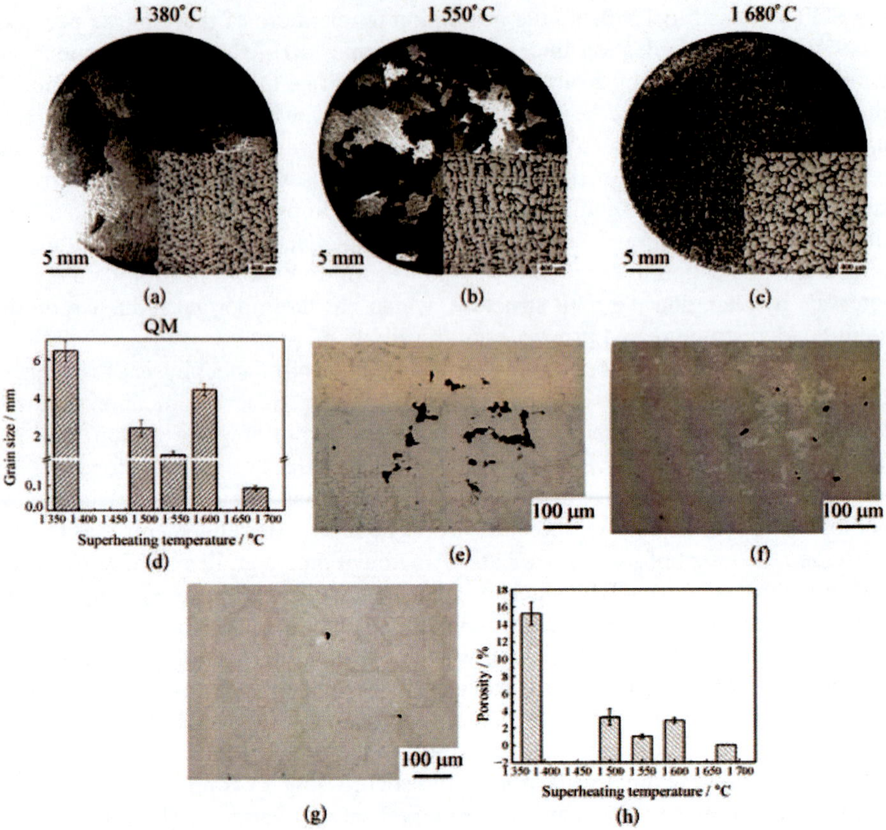

Fig. 5.2 Grain and dendritic structure **a–c**, shrinkage cavity **e–g** of K4169 superalloy after melt superheating treatment and their quantitative analysis results **d**, **h**

are a large number of carbide-based refractory particles in the initially melted nickel-base superalloy. These particles are thermodynamically unstable and will dissolve into the melt as the temperature increases. It has been reported that the carbides are not completely dissolved when the **SHT** is 1500 °C and the undercooling does not change significantly with increasing the carbon content [6]. Therefore, the carbide particles are unlikely to be the nucleation site of the γ phase and the undercooling change caused by the superheating treatment should be due to another reason.

Studies have shown that the structure factor of some superalloy melts (such as M963) has a pre-peak in the low-q-value region at lower superheating temperature [9]. The pre-peak disappears as the temperature increases; when the temperature decreases again, the pre-peak appears once again. The *ab initio* molecular dynamics (AIMD) simulations showed that when the concentration of Al in the Ni-base binary alloy $Ni_{1-x}Al_x$ is low, no obvious pre-peak is observed on the structure factor of the melt. With the gradual increase of the Al content, the pre-peak gradually appears on the structure factor and becomes very pronounced when the concentration of Al is 0.25 (x = 0.25) [10, 11]. Figure 5.3 shows the total and partial structure factors of molten $Ni_{1-x}Al_x$ with different Al concentrations at 1500 °C.

The bond angle distribution function (**BADF**) curves of molten $Ni_{1-x}Al_x$ are compared in Fig. 5.4. It can be seen that the **BADF**s of Ni–Ni-Al, Ni–Al-Ni, and Ni–Al-Al triples are similar to that of Ni–Ni-Ni, i.e., the peak on the left is stronger than the one on the right. However, the BADF of Al-Ni–Al is very different from those of Ni–Ni-Ni, Ni–Ni-Al, Ni–Al-Ni, and Ni–Al-Al. The BADF curve of Al-Ni–Al has a much significant peak at 105° and a shoulder or a relatively weak peak is present near 72°. As the Al concentration increases from 0.102 to 0.25, the intensity of the peak near 105° is only slightly weakened; the shoulder near 72° gradually becomes apparent until it becomes a peak. The BADF curve of Al-Al-Al is different from that of Ni–Ni-Ni and that of Al-Al-Al in pure Al melt. It has a strong peak around 100° and two shoulders around 60° and 140°.

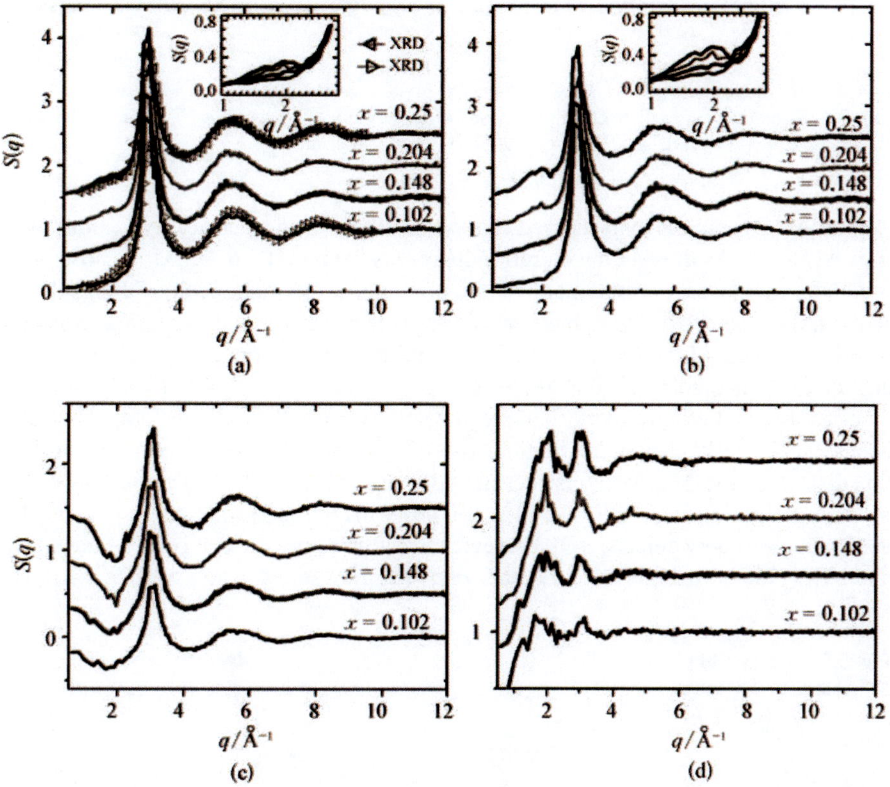

Fig. 5.3 Total **a** and partial **b**–**d** structure factors of $Ni_{1-x}Al_x$ at 1773 K. The left and right triangles in **a** are the XRD results from Ref. [8]; the insets in **a** and **b** are magnifications for S_{total} (q)and S_{NiNi} (q) with the q value ranging from1.0 to 2.8 $\AA^{-1}$

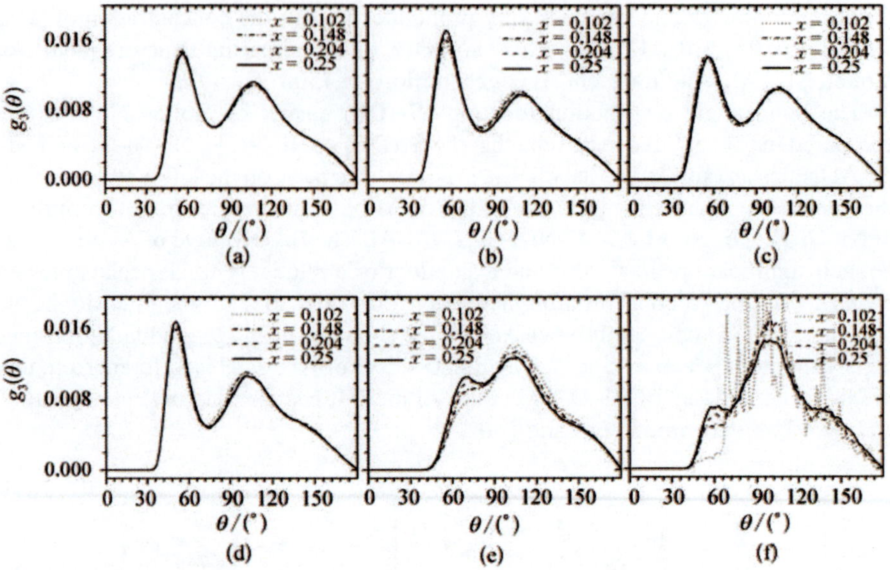

Fig. 5.4 The bond-angle distribution functions of molten $Ni_{1-x}Al_x$ for Ni–Ni–Ni **a**, Ni–Ni–Al **b**, Ni–Al–Ni **c**, Ni–Al–Al **d**, Al–Ni–Al **e**, and Al–Al–Al **f** triples at 1773 K

The AIMD simulations showed that Ni atoms have the possibility to bond with four Al atoms. As the Al concentration increases from 0.102 to 0.25, this possibility gradually increases. As stated above, the position of the main peak of the BADF of Al-Ni–Al is about 105°, which is very close to the bond angle of a methane molecule (CH4), i.e., 109.471°. Therefore, an Ni atom and its four bonded Al atoms have a chance to form a methane-like structure, as shown in Fig. 5.5a. When there is a large amount of such methane-like structures, they may form an Ni–Al bond network which is similar to the zinc-blende structure (see Fig. 5.5b). The Ni–Al bond network of this distorted zinc-blende-like structure will cause a pre-peak on the structure factor in the q value interval of 1.0–2.2 Å^{-1}. The AIMD simulations also revealed that the interaction between the unlike species in molten $Ni_{1-x}Al_x$, $Ni_{1-x}Ti_x$, $Ni_{1-x}Ta_x$, and $Ni_{1-x}Nb_x$ is stronger than the ones between the same species. Molten $Ni_{1-x}Ti_x$,

Fig. 5.5 Methane-like Ni–Al structure **a** and zinc-blende-like Ni–Al network **b**

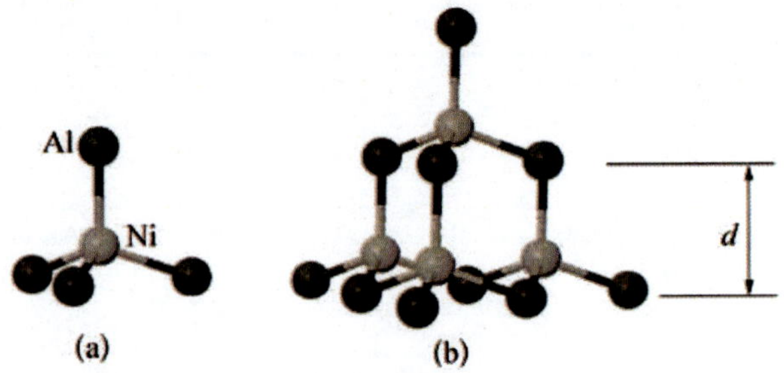

$Ni_{1-x}Ta_x$, and $Ni_{1-x}Nb_x$ exhibit similar characteristics to molten $Ni_{1-x}Al_x$. That is to say, as the concentrations of the solute elements gradually increase to 0.25, the similar zinc-blende-like Ni-M (M = Ti, Ta, Nb) bond networks will be formed in these melts too, as evidenced by the observable pre-peaks on their respective structure factors [12]. When the temperature is raised to 1973 K, the pre-peak on the structure factor of molten $Ni_{0.75}Al_{0.25}$ is still present.

Based on the AIMD simulations of molten $Ni_{1-x}Al_x$, $Ni_{1-x}Ti_x$, $Ni_{1-x}Ta_x$, and $Ni_{1-x}Nb_x$, it can be inferred that Ti, Ta, and Nb, though with relatively low contents in the superalloy melt, will play a similar role as the one of Al. The concentration of Al + Ti + Nb in the M963 alloy melt is about 16.28 at.%, and the structure factor of molten $Ni_{1-x}Al_x$ at this concentration should not have a significant pre-peak. However, at the lower **SHT**, the structure factor of molten M963 alloy indeed has a pre-peak [9]. It is conjectured that the solute elements such as Al, Ti, and Nb may aggregate in the melt, which leads to some solute-rich zones to be formed. With increasing the **SHT**, the pre-peak disappears, which indicates that the solute elements (Al + Ti + Nb) are well-dispersed in the melt. The pre-peak appears again with decreasing the **SHT**, which demonstrates that the solute-rich region appears again. It can be speculated that some other types of superalloy melts (such as K4169 melt) are also microscopically inhomogeneous, namely, there are solute-poor and solute-rich zones in the melt at lower superheating temperatures. During the solidification process, some of the solute-poor and/or solute-rich regions may be more susceptible to nucleation. However, it is still unclear at present why the nucleation undercooling of γ-phase in molten K4169 decreases again after 1600 and 1650 °C treatments (see Fig. 5.1c), which calls for further studies.

5.1.3 *Effect of Trace Elements on Fluidity and Solidification Structure*

5.1.3.1 The Influences of Trace Element Boron and Zirconium on the Solidification Characteristic Parameters, Microstructure, and Fluidity of the K4169 Alloy

Trace elements have significant effects on the solidification characteristics, microstructure, and performance of superalloys [13, 14, 15]. Studying these effects is critical for the control and optimization of alloy composition.

The solidification temperature interval is an important solidification parameter and is directly related to the fluidity of the alloy. The solidification characteristic temperatures of the K4169 alloy with different boron and zirconium contents are determined using the DSC experiments. The results are shown in Fig. 5.6 and Table 5.3. The γ phase is the matrix, which occupies a large volume fraction. Consequently, the initial melting temperature of the γ phase, being selected as the solidus temperature, can well characterize the fluidity.

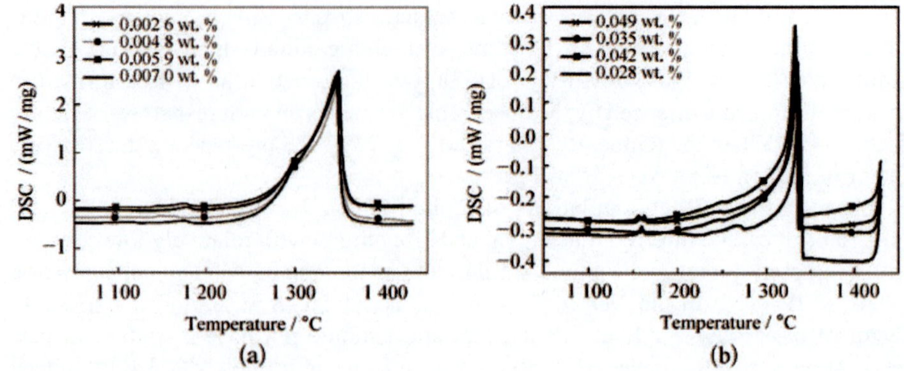

Fig. 5.6 The effects of boron **a** and zirconium **b** on the solidification process of K4169 by DSC

Table 5.3 The effect of boron on the solidification characteristic temperatures of K4169

B content (wt%)	0.0026	0.0048	0.0059	0.007
solidus temperature of the γ phase /°C	1295	1304	1316	1303
liquidus temperature /°C	1347	1349	1349	1346
solidification temperature interval/°C	52	45	33	43

According to Fig. 5.6 and Table 5.3, boron has little effect on the liquidus temperature and the solidus temperature of the γ phase rises at first and then decreases with increasing the boron content. When the boron content increases from 0.0026 to 0.0059 wt%, the temperature interval between the liquidus and solidus temperature of γ phase is reduced from 52 to 33 °C. When the boron content is 0.0059 wt%, the temperature interval is the smallest. When the boron content further rises to 0.007 wt%, the temperature interval increases once again to 43 °C. According to Fig. 5.6 and Table 5.4, zirconium has little effect on the liquidus temperature. The solidus temperature of γ phase increases firstly and then decreases. When the zirconium content increases from 0.028 to 0.042 wt%, the solidification temperature interval decreases from 52 to 31 °C. When the zirconium content is 0.042 wt%, the temperature interval achieves the smallest. When the zirconium content further grows to 0.049 wt%, the temperature interval is increased once again to 36 °C.

Table 5.4 The effect of zirconium on the solidification characteristic temperatures of K4169

Zr content (wt%)	0.028	0.035	0.042	0.049
solidus temperature of the γ phase /°C	1295	1307	1315	1310
liquidus temperature /°C	1347	1345	1346	1346
solidification temperature interval/°C	52	38	31	36

The study of Antony et al. [16] shows that, when 0.012 wt% boron is added to the superalloy Ni-10Co-9Cr-9 W-5.5A1-3.5Ta-1.5Mo-1.5Ti, the solidification temperature interval reduces from 79 to 72 °C. It is consistent with the results discussed previously. The wider the solidification temperature interval of the metal, the more developed the dendrite grain, and thus the smaller the volume fraction of the solid phase needed for the stopping of flow. Therefore, when the boron content is 0.0059 wt%, the superalloy has a good fluidity.

With the progress of technology, the aviation industry is developing toward to the high performance and low cost. Large-scale and thin-walled castings have great advantages in overall casting performance and are thus widely used. Improving the fluidity of superalloy melts is one of the effective methods to improve the casting quality of large-scale and complex thin-walled castings. Fluidity refers to the flowability of the molten metal itself, which is related to the chemical composition, temperature, impurity content, and physical properties of the molten metal. As an important solidification characteristic, the fluidity of liquid metal is directly related to the filling ability in the process of casting, which is especially important for the filling of large-scale and thin-walled castings.

Generally, the fluidity of the liquid alloy and its filling ability are characterized by two steps, i.e., casting it into a mold firstly and then measuring the streamline length after solidification or the area of the thin portion after filling. Most of the fluidity tests use gravity casting method, with only a small amount of tests using vacuum casting method. Most of the channels are straight rods, or curved into a certain shape. The common types of fluidity samples include radiation type (mainly used for rubber molds), spiral models (mainly used for metal molds), vacuum samples (by vacuum suction casting), U-shaped, stepped type, plate-shaped, disk type, and so on. During the fluidity measurement, the mold conditions, casting conditions, and channels should be same, so as to ensure the test results have good reproducibility and accuracy.

The spiral flow model (as shown in Fig. 5.7) is one type of popular models that have been used so far. It takes into account of the large-scale variation gradient and the

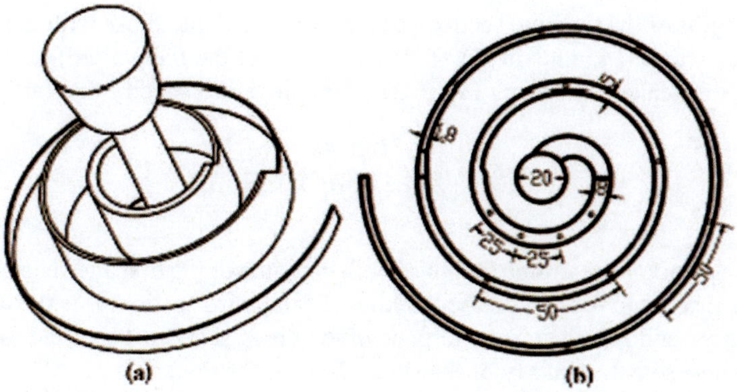

Fig. 5.7 The shape **a** and the size parameters (mm) **b** of the spiral flow model

thin wall of castings. The spiral in the spiral model is formed in an involute manner, which can be described by Eq. (5.1). The inside of the spiral plate is a cavity with a rectangular or trapezoidal cross-section. The flow channel consists of three sections. The area of the cross-section of the spiral channel in each section is determined by the large hole outflow theory: the ratio of the cross-sectional area of the first section to that of the second section is about 2:1. The cross-sectional area of the third section is variable.

$$\begin{cases} X = 10(\cos\varphi + \varphi\sin\varphi) \\ Y = 10(\sin\varphi - \varphi\cos\varphi)(\varphi \geq 0) \end{cases} \tag{5.1}$$

In order to increase the application scope of the fluidity testing, and reflect the fluidity of the alloy at different thicknesses, the length of each flow section can be determined as, $l_1 : l_2 : l_3 = 1:2:4$. Therefore, the length of the first section is 125 mm, the second section 250 mm, and the third section 500 mm.

The theoretical filling length of the alloy in the spiral model can be calculated via Eq. (5.2):

$$l = v\frac{F\rho_1}{P}\left[\frac{1}{\alpha_1} + \frac{x}{\lambda} + \frac{\sqrt{\pi \cdot \Delta x}}{b_2\sqrt{v}}\right]\frac{kL + C_1(T_p - T_k)}{T_L - T_m}, \tag{5.2}$$

where, l is the length of the streamline, v is the average flow velocity of the liquid metal in the cavity under the action of static head H, F is the area of the cross-section, P is the circumference of the cross-section, and ρ_l is the density of the liquid metal. In addition, the heat transfer coefficient of the mold is denoted as α_l, x is the thickness of the coating layer, λ is the thermal conductivity of the coating layer, b_2 is the heat storage coefficient of the mold, and k is the amount of solid phases at the front end of the flow when the flow is stopped. L is the latent heat of the alloy, C_l is the specific heat of the liquid metal, T_p is the casting temperature, T_k is the temperature when flow stops, T_L is the liquidus temperature, and T_m is the temperature of the mold shell.

The lengths of the first two sections can be converted into those with the thickness of the last section (l_1' and l_2' in Eq. (5.3)). The sum of the (converted) lengths of the three sections, calculated using Eq. (5.3), is employed to specify the fluidity.

$$l' = l_1' + l_2' + l_3 = \frac{l_1 P_1 F_3}{F_1 P_3} + \frac{l_2 P_2 F_3}{F_2 P_3} + l_3, \tag{5.3}$$

where, l' is the total streamline length after the lengths of the first and second sections being converted into those with the thickness of the third section, F is the area of the cross-section, and P is the circumference of the cross-section. The filled total length is used to describe the fluidity of the superalloy at this thickness.

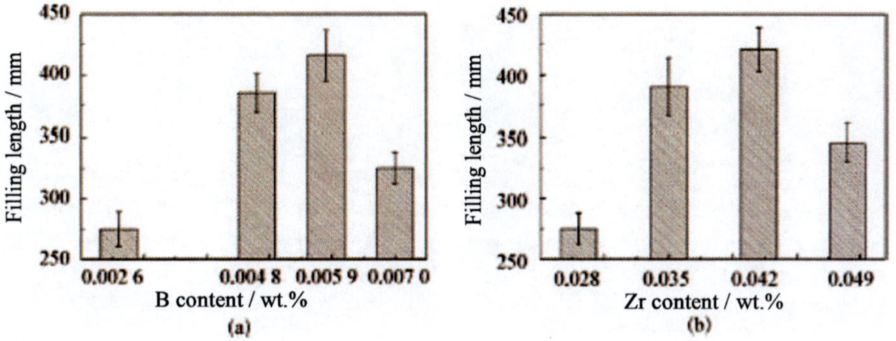

Fig. 5.8 The fluidity test results of the K4169 superalloy depending on the contents of boron **a** and zirconium **b**

This kind of mold model also has the advantages of low cost, convenient to manufacture, easy to use, and widely used etc. The melts of the superalloy K4169 with different boron or zirconium contents have been cast into the mold under vacuum with a same casting temperature of 1470 °C. Converting the lengths of the spiral streamlines of the first two sections (the first section with the thickness of 8 mm, the second one with the thickness of 5 mm) to those with the thickness of 1.8 mm using Eq. (5.3), the total length can be obtained, which reflects the fluidity of the superalloy K4169.

Figure 5.8 shows the measurements of the superalloy K4169 under different contents of boron or zirconium. According to these figures, with increasing the boron content, the fluidity of the alloy increases firstly and then decreases. When the boron content is in the range 0.0048–0.0059 wt%, the filling length is about 50% longer than that of the master alloy. When the boron content exceeds 0.0059 wt%, the filling length decreases once again, but is still higher than that of the master alloy. With rising the zirconium content, the filling length generally increases firstly and then decreases, too. When the zirconium content is between 0.035 and 0.049 wt%, the filling length is increased by 40–55% compared to the master alloy.

The effect of boron and zirconium on the fluidity is mainly due to the narrowing of the solidification temperature interval. When temperature interval is wide, the dendrites will be effectively developed. As a result, just with a small volume fraction of the solid phase, the flow will be stopped [17]. Therefore, when the boron content is 0.0059 wt% or when the zirconium content is 0.042 wt%, the alloy shows an enhanced fluidity.

The Laves phase is brittle, which promotes the formation and propagation of cracks in superalloys. In addition, a large amount of Nb will be consumed by the Laves phase in the K4169 alloy, resulting in a reduction in the amount of γ'' phase ($NbNi_3$) that is the main contributor to the strength. So, the Laves phase should be as few as possible in the superalloy. The effect of different boron contents on the morphology and content of the Laves phases are shown in Fig. 5.9. Laves phase in the master alloy has two kinds of forms, one is the eutectic Laves phase and the other is

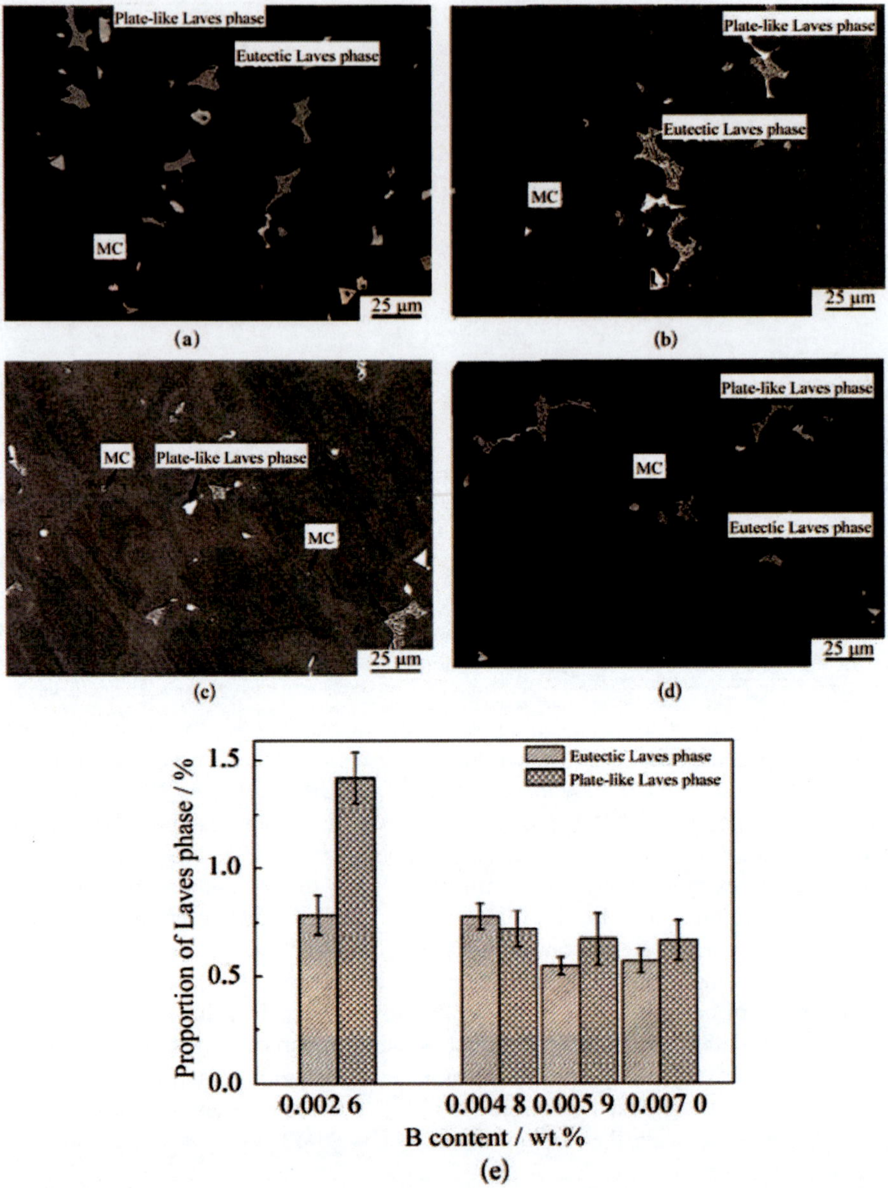

Fig. 5.9 The effect of boron content on the morphology **a**–**d** and content **e** of the Laves phase. 0.0026 wt% B **a**, 0.0048 wt% B **b**, 0.0059 wt% B **c**, 0.0070 wt% B **d**

the plate-like Laves phase. With the addition of boron, the morphology of the Laves phases did not change significantly, but their amount and size are reduced. With increasing the zirconium content, the amount of Laves phases is also significantly reduced (see Fig. 5.10). Since the interdendritic region, where the Laves phase is typically located, is the last solidified region of the K4169 alloy, these findings show

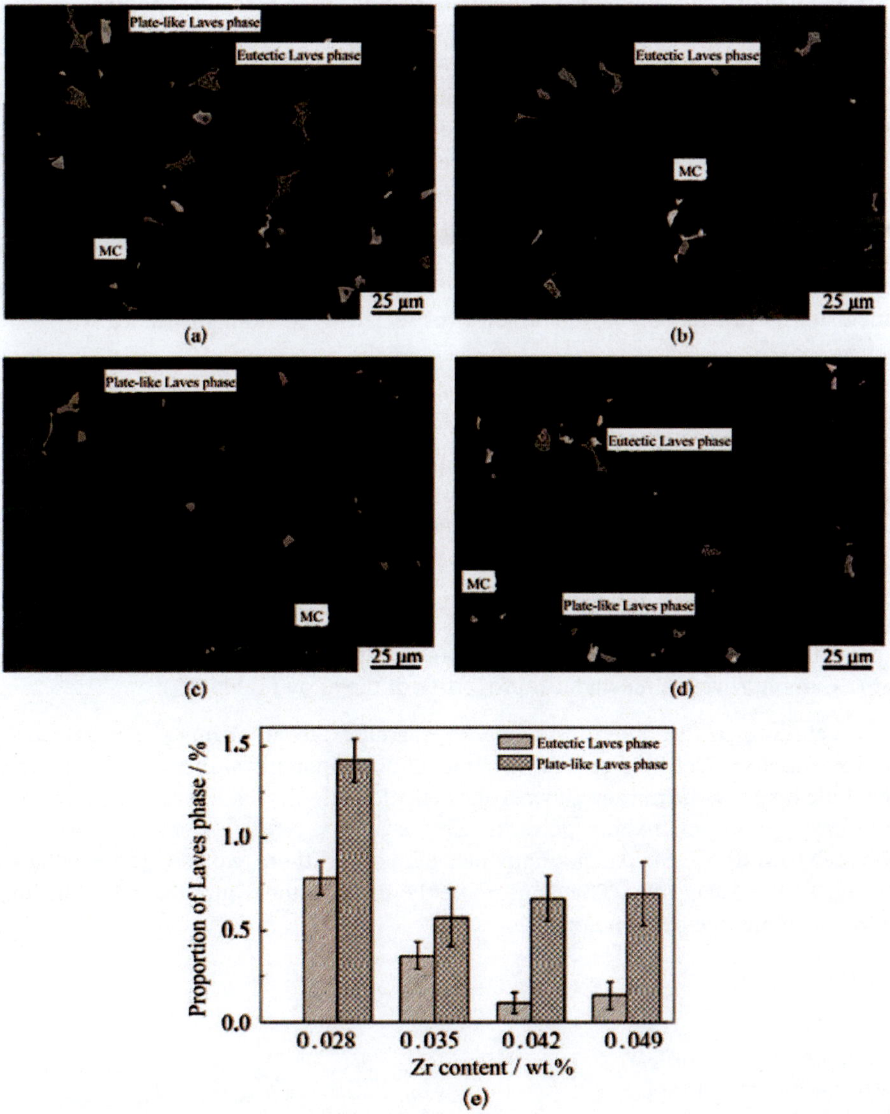

Fig. 5.10 The effect of zirconium content on the morphology **a**–**d** and content **e** of the Laves phase. 0.028 wt% Zr **a**, 0.035 wt% Zr **b**, 0.042 wt% Zr **c**, 0.049 wt% Zr **d**

that the addition of boron and zirconium will lead to a significant change in the solidification behavior of the last stage. It is generally believed that the enrichment of boron and zirconium at the grain boundaries reduces the diffusion rate at the grain boundaries and hinders the precipitation of the Laves phase.

5.1.3.2 Effect of Chemical Refinement on Solidified Grain Structure of K4169 Alloy

The chemical method refers to the addition of chemical elements or compounds into the molten metal or alloy, which can provide a large number of fine nucleation sites. These well dispersed and suspended nucleation sites will lead to great improvement of the nucleation rate and the remarkable refinement of grain size. These elements or compounds are refiners, which can be added directly or be formed from the chemical reacting of the added substances with the melt.

It is very important for the selection of chemical refiners without leading into new inclusions. For superalloys, the principle of selecting the refiner includes [18]:

(1) No harmful elements; the alloying elements of the refiner are same to the superalloy.
(2) In order to avoid the precipitation or floating of the refiner, the density difference between the refiner and the molten alloy should be as small as possible.
(3) The crystal lattice of the refiner particles should well match to that of the matrix, in order to reduce the interfacial energy between them, have good wettability between the two, and promote the new phase in its epitaxial growth when the alloy solidifies.
(4) Stable, and not easily dissolved in the melt or react with the alloying elements.
(5) Brittle, and the surface should be rough after crushing.
(6) Can improve the mechanical properties of the alloy (Table 5.5).

According to the above principles, researchers from *Northwestern Polytechnical University* selected several intermetallic compounds with the same component elements as the master alloy. After the calculation of lattice mismatch degree, the investigation of melting point, density, etc., two refiners were proposed, i.e., Co_3FeNb_2 and CrFeNb. All the component elements in these two refiners are included in the K4169 superalloy. They can refine the grain without introducing inclusions and worsening the performance.

Table 5.5 Crystal structure and physical parameters of the refiners [19]

Refiner	Crystal structure	Density (gcm^{-3})	Melting temperature (°C)
Co_3FeNb_2	Hexagonal	8.8	1500
CrFeNb	Hexagonal	8.2	>1650

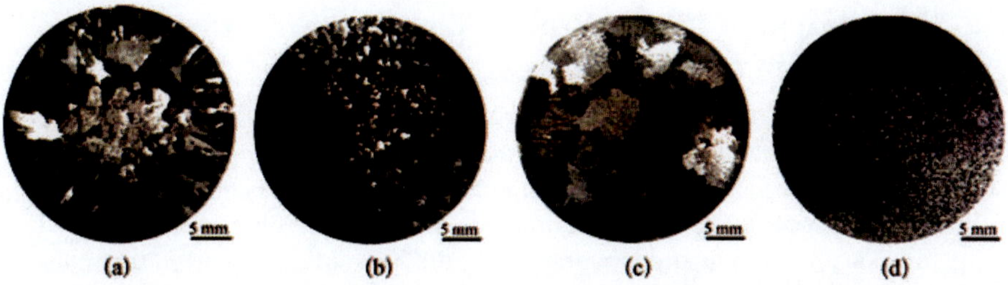

Fig. 5.11 The effect of the refiner on the grain structure of the K4169 alloy under different processing conditions. $T_p = 1380$ °C, $T_m = 900$ °C **a**, $T_p = 1380$ °C, $T_m = 900$ °C with refiner **b**, $T_p = 1380$ °C, $T_m = 1200$ °C **c**, $T_p = 1380$ °C, $T_m = 1290$ °C with refiner **d**

Figure 5.11 shows the effect of the refiner on the grain structure of the K4169 alloy under different processing conditions. For conventional processing methods, the solidification structure consists of equiaxed grains in the center and columnar grains on the edge (see Fig. 5.11a). After adding refiner, the grains all become equiaxed ones (see Fig. 5.11b) and the average grain size reduces from 4560 to 1230 μm. Under the **TCS** process condition without refiner, the equiaxed grains are relatively coarse (see Fig. 5.11c). After the addition of refiner, the grains are remarkably refined (see Fig. 5.11d), with the average grain size being reduced from 3340 to 126 μm. When the mismatch between the refiner and the substrate is less than 12%, the refiner has an obvious refining effect. The γ matrix of K4169 is an FCC structure with a lattice constant of 0.3586 nm. According to the two-dimensional lattice mismatch model, the mismatches between the refiners, Co_3FeNb_2 and CrFeNb, and the matrix are both less than 10%. For example, the mismatches between the (111) plane of γ phase and the (0001) planes of the refiners (Co_3FeNb_2 and CrFeNb) are 4.45% and 3.34%, respectively [19]. Acting as a nucleation site of the γ matrix, the refiner can increase the nucleation rate and refine the grain. In general, the number of refiner particles are 10^3 to 10^4 times of the number of grains. The number of particles is much more than the actual number of grains. The rest particles, not acting as nucleation, are enriched and segregated to the grain boundary during grain growth. They block the grain growth by hindering the migration of grain boundary and atomic diffusion, which also promotes grain refinement.

According to Fig. 5.11b–d, the refining efficiency when the mold is treated by thermal control processing method is higher than that by the normal processing method. In general, the alloy's casting structure depends on the nucleation and crystal growth conditions. The chilling effect, resulted from the large temperature difference between the melt and the mold shell, lowering the nucleation free energy, leads to the nucleation on the inner wall of the mold shell. When the temperature of the mold shell is raised to the solidus temperature of the alloy, the chilling effect is weakened and a large number of nucleation sites, no matter near or far from the edge, can initiate nucleation simultaneously, resulting in a significantly much finer grain structure as compared to that under the normal process [20].

5.2 Simulation of Filling Process Based on Structural Features of Large Castings

Hydraulic simulation has always been the main method of cavity filling process simulation. However, as the actual filling process of the cast is accompanied by the solidification process, the cavity section size, flow surface characteristics and melt properties vary with time and location. Since the hydraulic simulation separates the filling and solidification process, its simulation results may not correspond to the actual situation. As a result, the simulation method of low-melting alloy as flow medium and the finite element software simulation method have become the main methods for simulating the filling process of complex castings.

Among them, low-melting alloy with similar solidification characteristics, lower cost and lower melting point is used to simulate casting into transparent shell or actual shell, and the filling and solidification process is analyzed by actual observation or destructive detection. Computer simulation of casting process is a modern casting process research that has developed rapidly in recent decades. The filling and solidification process of casting are simulated using finite element software. It is of great significance to guide the design of casting process and improve the quality of castings. Because superalloy is easy to oxidize and because its melting temperature is high, casting production process test is difficult. Computer simulation of casting process can be used to reduce the number of technical tests, shorten the trial production period, and reduce the production cost.

5.2.1 Physical and Numerical Simulation of Filling Process of Gravity Casting Low Melting Alloy

5.2.1.1 Physical Simulation

The experimental mold is made of quartz glass, as shown in Fig. 5.12a, and ZL102 alloy is selected as the simulation medium. In the casting process, in order to reduce heat loss, the casting system is wrapped with aluminum silicate insulation. The alloy casting temperature is 800 °C and the mold temperature is 20 °C. The filling process of the alloy liquid is recorded by high-speed photography, and the relative temperature field change of the alloy liquid filling process is obtained by an infrared camera. The photographic equipment is the Phantom V12.1 high-speed camera of the American VRI company. The temperature field measurement is performed by the FLIR company's SC7300 medium-wave thermal imaging camera, as shown in Fig. 5.12b.

Figure 5.13 shows the fluid morphology at different times during the melt filling process of ZL102 aluminum alloy. It can be seen that the impinging jet from the gate is sprayed into the cavity after the alloy liquid enters the cavity, causing the liquid surface protrusion above the inner gate to form a center wave. Along with the

Fig. 5.12 Experimental model and apparatus (provided by Professor Yang Guangyu from Northwestern Polytechnical University) **a** quartz glass model **b** infrared camera

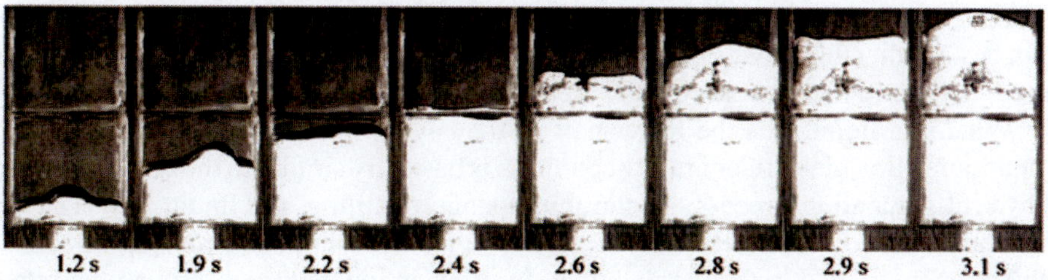

Fig. 5.13 Process of fluid morphology change with time during gravity filling process of aluminum alloy

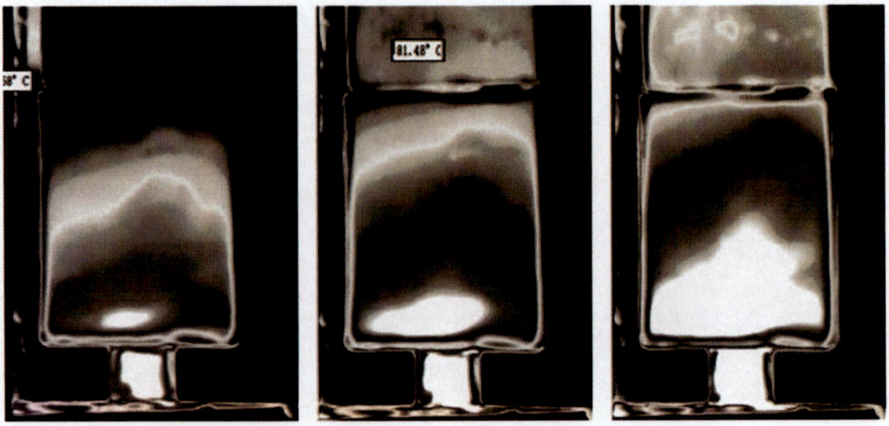

Fig. 5.14 The relative temperature field of gravity filling process in aluminum alloy

melt pouring, the filling gradually becomes smoother, the liquid level fluctuations decrease, and it tends to be stable.

Further observation of the location of the vertical submerged flow can be found that in the process of aluminum alloy melt filling liquid level raised position is not always is opposite to the sprue position, but takes place in a certain offset to the right. This is consistent with the numerical simulation results of the hydro-physical simulation process and the high-temperature alloy melt filling process.

Figure 5.14 shows the relative temperature field changes of various parts of the fluid during the gravity filling process of aluminum alloy melt. It can be seen that the temperature field variation of the alloy liquid filling process in the lower part of the model cavity is basically consistent with that of the numerical simulation result.

5.2.1.2 Numerical Simulation

Figure 5.15 shows the distribution of velocity field and temperature field at different stages of aluminum alloy filling process under the calculated gravity condition. It can be seen that the dimensions and boundary conditions of the numerical simulation model of the low-temperature alloy are consistent with the physical simulation.

When the liquid fills the lower half of the cavity, the filling process of ZL102 aluminum alloy melt under gravity condition is basically similar to the corresponding physical simulation process. At the initial stage of filling, the liquid flowing into the mold cavity under the action of gravity tends to maintain its moving direction and momentum due to the inertia effect, forming an undercurrent through the filled area. The undercurrent continuously exchanges momentum with the surrounding medium during the movement, and the result is that the undercurrent expands and the average velocity decreases gradually. Under the action of static pressure difference, the velocity direction of some fluid changes and flows in all directions. After

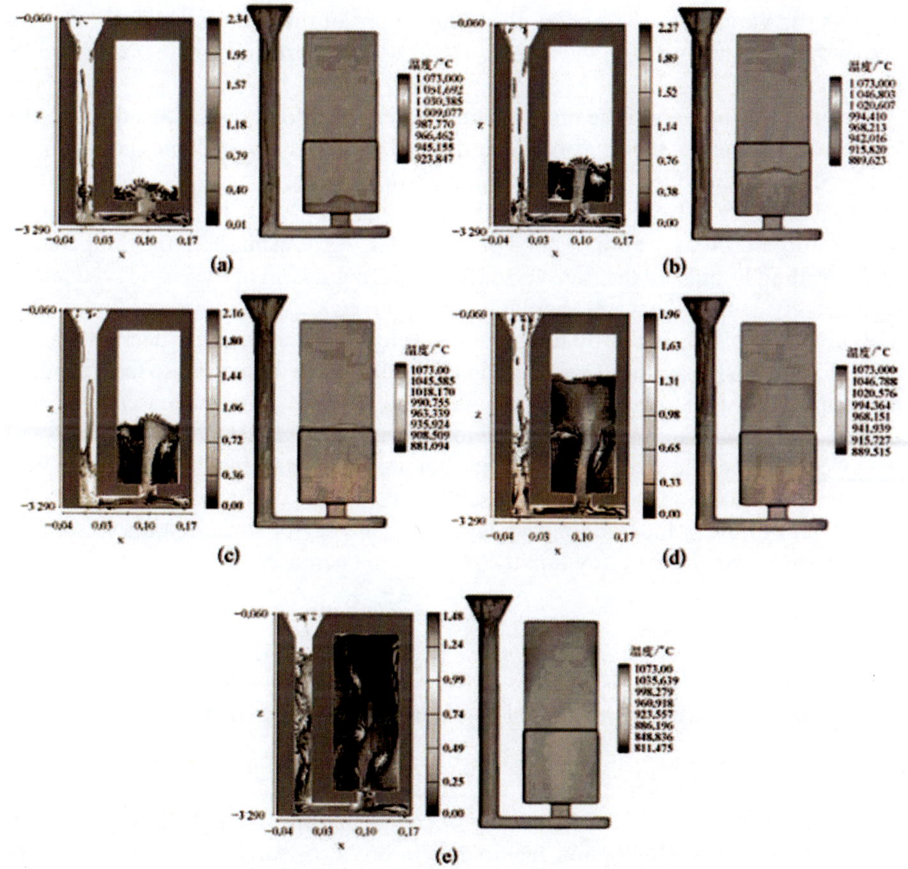

Fig. 5.15 Velocity field and temperature field of aluminum alloy filling process under gravity condition

encountering the mold wall, collision occurs. Due to momentum conservation, the direction of fluid movement changes again, forming an eddy current area. It can be seen from the figure that the central fluid always maintains a large flow velocity; when it reaches the liquid surface, its velocity still maintains a large value, so that the center of the free liquid surface is convex. In the whole process of mold filling, a wide central jet zone is formed horizontally, with vortexes on both sides, and the latter vortexes are obviously larger, and have not disappeared. Such a mold filling flow state keeps the central fluid at a high temperature, which not only increases the momentum exchange, but also increases the heat exchange in the high-temperature jet zone and surrounding vortexes. It can be seen from the temperature distribution in the figure that in the early stage of mold filling, the high-temperature jet zone of the casting is not particularly obvious, that is, the transverse temperature gradient remains at a small level. As the filling process continues, especially after passing through the variable cross section, the temperature of the central fluid is obviously higher than the surrounding, and there is a large temperature gradient in the transverse direction.

Due to the mutation of the cross section, when the alloy liquid flows through the variable cross section, the area of the central high-speed jet area becomes larger due to the continuity of the fluid. Subsequently, due to the increase of melt flow, momentum and heat exchange always exist along the way, and the heat dissipation of the high-temperature liquid in the thin-walled part is faster. As a result, both the temperature of the leading liquid and the transverse temperature gradient of the casting decreases significantly. Unlike the water simulation, thermal physical parameters of liquid alloy melt are greatly influenced by temperature. With melt temperature decreasing, the alloy liquid density, surface tension and viscous force are increased, so the impact of the coupling temperature field after the jet encountered resistance increases gradually, the free surface gradually leveled off, velocity decrease again. Due to the increase of the viscous force, the size of the velocity boundary layer formed by the cavity wall increases more obviously. At the later stage of the filling of the thin-walled part, the alloy liquid in the velocity boundary layer near the cavity surface starts to solidify as the temperature of the alloy liquid continues to decrease.

5.2.2 *Mold Filling Experiment and Numerical Simulation of Gravity Casting K4169 Alloy Feature Parts*

Through numerical simulation and experimental simulation of low melting point alloys, we have a certain understanding of the filling process of thin-walled specimens and variable cross-section specimens. In order to verify the characteristics of the filling process of high-temperature alloy thin-walled samples, the following is a further experimental analysis of the actual pouring process of high-temperature alloys.

5.2.2.1 Numerical Simulation

1. Influences of pouring temperature on filling rate

The pouring temperature has a great influence on the filling process. Higher pouring temperature, better fluidity of alloy liquid and filling. Filling rate refers to the volume ratio of filling liquid and cavity. In order to investigate the effect of pouring temperature on filling rate, the gate area is 610.86 mm^2 (the following are the same). The pouring flow rate is 2 kg/s, the heat transfer coefficient is 1000 W m^{-2} K^{-1}, and the casting temperature is 800 °C. The initial pouring temperatures were 1380, 1450, and 1550 °C respectively, then the numerical simulation was carried out.

According to the theory of casting formation [21], it is difficult to continue filling when the solidification fraction of liquid alloy flow head reaches 20%. Figure 5.16 shows the effects of pouring temperature on filling of K4169 alloy sheet. The left side of each pouring system is 1.5 mm sheet and the right side is 1 mm sheet. When the solidification fraction of 20% appears at the front end of K4169 alloy solution (according to Fig. 3.3, the corresponding temperature of K4169 alloy solution with 20% solidification fraction is 1350 °C), the filling area at 1450 °C is obviously larger than that at 1380 °C, which indicates that the bottom-pouring melting die is made. In the process of casting, the increase of pouring temperature of alloy liquid is beneficial to filling of alloy melt. According to the calculation of the relationship between the viscosity and temperature of K4169 alloy (see Fig. 3.6), it is found that the increase of temperature is beneficial to the decrease of the viscosity of the alloy liquid and the filling of the alloy liquid. These two aspects prove that increasing pouring temperature is beneficial to filling K4169 alloy liquid. In addition, it can be clearly seen in Fig. 5.16a that the filling of 1.5 mm sheet metal melt is smoother and the filling area is larger than that of 1 mm sheet metal. Figure 5.16c is the filling ratio of sheet metal with the pouring temperature of 1550 °C. It can be seen that when the pouring temperature reaches 1550 °C, the alloy melt can completely fill the whole sheet casting. From Fig. 5.16a, it can be seen that the main reason for the difficulty of filling K4169 alloy is that there are some solidification fraction alloy liquid nodes at the front end of the alloy liquid. According to the calculation, the temperature of the alloy liquid corresponding to 20% solidification fraction is 1350 °C.

2. The influence of shell temperature on filling rate

The numerical simulation was carried out under the conditions of lower pouring temperature of 1380 °C, heat transfer coefficient of 1000 W m^{-2} $K\cdot^{-1}$, pouring flow rate of 2 kg/s, and shell temperature of 700, 800, and 900 °C, respectively.

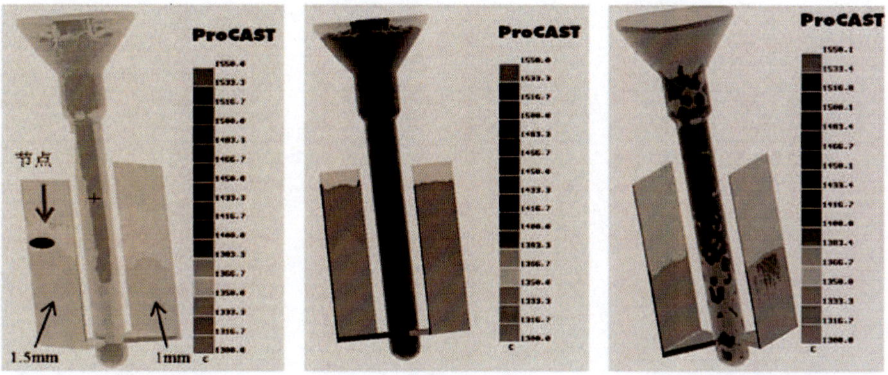

(a) (b) (c)

Fig. 5.16 The influence of pouring temperature on filling rate of K4169 sheet casting **a** 1380 °C **b** 1450 °C **c** 1550 °C

Figure 5.19 shows the effect of shell temperature on the temperature distribution of liquid filling of K4169 alloy. It can be seen from Fig. 5.17 that when the temperature of K4169 alloy liquid is 1350 °C, the solidification fraction of the alloy liquid reaches 20%, and filling becomes difficult. Figure 5.19 shows that for 1 mm thin plate (right), when the shell temperature is 700 °C, the middle temperature of thin plate castings has dropped to 1350 °C, which is easy to form defects such as insufficient pouring. With the increase of shell temperature, the area of liquid alloy in thin-walled parts below 1350 °C decreases. When the shell temperature is 800 °C, there is only a small area of low temperature at the front of the liquid alloy during the filling process. When the shell temperature is 900 °C, there is no temperature lower than 1350 °C in the front of the liquid alloy during the filling process, which makes it easier to complete the filling process. It can be seen that in the process of bottom injection investment casting, the increase of shell temperature is beneficial to the increase of filling rate (Fig. 5.18).

3. Summary

The effects of casting parameters such as pouring temperature and shell temperature on the filling and solidification process of K4169 alloy melt in bottom casting were studied using numerical simulations. The results are as follows:

(1) The increase of pouring temperature is beneficial to the filling of K4169 alloy melt. When the pouring temperature is below 1550 °C, the relationship between pouring temperature and filling ratio is linear. The increase of shell temperature is beneficial to the filling of liquid alloy. When the shell temperature is below 900 °C, the relationship between shell temperature and filling rate is linear.
(2) According to the simulation results, the optimum technological parameters of thin-walled castings with good filling were determined. When the casting speed was 2 kg/s and 3 kg/s respectively, the casting temperature of thin-walled castings with complete filling was 1550 °C or higher.

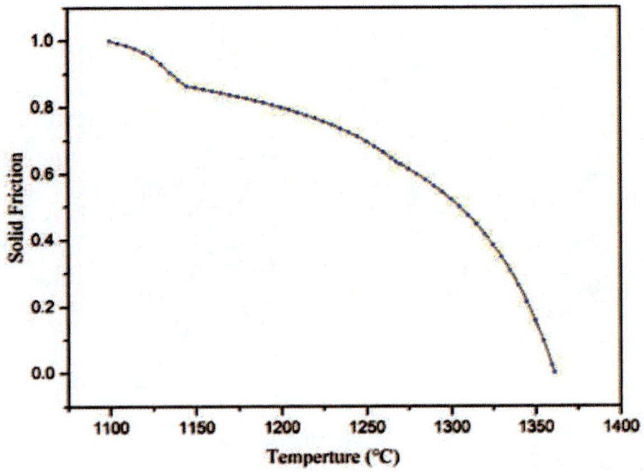

Fig. 5.17 The relationship of K4169 alloy solid fraction and temperature

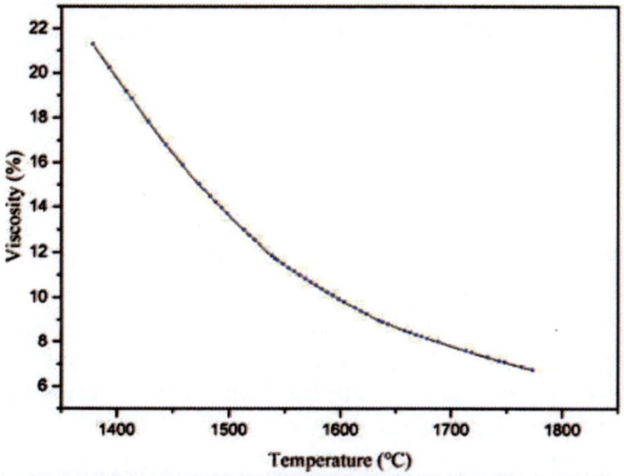

Fig. 5.18 The relationship of K4169 alloy viscosity and temperature

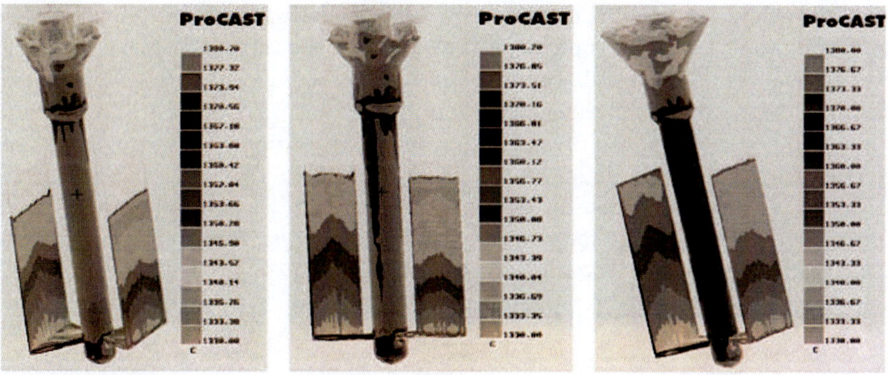

(a) (b) (c)

Fig. 5.19 The effect of shell temperature on the temperature distribution of liquid filling of K4169 alloy. **a** 700 °C, **b** 800 °C, **c** 900 °C.

(3) For thin-walled castings with different thicknesses, casting parameters have different effects on the filling rate and solidification characteristics. When the thickness of thin-walled castings is 1 mm, the influence of casting parameters on the filling of thin-walled castings is obvious. In the solidification process of the castings, because the castings are thinner and the solidification speed is too fast, the solidification characteristics of the castings have no obvious change with the change of process parameters. The 1.5 mm thin-walled castings can obtain excellent finished products by optimizing casting parameters. It shows that under

gravity conditions, the filling and solidification process of large complex thin-walled castings should be strictly controlled, but the forming of thinner thin-walled castings is still not ideal.

5.2.2.2 Experimental Analysis of Thin-Walled Features

1. Pouring results of thin-walled feature parts of superalloy

The filling, solidification cooling rate, and solidification structure of thin-walled feature parts of K4169 superalloy with different casting temperature and shell preheating temperature and different thickness and size were studied using bottom-casting investment casting process.

Superalloy K4169 is chosen as the sample, of which the size of thin-walled plate are 2 mm × 1.5 mm × 1 mm. The bottom pouring system is adopted in the investment casting.

There are two steps in the manufacture of shell, i.e., wax pattern preparation and shell preparation. According to the need of casting system, the characteristic paraffin parts are pressed and then bonded to the casting shape module. After wax mould preparation, three single-hole ceramic blind pipes are bonded on the wax sheet to facilitate the insertion of thermocouples during the pouring experiment. Temperature monitoring of shell and K4169 melt filling and solidification process are carried out in real time, as shown in Fig. 5.20.

In the experiment, the pouring temperature is 1550 °C, the preheating temperature of the shell is 900 °C, and the pouring flow rate is 2 kg/s. The shape of 2 mm thin plate castings obtained by pouring is shown in Fig. 5.21.

(1) The cooling rate curve of the shell.

Fig. 5.20 The positions of thermocouples

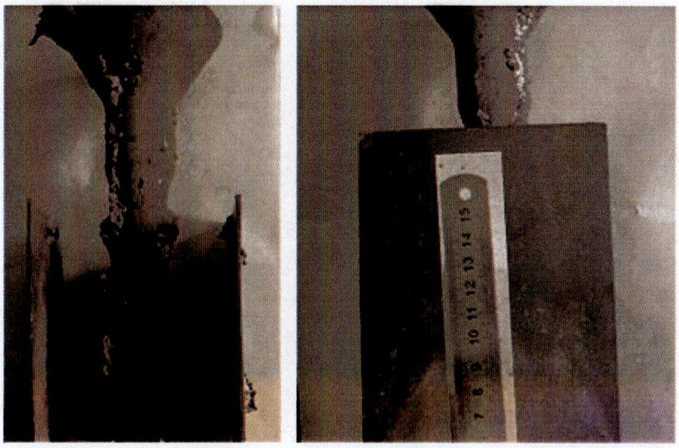

Fig. 5.21 The shape of 2 mm thin plate

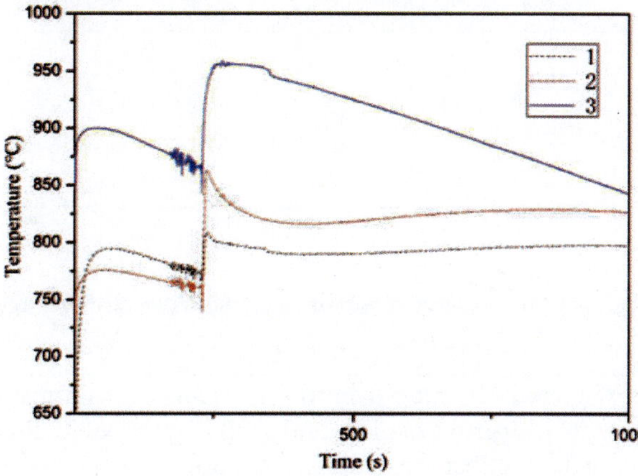

Fig. 5.22 The temperature cooling curves at positions 1, 2 and 3 corresponding to the shell in Fig. 5.22

Under the above pouring conditions, the temperature curves of the shell are recorded by the black box furnace temperature tracker as shown in Fig. 5.22. The temperature cooling curves at positions 1, 2, and 3 corresponding to the shell in Fig. 5.22 are recorded, respectively, in the channels 1, 2, and 3 shown in the figure. The first peak of the three curves is the initial temperature of the shell after it is removed from the resistance furnace, which is about 900 °C. After 250 s, three temperature curves increase sharply, indicating that the melt fills into the corresponding position of the thermocouple in the shell at this moment. Through the order of steep increase of curve and temperature, the sheet metal is filled from bottom to top. In order to verify the regularity of temperature change, the above process was simulated using the above parameters, and the temperature distribution was similar to the actual process.

In addition, the filling rate of K4169 at the thin plate is estimated to be about 0.1 m/s by the time difference between the steep increasing points of curve 1 and curve 2 and the distance between positions 1 and 2 on the transverse shell.

The cooling curves of the shell at positions 1, 2, and 3 obtained in the numerical simulation, as shown in Fig. 5.23, are in good agreement with the cooling curves

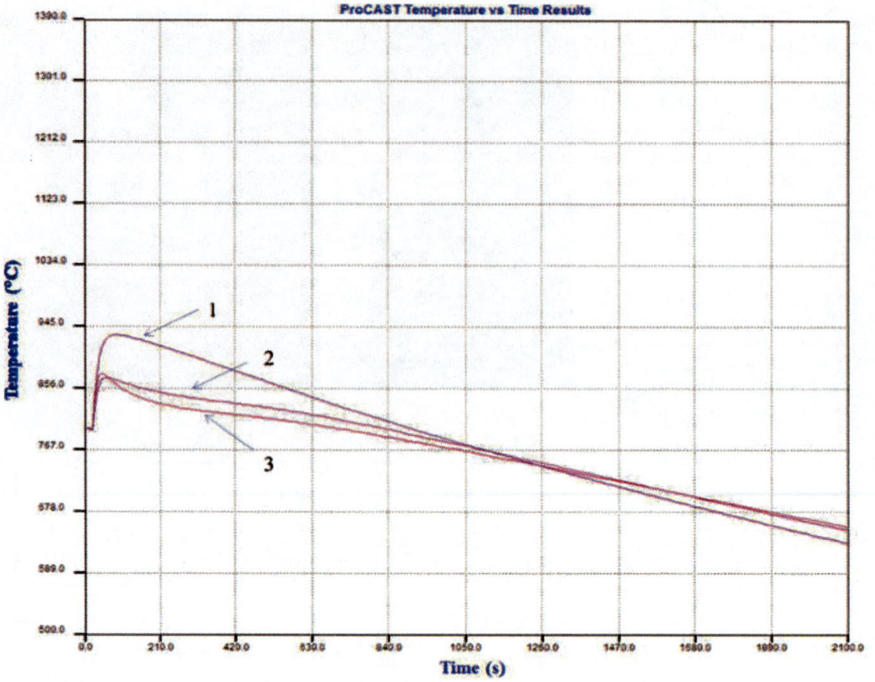

Fig. 5.23 Temperature curves of medium shell at positions 1, 2 and 3 are numerically simulated

of the shell obtained from the experiment. The results show that the selection of thermal properties of alloy and shell is reasonable. Procast numerical simulation can accurately reflect the cooling conditions of shell and castings in the casting process.

(2) The change of solidification structure in different parts of the gating system.

The grain size of 1.5 mm thin plate is analyzed for different parts of the casting system, i.e., main runner, inner runner corner, bottom of thin plate, top of thin plate, and cold compartment in casting, as shown in Figs. 5.24, 5.25, 5.26 and 5.27.

It can be seen from Fig. 5.24 that the grain size formed at the columnar main runner is larger. Because it is close to the gate, the high-temperature alloy liquid passes through for a long time, has a larger diameter, and solidifies quickly, which is beneficial to the growth of dendrites. Figure 5.25 is a picture of grain structure at the corner of 5 mm fan-shaped inner runner. It can be clearly seen that at 5 mm thickness plate, the grain is still relatively large and about 2–3 large grains are formed in the cross-section. At the transition point from 5 mm thickness plate to 1.5 mm thickness plate, the grain size decreases gradually.

Figure 5.26a, b show the grain structure at, respectively, the bottom and top of a 1.5 mm thin plate. It is shown that the grain at the bottom of the thin plate is larger than that at the top. It shows that the cooling speed of the top of the thin plate is faster when the filling process of the thin plate is filled from the bottom to the top in turn.

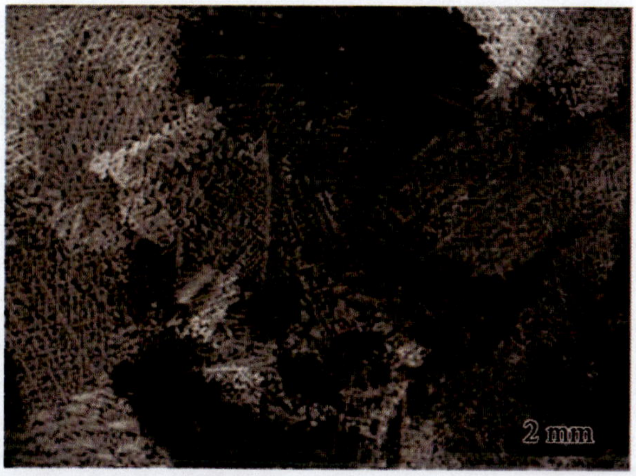

Fig. 5.24 Grain structure at main runner

Fig. 5.25 Grain structure at the corner of inner runner

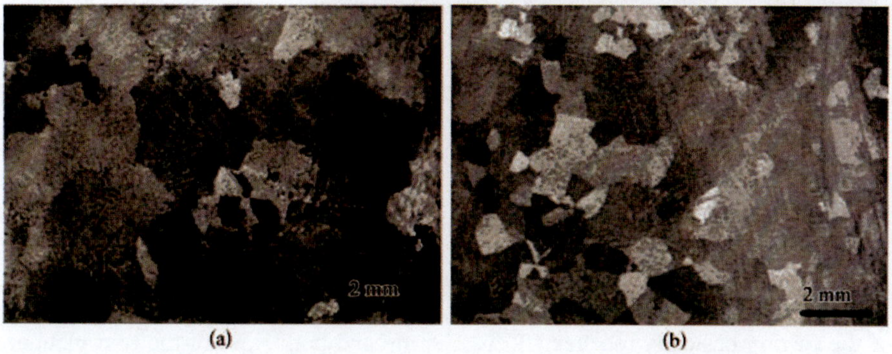

Fig. 5.26 Grain structure at the bottom and top of a 1.5 mm thin plate, respectively

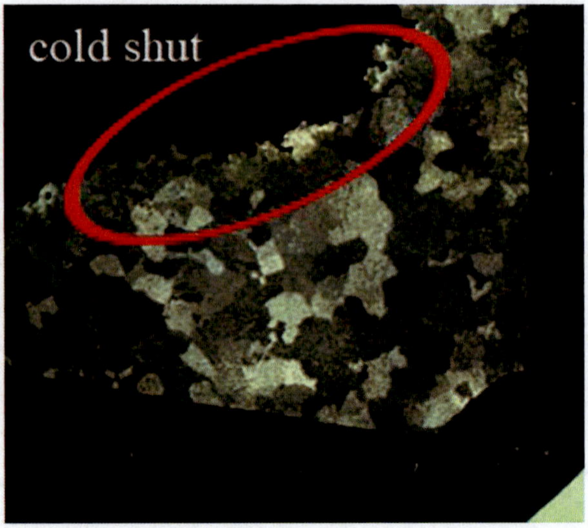

Fig. 5.27 Grain size at the front end of the cold barrier

The continuous heat conduction caused by melt passing near the inner runner at the bottom of the sheet is beneficial to the growth of dendrites, resulting in a larger grain size.

For the incomplete filling parts of 1.5 mm thick sheet, the grain structure at the defect of cold barrier is analyzed, as shown in Fig. 5.27. The grain size at the front end of the cold barrier is smaller, which indicates that the solidification rate of the melt is too fast. The melt continuously dissipates heat after flowing through the cavity, the temperature decreases drastically, the viscosity increases, the resistance caused by surface tension increases, and the flow eventually stops. According to the theory of casting formation [21], it is difficult to continue filling when the solid fraction of liquid–liquid flow head reaches 20%. Compared with Figs. 5.26b and 5.27, it can be seen that the grain size of the cold compartment is close to the top edge of the thin plate.

2. The effect of wall thickness on filling rate and solidification structure of sheet metal.

(1) The influence of thickness on filling rate of sheet specimen.

In the experiment, the castings obtained under the conditions of melt temperature of 1550 °C and shell preheating temperature of 900 °C are shown in Fig. 5.28. The thin plates in the figure from left to right correspond to thicknesses of 2 mm, 1.5 mm, and 1 mm, respectively. For 2 mm thin plate, as shown in Fig. 5.28a, the filling rate of the casting basically reaches 100%. The top and both sides of the thin plate are fully filled. No obvious defects such as cold barrier and shrinkage holes are observed on the surface of the casting, and the surface is smooth. For 1.5 mm and 1 mm thin sheets, as shown, respectively, in Fig. 5.28b, c, the filling integrity of casting is gradually improved. The filling rate of 1.5 mm thin sheet is about 90%, and that of 1 mm thin sheet is only about 30%. The surface of 1.5 mm and 1 mm thin plate is not smooth and there are many cold insulation defects at the edge of the casting.

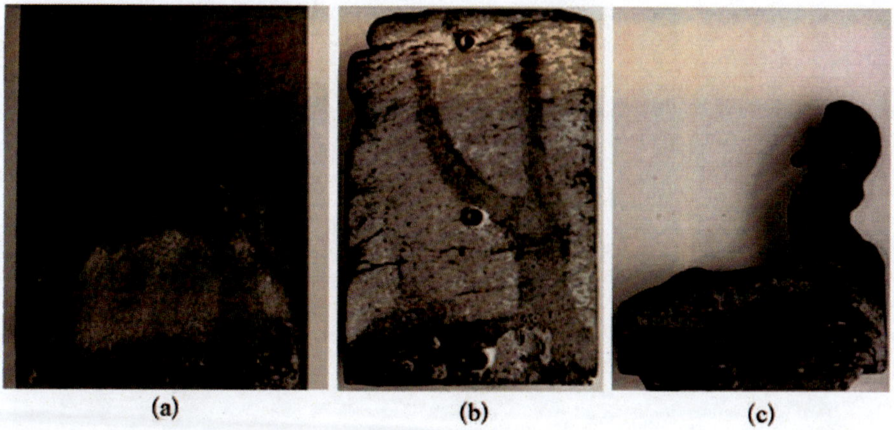

Fig. 5.28 Filling rate of sheet with different thickness and size

This shows that the thickness of the mold wall becomes smaller, and defects such as insufficient pouring and cold isolation are easy to occur.

(2) Variation of grain size of sheet metal with different thickness.

The metallographic structure pictures of 2 mm, 1.5 mm, and 1 mm thin plate castings, which correspond to the position 2 of the thin plate shown in Fig. 5.20, are shown in Fig. 5.29. The typical dendrite structure in the solidified structure of the sheet can be clearly seen in the figure. The dendrite distribution in the thin-walled parts is uniform. Dendrites grow from the outside to the inside. It can be inferred that the dendrite growth mode of thin-walled parts is from both sides to the center. Figure 5.29c shows that the length of the main axis of the branch in a 1 mm thin plate accounts for more than 60% of the thickness. The results are in good agreement with those observed in Al-20%Cu alloy sheet under synchrotron radiation. The dendrite growth mode of the sheet is from both sides to the center.

For the same condition, the grain structure obtained by cross-sectioning the thin plate with different thickness and size are shown in Fig. 5.30. It can be seen from the figure that the grain size of 1 mm thin plate is smaller. Because of the high cooling

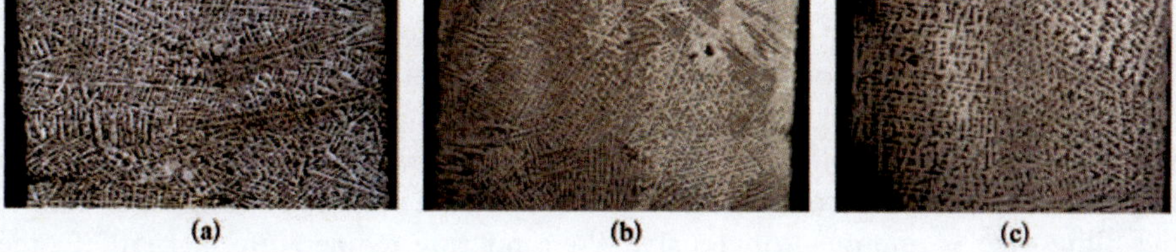

Fig. 5.29 Metallographic structure of sheet with different thickness and size. **a** 2 mm **b** 1.5 mm **c** 1 mm

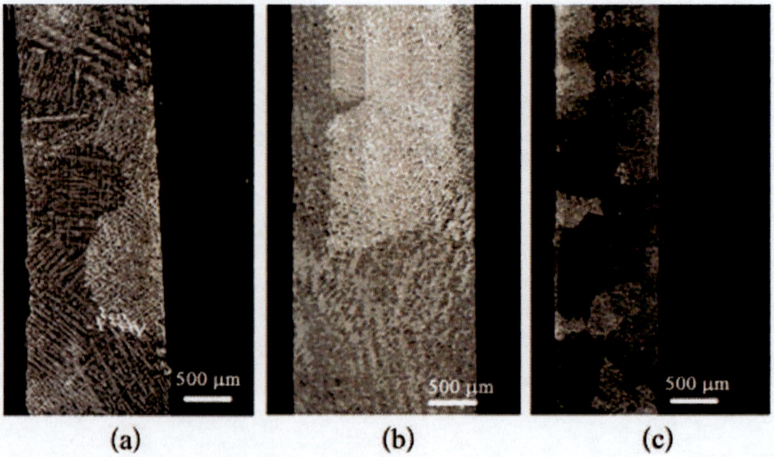

Fig. 5.30 Grain structure of sheet metal with different thickness and size. **a** 1.5 mm **b** 2 mm **c** 1 mm

rate, the nuclei grow rapidly on the surface of the plate. The average number of grains on the cross-section is about 2–3, as shown in Fig. 5.30c. With the increase of the thickness and size of the sheet, the grain size becomes larger. As shown in Fig. 5.30a, the average number of grains on the cross-section of 2 mm sheet is about 1–2. It can be predicted that the cross-section of a single grain may occur under appropriate conditions, whether it is 1 mm or thicker wall thickness.

3. The influence of pouring process parameters on filling rate and solidification structure of thin-walled plate.

The main factors affecting the filling rate of melt and the solidification structure of thin-walled plate in bottom casting of K4169 alloy are pouring temperature, pouring speed, and shell temperature. When the gate position and gate size are designed, the pouring speed becomes quantitative. Therefore, the pouring temperature and shell temperature are studied through experiments. In this study, 1.5 mm thin-walled castings were selected as the research object.

(1) The influence of pouring temperature on filling rate and solidification structure of thin-walled plate.

a. Filling rate of thin-walled plate.

The characteristic parts of 1.5 mm K4169 alloy sheet were selected. The pouring speed was 2 kg/s, the shell temperature was 800 C, and the pouring temperature was 1380, 1450, and 1550 °C, respectively. The experimental results were shown in Fig. 5.31. It can be seen from the figure that with the increase of pouring temperature, the filling rate is improved obviously. When pouring temperature is 1550 °C, the filling rate is 100%, which is consistent with the simulation results. In the process of numerical simulation, when the solidification fraction of K4169 alloy liquid is 20%,

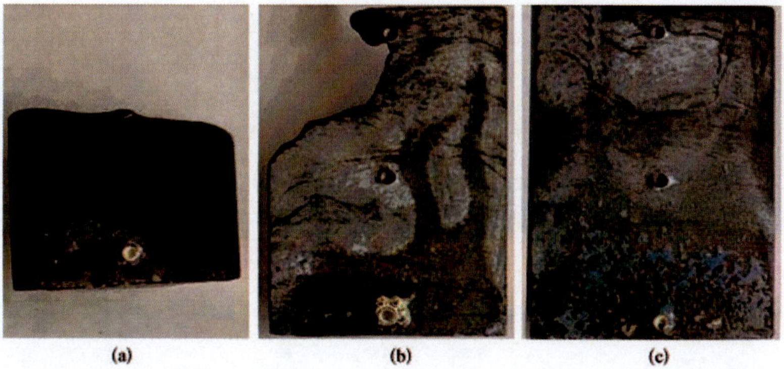

Fig. 5.31 Effect of pouring temperature on filling rate of K4169 alloy thin-walled parts

it just means that the joint here is easy to solidify first, which hinders the filling of alloy liquid, but the simulation can still complete the filling. Figure 5.32 shows the relationship between pouring temperature and filling rate of K4169 sheet casting. It can be seen from the figure that the relationship between pouring temperature and filling rate is basically linear. The higher the pouring temperature is, the larger the filling rate is.

b. Solidified structure of thin-walled plate.

For thin-walled castings under medium casting condition, the feature points on the castings (located at 2 midpoint of Fig. 5.20) are selected to obtain the metallographic structure as shown in Fig. 5.33. It can be seen from the figure that the grain size of the castings at 1550 °C is much larger than that of the castings at 1380 °C and 1450 °C. With the increase of pouring temperature, dendrite growth and grain growth process in castings are advantageous and coarse grains are easy to form. At the same

Fig. 5.32 The relationship between pouring temperature and filling rate of K4169 alloy sheet casting

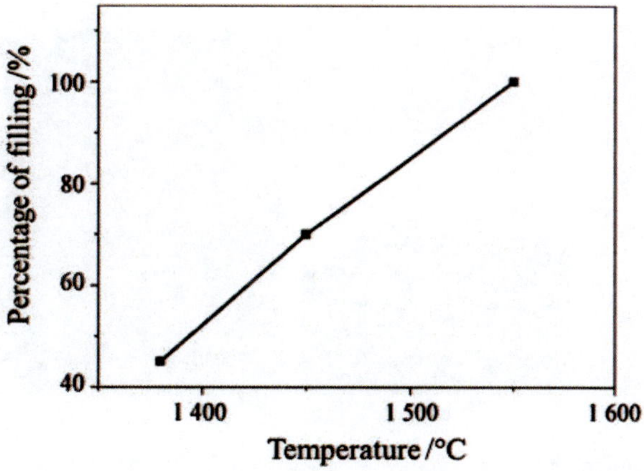

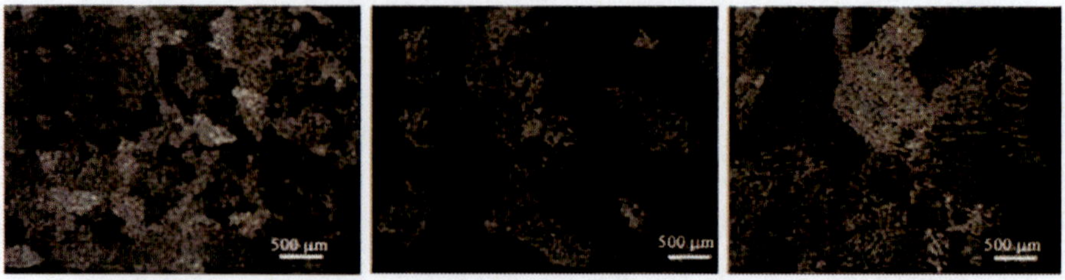

(a) (b) (c)

Fig. 5.33 Metallographic structure of thin-walled castings at different pouring temperatures

time, increasing pouring temperature to improve filling rate is not conducive to grain refinement.

(2) The influence of shell temperature on filling rate and solidification structure of thin-walled plate.

a. Filling rate of thin-walled plate

The characteristic parts of 1.5 mm K4169 alloy sheet were selected. The pouring flow rate was 2 kg/s, the pouring temperature was 1450 °C, the shell temperature was 700, 800, and 900 °C, respectively. The experimental results were shown in Fig. 5.34. It can be seen from the figure that with the increase of shell temperature, the filling rate is obviously improved. When the pouring temperature is 900 °C, the filling rate reaches 100%. With the increase of the shell temperature, the solidification time of alloy liquid at the main runner and the inner runner is prolonged, and the duration of the feeding channel becomes longer, which is beneficial to the feeding of thin-walled castings and reduces the formation of shrinkage holes in castings. This is basically consistent with the simulation results, except that the pouring temperature needed to reach 100% filling rate in numerical simulation is only 1380 °C but higher pouring temperature is actually needed in experiments, which may be caused by deviation between the

(a) (b) (c)

Fig. 5.34 The effect of shell temperature on the filling rate

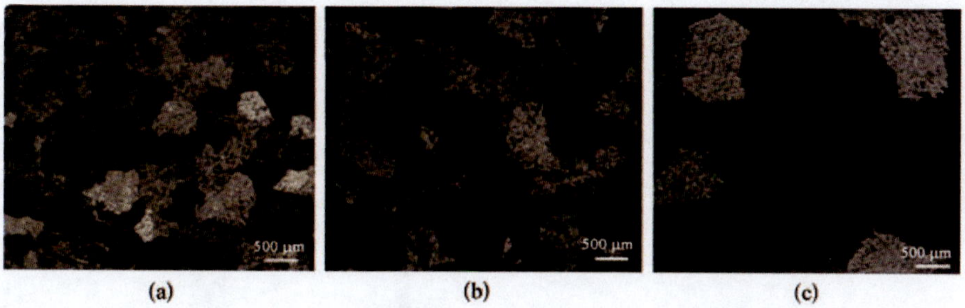

Fig. 5.35 Metallographic structure of thin-walled castings at different shell temperature

set parameters of shell thermal conductivity, ambient temperature, specific heat of melt, and viscosity.

b. Solidification structure of thin-walled plate

For thin-walled castings under medium casting condition, the feature points on the castings (located at 2 midpoint of Fig. 5.20) are selected to obtain the metallographic structure as shown in Fig. 5.35. It can be clearly seen from the figure that the grain size of the castings with the 900 °C preheating temperature of the shell is much larger than that of the castings with the casting temperature of 700–800 °C. With the increase of the shell preheating temperature, dendrite growth and grain growth processes in castings are facilitated and coarse grains are thus easy to form. It shows that in the casting of high temperature alloy thin-walled castings, reducing the preheating temperature of the shell is beneficial to grain refinement, but at the same time, incomplete filling of the castings is easy to occur.

4. Filling and solidification process analysis of thin-walled castings.
(1) Analysis of surface tension and flow rate of liquid alloy in filling process of thin-walled castings.

The stress analysis of the front end of the alloy liquid is carried out. As shown in Fig. 5.36, the Laplace force produced by the surface tension σ is $P_L = 4\sigma/\delta$ (δ is the wall thickness of the casting) and ΔP is the differential pressure formed by the alloy hydraulic head at the gate. In the filling process, if the resistance and gravity of the casting and the alloy liquid are neglected, the resultant force of the front end of the alloy liquid is $P = \Delta P - 4\sigma/\delta$. For the front end of the alloy liquid at the same height, ΔP and σ are the same and the driving force of the alloy liquid filling is mainly affected by the wall thickness. The smaller the wall thickness is, the smaller the driving force is, and the more difficult the filling of liquid alloy is. This can explain the filling rate of thin plates with different thickness sizes in Fig. 5.28. At the same time, it can be concluded that increasing the pressure difference in the filling process of liquid alloy is conducive to improving filling rate, which can be provided by the counter-gravity pressure regulating casting technology. In addition, the formula of liquid flow rate between thin-walled plates is given as $Q = \frac{W\delta^3}{12\eta}\frac{\partial P}{\partial x}$ where W is the

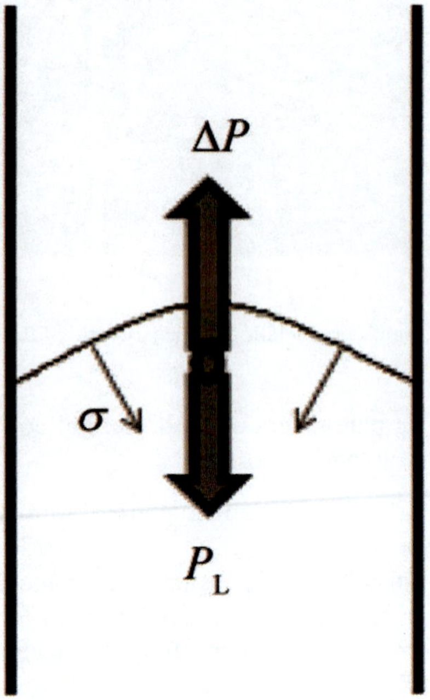

Fig. 5.36 The force diagram of the front end of alloy liquid

width of thin-walled plates, δ is the thickness of castings, η is the viscosity of alloy liquids, and $\frac{\partial P}{\partial x}$ is the pressure difference at the front end of alloy liquids [22]. Based on this formula, it can be concluded that the thicker the thickness of thin plate, the larger the filling flow rate of liquid alloy. Because the cooling speed of the sheet is relatively high, especially the thickness and size of the sheet are too small, as shown in Fig. 5.28b, it is easy for the alloy liquid to form laminated cold insulation defects.

(2) Analysis of heat transfer during solidification of thin-walled castings.

Because the thermal conductivity of ceramic mold is much smaller than that of K4169 alloy castings, and the wall thickness of the castings is larger than that of the castings, the thermal resistance of the castings is much larger than that of the castings, and the temperature gradient of the castings in solidification heat transfer is much larger than that of the castings, which exceeds one or two orders of magnitude. The temperature gradient in thin-walled castings can be neglected and the heat transfer during solidification is mainly controlled by the thermal resistance of the casting. According to the temperature distribution in the casting mold, the relationship between solidification layer thickness M and time t is obtained by Fourier's law [22]:

$$M = C\sqrt{t}$$

where C is the solidification constant. According to this formula, it can be seen that the solidification thickness of thin-walled castings is proportional to the square root of solidification time. The solidification constant C is affected by the physical properties of the castings and metals, as well as the pouring temperature and the initial temperature of the castings. For thin-walled castings, the smaller the wall thickness is, the shorter the solidification time of castings, the shorter the filling time and feeding time of liquid alloy, and more likely it is to cause insufficient pouring.

In conclusion, the main factors affecting the final filling volume of alloy liquid in thin-walled castings are the thickness of thin-walled plate, filling speed, pressure difference, pouring temperature, and mold temperature. To better facilitate the filling of thin-wall casting alloy liquid per unit time, it is desired to have a larger thickness of the thin plate, a faster filling speed, a greater pressure difference, and a higher pouring temperature. Aiming at precise control of pouring temperature, it has become mature in actual production. However, under the gravity condition, the pressure difference is not easy to control, and under the impact of the alloy liquid in the pouring process, the pressure is easy to fluctuate and the filling is unstable. In anti-gravity castings, the filling pressure difference of the castings can be accurately controlled by adjusting the gas pressure on the surface of the alloy liquid, so that the melt can be filled smoothly, which is beneficial to the complete filling of thin-walled castings. In addition, the pressure difference can be adjusted continuously during the solidification process of the castings in pressure-regulated casting, which in theory benefits the melt feeding in thin-walled castings and improves the solidification characteristics of the castings.

(3) Summary

The effects of different pouring temperature and shell preheating temperature on filling, solidification cooling rate, and solidification structure of thin-walled feature parts of K4169 high-temperature alloys with different thicknesses and sizes were investigated by means of bottom-pouring investment casting under gravity conditions. The results were then compared with the numerical simulations.

a. By adding thermocouples to different positions of the shell on the filling path of the alloy liquid, not only the cooling condition of the shell and the casting can be reflected, but also the filling speed of the alloy liquid can be calculated. The cooling curve of the shell measured in the experiment is basically the same as that obtained in the numerical simulation, which shows that the numerical simulation can accurately reflect the cooling situation of the shell and the casting points during the casting process.
b. Through comparative analysis of filling rate of thin-walled castings with different thicknesses, it is found that when pouring at 1550 °C with a shell preheating temperature of 900 °C, if the thickness of thin-walled plate is too small, it is easy to solidify in some parts of the casting during filling and solidification, which is not conducive to the complete filling of the castings.
c. The measurement results show that the pouring temperature and filling rate of thin-walled samples basically hold a linear relationship. The higher the pouring

temperature is, the higher the filling rate is. With the increase of the shell temperature, the filling rate is also significantly improved.

d. The experimental results show that grain refinement and filling integrity are a pair of technical contradictions. Therefore, systematic control of the pouring temperature and shell temperature is the key to the casting process of high temperature thin-walled castings.

5.3 Analysis of Solidification Microstructure Evolution of Large Castings Using Synchrotron Radiation

The difficulty in filling and the shrinkage of thin-wall structures during gravity casting have been long-lasting problems in the casting process. Although various pressure casting methods have been successfully applied to mass production of thin-wall castings of light alloys [23, 24], the large complex thin-wall castings of high-temperature alloy are still mainly produced by the traditional gravity casting. In general, the flowability of liquid metal can be increased by increasing the pouring temperature and shell temperature, and the filling ability of thin-wall castings can be improved to a certain extent by extending the solidification time. However, casting defects such as coarse grains, shrinkage, porosity, and hot cracking, which can significantly reduce the mechanical properties of castings, are more likely to be induced if using the above methods. In previous studies, the effects of wall thickness, pouring temperature and cooling rate on the microstructure of thin-wall castings have been fully explored. The wall thickness has a great influence on the cooling rate. The cooling rate can be sharply increased when reducing the wall thickness [25, 26], which reduces the spacing of dendrite arms and the grain size. Studies have shown that the amount of non-equilibrium eutectic will increase when the cooling rate is in the range of 0–1 K/s, while it will decrease when the cooling rate is in the range of 1–10 K/s [27]. The grains are finer and more uniform at low superheat and grow nonlinearly with increasing the temperature at medium superheat, which eventually changes to a linear relationship with the temperature at high superheat [28].

Many methods are involved in the study of solidification structure on microscale. The common methods include traditional metallographic experiment, transparent organic alloy simulation, rapid cooling technology, computer simulation and in situ synchrotron radiation X-ray imaging. Synchrotron radiation is a pulse light which can be accurately controlled. It possesses many excellent properties such as continuous spectrum from far infrared to X-ray, high intensity, high collimation, and high polarization. Compared with other light sources, it has great advantages in many frontier researches. Synchrotron radiation X-ray imaging is widely used in the study of microstructure evolution during the solidification of metals since the appearance of the third generation synchrotron radiation light source. Its advantages of in situ, real-time, continuous fast imaging make it particularly suitable for the studies of dynamics evolution. At present, researches on the alloy structure using synchrotron radiation

X-ray imaging are mainly focused on dendrite morphology and its growth kinetics, solidification interface morphology, and its growth kinetics, phases in alloys, grains, particle morphology and its growth kinetics, defect morphology and its evolution dynamics, and so on.

The synchrotron radiation facilities in China include the first generation synchrotron radiation source in Beijing, the second generation synchrotron radiation source in Hefei, and the third generation synchrotron radiation source in Shanghai.

Shanghai Synchrotron Radiation Facility (SSRF) was built and launched in 2009. Its electron beam energy is 3.5Gev. It ranks the fourth in the world, following the top three of Spring 8 in Japan (8Gev), APS (7Gev) in the United States, and ESRF (6Gev) in the European Union. It can accommodate more than 60 line stations. SSRF X-ray Imaging and Biomedical Application Beamline (BL13W1) is one of the seven line stations built in the first batch. Imaging methods of microscopic computer tomography, in-line phase contrast imaging and X-ray fluorescence imaging could be provided for scientific researches in the fields of materials, paleontology, archaeology, geophysics and so on.

5.3.1 Selection of Model Alloy and Preparation of Samples for Synchrotron Radiation X-ray Imaging

For superalloys, owing to their high melting point, there are significantly strict requirements for the experimental equipment. On the other hand, the composition contrast of nickel-base alloys is small in solidification process. In order to obtain a clear picture of the alloy structure, the incident light needs higher radiation intensity and brightness. The current light source is difficult to achieve a short exposure time, and a large time interval is unacceptable for the study of solidification process. Therefore, a suitable model alloy is needed to replace K4169 alloy for the synchrotron radiation X-ray imaging experiments.

The hypoeutectic Al-Cu alloy can be commonly used as the model alloy, due to the following reasons.

(1) The melting point of the hypoeutectic Al-Cu alloy is lower, as shown in Fig. 5.37, the temperature at the liquidus is in the range of 548–660 °C.
(2) The atomic weight of Cu is more than twice that of Al, so this alloy has a good composition contrast.
(3) Both of the Al-Cu alloy and K4169 alloy are eutectic alloy. Above the liquidus, the hypoeutectic Al-Cu alloy has only two phases of $L + \alpha$, and the K4169 alloy has two phases of $L + \gamma$. They have high similarity in solidification structure and dendrite growth morphology, as shown in Figs. 5.37 and 5.38.

According to the features of typical thin-walled castings, two sample geometries with wall thicknesses of 0.5 mm and widths of 2 mm and 5 mm were prepared. As shown in Fig. 5.39, the Al-20 wt% Cu (melting point $Tm = 600$ °C) slice was sandwiched by two Al2O3 sheets and sealed with a mixture of silica sol and Al2O3

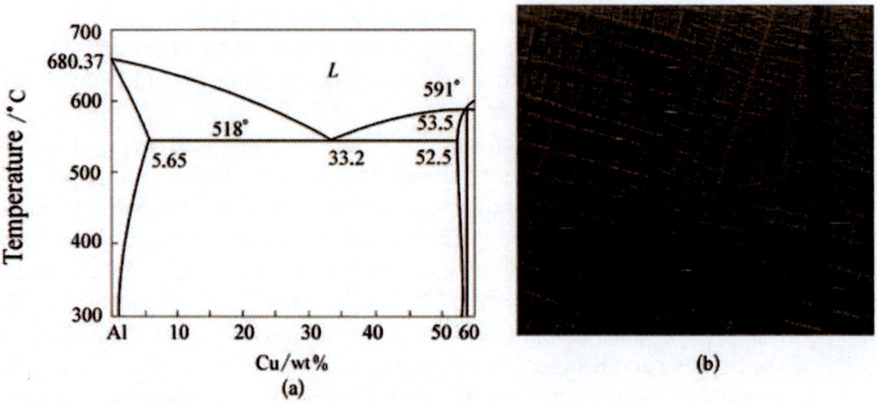

Fig. 5.37 Phase diagram and solidification microstructure of the model alloy **a** Phase diagram of Al-Cu alloy; **b** microstructure of the Al-20wt%Cu alloy

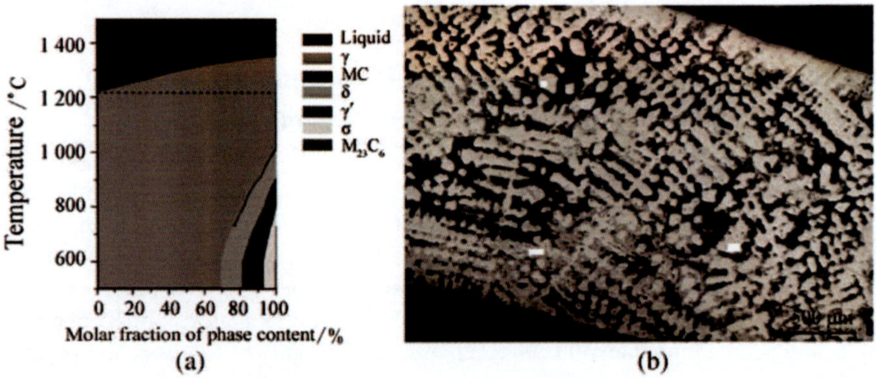

Fig. 5.38 Phase composition of K4169 alloy based on phase diagram by thermodynamic calculation and thin-wall solidification structure. **a** Phase composition of K4169 alloy based on phase diagram by thermodynamic calculation **b** thin-wall solidification structure

powders, which was then mounted on a U-tape quartz tube. The two quartz tube branches were set along the two sides of the sample to permit cooling by passing Ar gas through the tube. The X-ray radiography experiments were carried out in a self-made vacuum resistance heating furnace. Before freezing, the samples were heated to different beginning temperatures (*BT*) of 680, 700, 750, 800, and 850 °C, respectively, in order to simulate different pouring temperatures and mold temperatures. After holding for 15 min at the BT to homogenize the melt, high-purity Ar gas was introduced to the quartz tube from the bottom inlet at a constant flow rate of 8.51 m^3 h^{-1} at 0.1 MPa pressure. The cooling rate was measured by an embedded K-type thermocouple, as shown in Fig. 5.39.

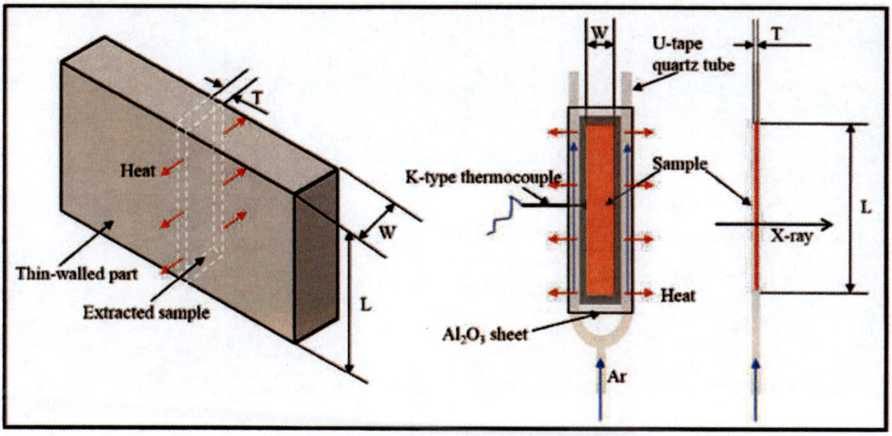

Fig. 5.39 Schematic of the thin-walled sample and the cooling method

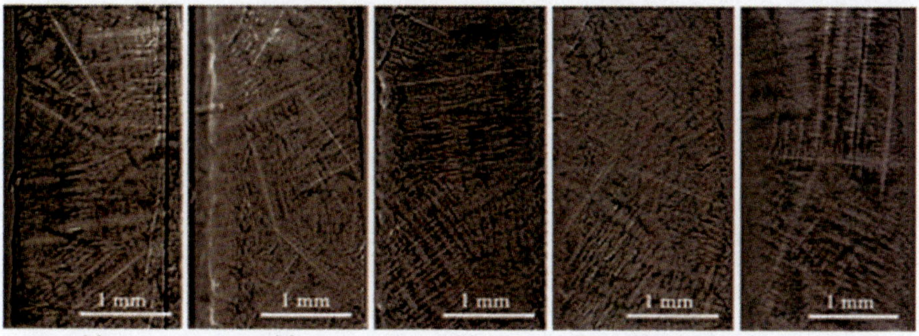

Fig. 5.40 The microstructure of the 2 mm sample under different beginning temperatures

The synchrotron X-ray phase contrast imaging experiments were performed at BL13W of SSRF. The incident monochromatic X-ray energy used in all experiments was 25 keV. The spatial resolution of all images was $7.4 \times 7.4\ \mu m^2\ pixel^{-1}$. The exposure time for each frame was 0.5 s.

5.3.2 *Dendrite Growth Kinetics*

5.3.2.1 Dendrite Morphology at Thin-Walled Structure

Figure 5.40 shows the microstructure of the 2 mm sample, which consists of elongated equiaxed grains. The dendritic arms intersect to form a dense network; the eutectic phase is separated into a capillary network within the interdendritic spaces.

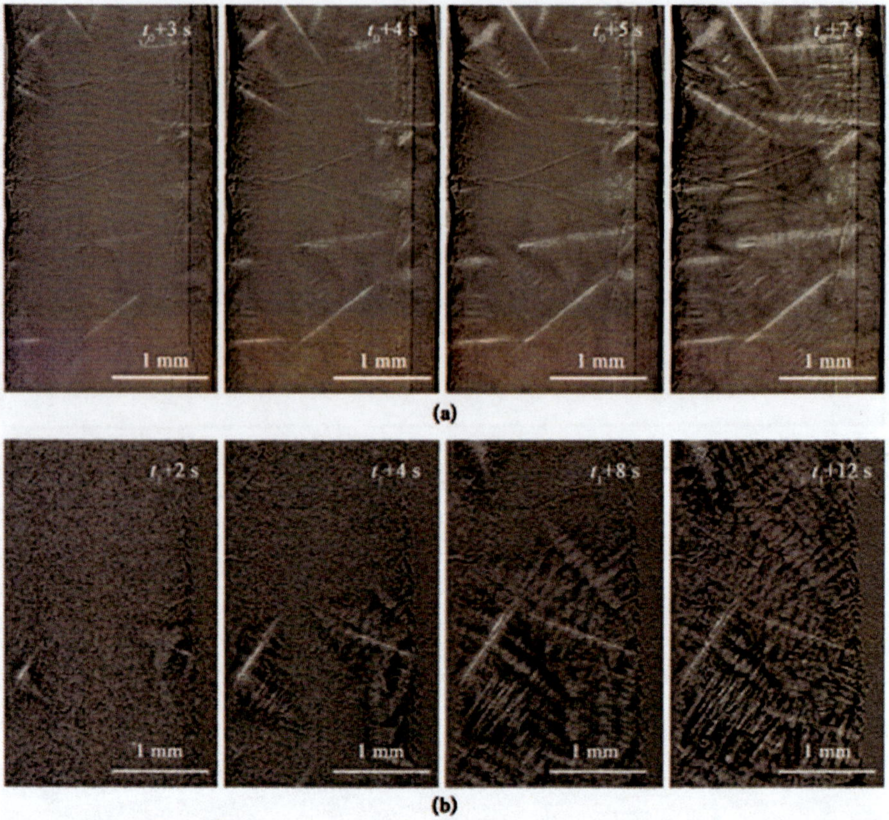

Fig. 5.41 Sequence of images showing the microstructural evolution of the 2 mm sample at beginning temperatures of 680 °C or 800 °C

As shown in Fig. 5.41a, when solidification begins at 680 °C, equiaxed grains nucleate at the chilled edges immediately after Ar gas is blown into the U-tape quartz tube. Dendrite arms grow rapidly into elongated equiaxed grains towards the center. Dendrite networks are formed in a few seconds, and the composition of the residual interdendritic liquid approaches the eutectic point. Capillary channels of characteristic sizes of 1–10 μm are formed among the dense dendritic network. Under the action of surface tension, this dense dendritic network prevents the melt from flowing, which can then imped melt feeding from the neighboring areas. During the solidification process, no obvious dendrite coarsening is observed. The solidification structure is composed of elongated equiaxed grains.

When the initial solidification temperature increases to 800 °C, as shown in Fig. 5.41b, the number of dendrites decreases significantly when grains nucleate on the side wall. The dendrite arms of equiaxed grains extend from the sidewall to the center to form elongated equiaxed grains. Generally speaking, the sidewall

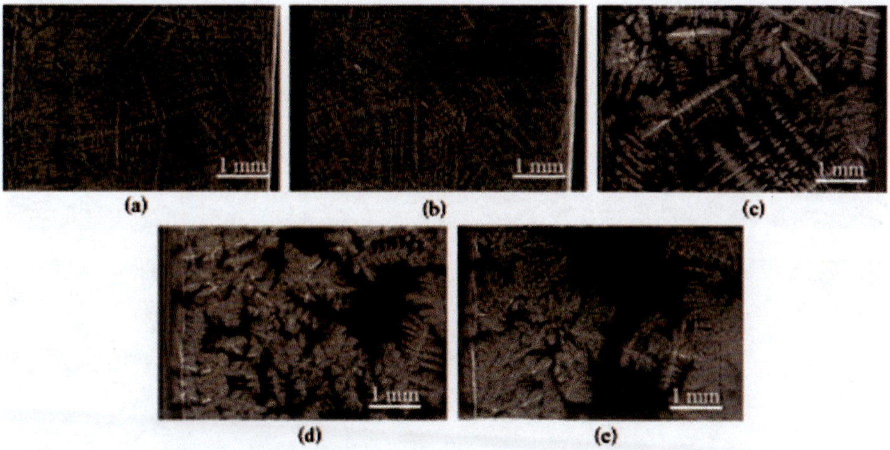

Fig. 5.42 The microstructure of the 5 mm sample under different beginning temperatures. **a** 680 °C; **b** 700 °C; **c** 750 °C; **d** 800 °C; **e** 850 °C

cooling has no effect on the central solidification for the 2 mm wall thickness specimen. The starting temperature varies from 680 to 850 °C and the density of the dendrite network does not change much.

At any beginning temperature, the structure of 5 mm thin-walled specimens is completely composed of equiaxed grains (see Fig. 5.42). When solidification begins at 680 °C (see Fig. 5.43a), a large number of fine equiaxed grains appear in a short time from the sidewall to the center. Dendrite networks are so dense that the mush zone is not large enough for feeding. When solidification begins at less than 750 °C, the solidification structure is dense equiaxed grains (see Fig. 5.42a–b).

However, when solidification begins at more than 750 °C (see Fig. 5.43b), small equiaxed grains formed in the central region of the sample float up under buoyancy. Due to wall thickness, the cooling rate of specimen decreases, leading to a decrease in the density of dendrite network and an increase in the dendrite gap. Obvious eutectic zone can be seen (see Fig. 5.42c). When solidification begins at 800 °C, the dendrite network becomes loose and large eutectic zones appear between the dendrites (see Fig. 5.42d). When solidification begins at 850 °C, a large eutectic zone appears in the center of the middle-lower part of the sample (see Fig. 5.42e). The solidification structure of the 5 mm thin-walled specimens is greatly affected by the beginning temperature: with the increase of the beginning temperature, the dendrite network becomes loose gradually, and the eutectic structure gradually aggregates and join into a large area in the center.

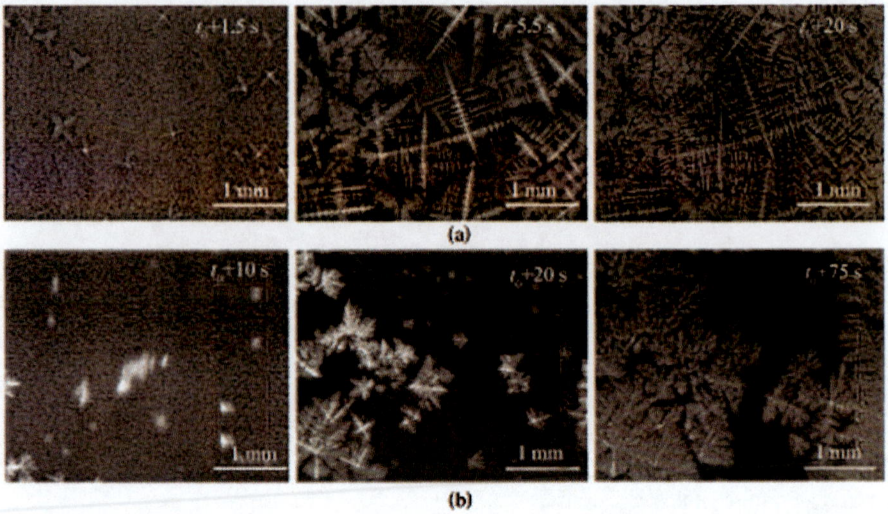

Fig. 5.43 Sequence of images showing the microstructural evolution of the 5 mm sample at a beginning temperature of 680 °C or 850 °C

The coarsening of dendrites can be caused by a slow solidification rate. Nucleation still starts from the sidewall, and the dendrite growth toward the center is delayed by the high temperature melt in the center. As mentioned previously, unlike the 2 mm thin-walled specimens, the solidification structure of the 5 mm thin-walled specimens is more sensitive to the initial melt temperature.

5.3.2.2 Thermal and Solute Fields

The microstructure morphology was found to mainly depend on the distribution of thermal and solute fields. As mentioned above, the solidification microstructure of the 2 mm samples contains elongated grains. In contrast, the solidification microstructure of the 5 mm samples contains normally shaped equiaxed grains.

Samples are extracted from the thin-walled structures, in which the heat dissipation occurring in the experiments along the direction perpendicular to the samples is presumed to be equal to zero. With the vacuum environment of the experimental setup, convective heat transfer on the samples' surfaces can be set to zero. The Ar gas cooling on both sides of the samples similarly renders the radiative heat transfer to the sample surface negligible.

In general, the profile of the eutectic front is considered to be an isotherm because eutectic has a fixed melting point. In our experiments, the shape of the eutectic front is elucidated by the image subtraction method, as shown in Fig. 5.44a and b, outlining the isothermal line of the cross-section of sample. The concave eutectic profile indicates that the temperature at the center exceeds that of the sides, and a

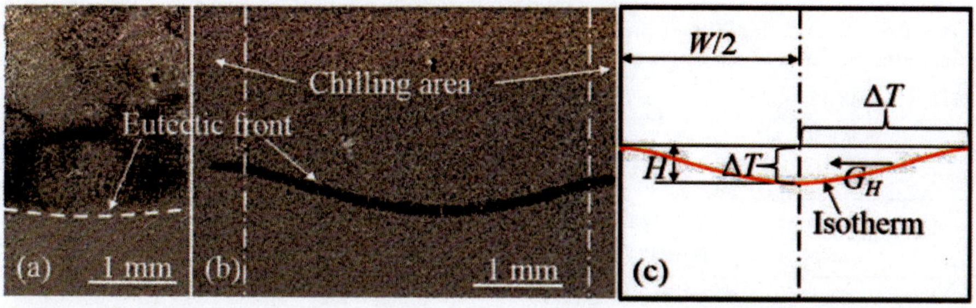

Fig. 5.44 The eutectic front of the two geometries of thin-walled samples: **a** 2 mm sample; **b** 5 mm sample; **c** Schematic of symmetrized isotherm

horizontal thermal gradient (*GH* > 0) exists from the side to the center. As the slope along the isotherm varies continuously, the horizontal temperature gradient at each point is different.

The effect of wall thickness on the temperature gradient of sidewall/center and solidification time difference can be visually displayed by analyzing the cross-section of the sample. If the solidification velocity of the solid phase is *Vs*, the concave depth of isotherm is *H*, the cooling rate is *Rc*, and the wall thickness is *W*, then the temperature difference from the surface to center (ΔT) is:

$$\Delta T = \frac{R_c H}{V_s} \tag{5.4}$$

The average horizontal temperature gradient G_H is

$$G_H = \frac{2\Delta T}{W} \tag{5.5}$$

The difference t_d in solidification time between the sidewall melt and the center melt is

$$t_d = \frac{\Delta T}{R_c} \tag{5.6}$$

Table 5.6 shows that *GH* increases with increasing *BT*. Moreover, thinner walls correlate to smaller *GH* values.

For the 2 mm sample, the *Rc* is relatively large, the G_H is very small, and the *td* is approximately 0.016–0.054 s, implying that the chilling effect influences the entire section across the sample. In addition, the growth rate of the primary dendritic arms with priority exceeds 1 mm s^{-1}, indicating that the quick formation of the dendritic network can form quickly and the dendrite network formation time is small (DNFT).

For the 5 mm sample, G_H increases as R_c decreases. For *BT* of 800 °C or less, the *td* is approximately 1.172–4.568 s. The melt near the sides solidifies first, and the dendrites grow stepwise towards the central region (the growth rate of the dendritic

Table 5.6 *Rc*, ΔT, and *GH* of different beginning temperatures (*BT*) for the two geometries of thin-walled samples

BT/ °C	W = 2 mm			W = 5 mm		
	R_c/ K s^{-1}	ΔT/ K	G_H/ K mm^{-1}	R_c/ K s^{-1}	ΔT/ K	G_H/ K mm^{-1}
680	2.48	0.0405	0.0405	1.68	1.9691	0.7876
700	2.08	0.0416	0.0416	1.04	2.1956	0.8782
750	1.49	0.0433	0.0433	0.69	3.1516	1.2606
800	1.16	0.0458	0.0458	0.50	3.8519	1.5408
850	0.87	0.0472	0.0472	0.33	5.1458	2.0583

arms with priority is approximately 200 μm s^{-1}). Newly formed equiaxed dendrites appear in the supercooled melt in front of the dendrite tips and the final solidified microstructure consists of coarse equiaxed grains. As the *BT* reaches 850 °C, *Rc* decreases dramatically, G_H increases dramatically, and *td* reaches 15.593 s. The newly formed equiaxed grains in the center float up as the structure solidifies because of their lower density. Cu becomes concentrated after the formation and floating of the equiaxed dendrites and during the dendrite coarsening process, which leads to the formation of a bulk eutectic area at the center.

5.3.2.3 Influence of Wall Thickness on Filling and Feeding Abilities

The limiting factor in the production of thin-walled castings is the complete filling and feeding of the thin-walled features. Increasing the pouring temperature and mold temperature is a technique that has been widely used in the industry. The dense dendritic network forming seriously impedes the melt flow and causes tiny intercrystalline voids. With a wall thickness of 2 mm, simply increasing the pouring temperature and mold temperature coarsens the microstructure and causes the aggregation of the eutectic phase, both of which degrade the performance of cast alloys. Here, the solidification process of the thin-walled structures is investigated deeply to identify the factors which most greatly affect the filling and feeding process.

(1) Filling pressure and feeding pressure

The molten alloy is well known to be subjected to the viscous force *F* and Laplace pressure *P* imposed by the mold's inner surface during the pouring and filling processes,

$$F = \eta S \frac{dv}{dx} \tag{5.7}$$

where η is the kinematic viscosity coefficient of the melt, *S* is the contact area between two flow layers, and *dv/dx* is the velocity gradient. η is closely associated with the temperature; it decreases with increasing the melt temperature. F is independent of the wall thickness.

$$P = \frac{4\sigma}{W} \tag{5.8}$$

where σ is the surface tension coefficient of the melt, which is determined by the wetting relationship of the molten alloy and the mold wall. W is the wall thickness of the casting, as established in the Introduction. P increases with decreasing W. For thin-walled castings, properly increasing the pouring temperature, improving the quality of the mold cavity's surface, and imposing pressure can all facilitate the filling process.

Upon the formation of dendrites, the space left for the melt decreases and the resistance force increases accordingly. As the dendrites impinge upon each other, the interdendritic space forms a capillary network. Under this condition, the flow of the melt depends on the pressure gradient in the melt; the relationship between these properties can be presented by the following equations.

When the molten alloy flows in the interdendritic space driven by a pressure P, the maximum flow velocity $V1$ can be expressed using Darcy's law:

$$V_1 = \frac{k}{\eta f_l} \nabla P \tag{5.9}$$

where k is the permeability of the melt, fl is the liquid fraction, and ∇P is the pressure gradient. The velocity component of the molten alloy's solidification shrinkage is assumed to be $V2$, defined as:

$$V_2 = \frac{R_c}{G_H} \beta \tag{5.10}$$

where β is the solidification shrinkage rate. This represents the minimum velocity of the melt necessary to accommodate solidification shrinkage. According to (5.8) and (5.9), if $V1 \geq V2$, microshrinkage will not form.

The microshrinkage-free conditions can be obtained by combining Eqs. (5.8) and (5.9), which leads to

$$\nabla P \geq \frac{\eta f_l R_c}{k G_H} \beta \tag{5.11}$$

The relative ratios of ∇P were calculated according to Table 5.6 for the two sample geometries. For the 2 mm sample, the ratios of ∇P were:

$$\nabla P_{680\,^\circ\mathrm{C}} : \nabla P_{700\,^\circ\mathrm{C}} : \nabla P_{750\,^\circ\mathrm{C}} : \nabla P_{800\,^\circ\mathrm{C}} : \nabla P_{850\,^\circ\mathrm{C}} = 1 : 0.817 : 0.562 : 0.414 : 0.301$$

By comparison, the ratios of ∇P for the 5 mm sample were:

$$\nabla P_{680^\circ C} : \nabla P_{700^\circ C} : \nabla P_{750^\circ C} : \nabla P_{800^\circ C} : \nabla P_{850^\circ C} = 1 : 0.556 : 0.257 : 0.152 : 0.075$$

The ratios between the feeding pressure of the 2 mm sample and the feeding pressure of the 5 mm sample were:

$$\begin{aligned}
\nabla P_{2mm} : \nabla P_{5mm|680^\circ C} &= 28.707 \\
\nabla P_{2mm} : \nabla P_{5mm|700^\circ C} &= 42.221 \\
\nabla P_{2mm} : \nabla P_{5mm|750^\circ C} &= 62.868 \\
\nabla P_{2mm} : \nabla P_{5mm|800^\circ C} &= 78.049 \\
\nabla P_{2mm} : \nabla P_{5mm|850^\circ C} &= 114.967
\end{aligned}$$

Increasing the *BT* can reduce the feeding pressure to some extent, making the feeding process easier for thicker-walled structures. Decreasing the wall thickness is vital for the success of the filling and feeding processes; this effect may increase by orders of magnitude with minimal decreases in wall thickness.

(2) Filling time and feeding time

Filling and feeding are the two key stages for a successful casting. The filling ability of a certain alloy at the thinnest part of the mold relates simultaneously to the pouring temperature, mold temperature, and mold inner-surface properties. As all the mold-related factors are defined in a given casting, the factor left to manipulate is the pouring temperature. For high-quality castings, feeding must be specifically considered in the design of the mold's gating system. With hypoeutectic alloys, the solidification process can be divided into two stages, i.e., dendrite formation and eutectic phase solidification. Image analysis shows that the casting molding of thin-walled castings can be divided into effective filling stage and eutectic phase solidification stage using the formation of the dendritic network as a dividing point. The stage before dendritic network formation is the effective filling stage. The dendrite network formation time (DNFT) is equivalent to the effective filling time (for thin-walled castings). Then the stage from the dendritic network formation to the "eutectic phase solidification" is defined as the feeding period. Therefore, the feeding time (FT) is defined as the time interval after DNFT and prior to freezing.

Figure 5.45a shows the DNFT of the two geometries of thin-walled samples at different *BT* values. The experimental data can be described through fitting into the exponential function $y = ax^n$. The magnitude of the exponent n reflects the sensitivity of the DNFT to *BT*. When *BT* increases from 660 to 850 °C, the DNFT of 2 mm samples increases by only a factor of 2.14; in contrast, the DNFT of 5 mm samples increases by a factor of 6.36. Compared to the 5 mm sample geometry, the DNFT of the 2 mm sample is less sensitive to *BT*.

Under the same conditions, the FT of the 2 mm sample increases by a factor of 3.21 as *BT* increases, and by a factor of 4.5 for the 5 mm sample, as shown in

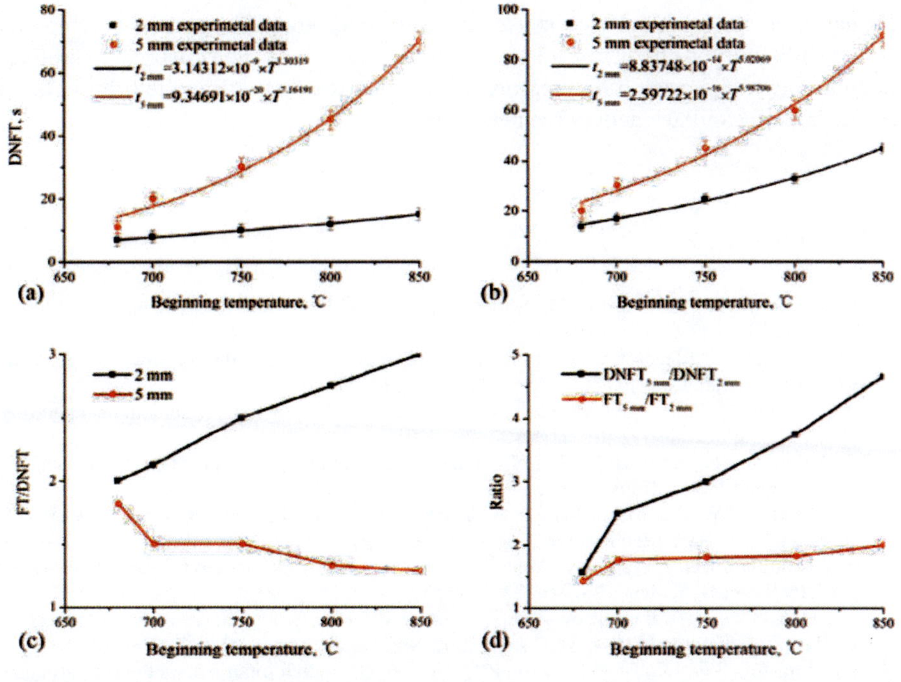

Fig. 5.45 The different solidification stages' dependence on the beginning temperature for the two geometries of thin-walled samples. **a** Dendrite network forming time (DNFT); **b** feeding time (FT); **c** ratios of FT/DNFT; and **d** ratios of DNFT5 mm/DNFT2 mm and FT5 mm/FT2 mm

Fig. 5.45b. For the 2 mm sample, in general, the sensitivity of FT to *BT* is stronger than that of DNFT. For the 5 mm sample, the sensitivity of DNFT to *BT* is stronger than that of FT.

This characteristics can be presented using the ratio of FT/DNFT for each sample geometry at each *BT*, as shown in Fig. 5.45c. The ratio of FT/DNFT of the 2 mm sample increases with increasing *TB*, but the ratio of the 5 mm sample decreases with increasing *BT*. This indicates that even infinite increase in the pouring temperature for a 2 mm thin-walled part will not improve the filling ability.

Figure 5.45d shows that the ratio of DNFT between the two geometries of samples increases rapidly with increasing *BT*. However, the ratio of FT between the two sample geometries remains constant with increasing *BT*. This indicates that the filling time is sensitive to the wall thickness of castings and decreasing the wall thickness will sharply reduce the filling ability.

As a result, increasing the *BT* alone cannot sufficiently improve the filling ability of thin-walled castings. Limited by the heat resistance of the mold and the refinement of the microstructure, the pre-heating temperature of the mold cannot be raised arbitrarily. Castings with thin-walled parts would be of better quality if filled during the "melt" stage. In other words, thin features require rapid filling. Increasing the pouring temperature to promote the completion of the filling process before the formation of dendritic network is useful to gravity casting. However, achieving this condition is quite difficult for a large and complex thin-walled casting. Therefore, maintaining a high hydraulic pressure to facilitate the flow of the melt through the interdendritic capillary network is necessary.

References

1. G.A. Knorovsky, M.J. Cieslak, T.J. Headley et al., Inconel 718: A solidification diagram. Metall. Trans. A **20**, 2149–2158 (1989)
2. M.J. Cieslak, T.J. Headley, G.A. Knorovsky et al., A comparison of the solidification behavior of Incoloy 909 and Inconel718. Metall. Trans. A **21**, 479–488 (1990)
3. Y.H. Xiong, A.M. Yang, P.J. Li et al., Investigation on grain structure refinement of superalloy K4169. J. Aeronautical Mater. **21**(3), 24–28 (2001)
4. T. Antonsson, H. Fredriksson, The effect of cooling rate on the solidification of Inconel718. Metallurgical Mater. Trans. B **36**, 85–96 (2005)
5. G.L. Wang, C.W. Wu, M.C. Zhang et al., The microanalysis of precipitation behavior of α-Cr in Inconel 718. Rare Metal Mater. Eng. **31**(1), 37–40 (2002)
6. E.V. Kolotukhin, G.V. Tyagunov, Crystallization of superalloys with various contents of carbon. J. Mater. Process. Technol. **53**, 219–227 (1995)
7. L. Liu, B.L. Zhen, A. Banerji, W. Reif, F. Sommer, Effect of melt homogenization temperature on the cast structures of IN 738 LC superalloy. Scr. Metall. Mater. **30**, 593–598 (1994)
8. F.S. Yin, X.F. Sun, J.G. Li, H.R. Guan, Z.Q. Hu, Effects of melt treatment on the cast structure of M963 superalloy. Scripta Mater. **48**, 425–429 (2003)
9. F.S. Yin, X.F. Sun, H.R. Guan, Z.Q. Hu, Effect of thermal history on the liquid structure of a cast nickel-base superalloy M963. J. Alloy. Compd. **364**, 225–228 (2004)
10. O.S. Roik, O.V. Samsonnikov, V.P. Kazimirov et al., Short and medium-range order in liquid binary Al-Ni and Al-Co alloys. J. Mol. Liq. **145**, 129–134 (2009)
11. J.B. Ma, S.H. Chen, Y.B. Dai et al., The local structure of molten Ni_{1-x} Al_x: An ab initio molecular dynamics study. J. Non-Cryst. Solids **425**, 11–19 (2015)
12. J.B. Ma. The local structure of molten Ni-based alloy: An ab initio molecular dynamics study[D]. (Shanghai Jiao Tong University, 2015)
13. P.J. Zhou, J.J. Yu, X.F. Sun, H.R. Guan, Z.Q. Hu, The role of boron on a conventional nickel-based superalloy. Mater. Sci. Eng. A **491**, 159–163 (2008)
14. Y.L. Tsai, S.F. Wang, H.Y. Bor, Y.F. Hu, Effects of Zr addition on the microstructure and mechanical behavior of a fine-grained nickel-based superalloy at elevated temperatures. Mater. Sci. Eng. A **607**, 294–301 (2014)
15. J. Zhang, R.F. Singer, Effect of Zr and B on castability of Ni-based superalloy IN792. Metallurgical Mater. Trans. A **35**, 1337–1342 (2004)
16. K.C. Antony, J.F. Radavich, Solute effects of boron and zirconium on microporosity. Proceedings of the third international symposium on superalloy 1976, (Claitor Publishing, Warrendale, PA, 1976) pp. 137–146
17. W. Zhou, L. Liu, Z.Q. Jie et al., Effect of element B on fluidity and microporosity of K4169 superalloy. Rare Metal Mater. Eng. **43**(12), 3082–3088 (2014)
18. L. Liu, T.W. Huang, Y.H. Xiong, A.M. Yang, Z.L. Zhao, R. Zhang, J.S. Li, Grain refinement of superalloy k4169 by addition of refiners: cast structure and refinement mechanisms. Mater. Sci. Eng. A **394**, 1–8 (2005)
19. F. Wang, J. Zhang, T.W. Huang, L. Liu, H.Z. Fu, Preparation of inoculants used in superalloy and analysis of the atomic matching models. J. Mater. Sci. Technol. **29**, 387–392 (2013)
20. Z.Q. Jie, J. Zhang, T.W. Huang, H.J. Su, Y.F. Li, L Liu, W.C. Yang, H.Z. Fu, Enhanced grain refinement and porosity control of the polycrystalline superalloy by a modified thermally-controlled solidification. (Advanced Engineering Materials, 2016) pp. 1–7

21. G.Y. An, *Theory of Casting Forming [M]* (China Machine Press, Beijing, 1990).
22. B.N. Lin, Z.J. Wei, *Transfer Principles of Metal Hot Forming [M]* (Harbin Institute of Technology Press, Harbin, 2000).
23. M.R. Ghomashchi, A. Vikhrov, Squeeze casting: an overview. J. Mater. Process. Technol. **101**, 1–9 (2000)
24. M. Wang, J.M. Zeng, W.D. Huang, Adjusted pressure casting technique for precision casting of large-scale complicated & thin-wall components. Foundry Technol. **25**(5), 353–358 (2004)
25. S.W. Kim, G. Durrant, J.H. Lee, B. Cantor, The effect of die geometry on the microstructure of indirect squeeze cast and gravity die cast 7050 (Al-6.2Zn-2.3Cu-2.3Mg) wrought Al alloy. J. Mater. Sci. **34**, 1873–1883
26. K.M. Pedersen, J.H. Hattel, N. Tiedje, Numerical modelling of thin-walled hypereutectic ductile cast iron parts. Acta Mater. **54**, 5103–5114 (2006)
27. D. Eskin, Q. Du, D. Ruvalcaba, L. Katgerman, Experimental study of structure formation in binary Al-Cu alloys at different cooling rates. Mater. Sci. Eng. A **405**, 1–10 (2005)
28. J.B. Ferguson, T.K. Meysam, C.M. John, K.R. Pradeep, K. Cho, C.S. Kim, Predicting the effect of pouring temperature on the crystallite density, remelting, and crystal growth kinetics in the solidification of aluminum alloys. Metallurgical Mater. Trans. B **45**, 1407–1417 (2014)

Chapter 6
Prediction and Control of Casting Defects in Large Castings

Jun Wang, Haiyan Gao, Xinhua Tang, Maodong Kang, and Yang Zhou

There are several characteristics of casting components with large scales: large shape size, complex structures, large wall thickness variations, many locations of varying sections and large portion of thin-wall area. In the process of casting, shrinkage cavity, shrinkage porosity, inclusion, cold separation, and other casting defects are easy to form, which seriously reduce the quality of castings and even lead to the waste of castings, limiting the development of high-performance aero-engine. The key to the development of large scale aero-engine castings contains the detection technology of casting defects, the quantitative prediction of casting defects, and the control and repair technologies of casting defects in large castings.

6.1 Detection Methods for Defects of Large Castings

6.1.1 Requirements of Castings Pre-treatment

X-ray detection method is mainly used to examine the existence of cracks, holes, inclusions, and other defects in the casting. Before radiographic examinations, the casting must undergo necessary pre-treatment. First, the sprue and riser of casting needs to be cut off and polished. Secondly, the shell material on the surface of castings must be removed to avoid their interference on the radiographic inspection and identification. Finally, the surface quality of castings must be acceptable through visual inspection.

6.1.2 Selection of X-ray Detection Technology

Considering conventional methods to perform nondestructive testing for large castings, the tube voltage should be maintained as low as possible to reduce X-ray

B. Sun et al., *Precision Forming Technology of Large Superalloy Castings for Aircraft Engines*, https://doi.org/10.1007/978-981-33-6220-8_6

energy and meanwhile high enough to keep X-ray penetrating. Under this condition, the absorption capacity of the irradiated casting is enhanced, which can increase the black and white contrast of the defective and non-defective parts, thus improving the sensitivity of the negative film. The contrast of X-ray negative film is related to the type of film. The film commonly used for conventional industrial testing is AGFAC7 film manufactured by the GE company. The researchers compared the difference in the effect of various types of films on the nondestructive testing by selecting AGFAC7 and AGFAD3 specifications (AGFAC7:G3, medium grain size film and AGFAD3:G1, particle size film). For thin-wall parts, high sensitivity AGFAD3 negative film is used to improve the resolution and reduce the omission ratio, as shown in Fig. 6.1a. As the thickness of components increasing, the difficulty in detecting minor defects increases, and the difference in detection ability between the two types of films decreases, as shown in Fig. 6.1b. The selection principles of negative film of large casting are as follows. The highly sensitive negative film is used for thin-wall

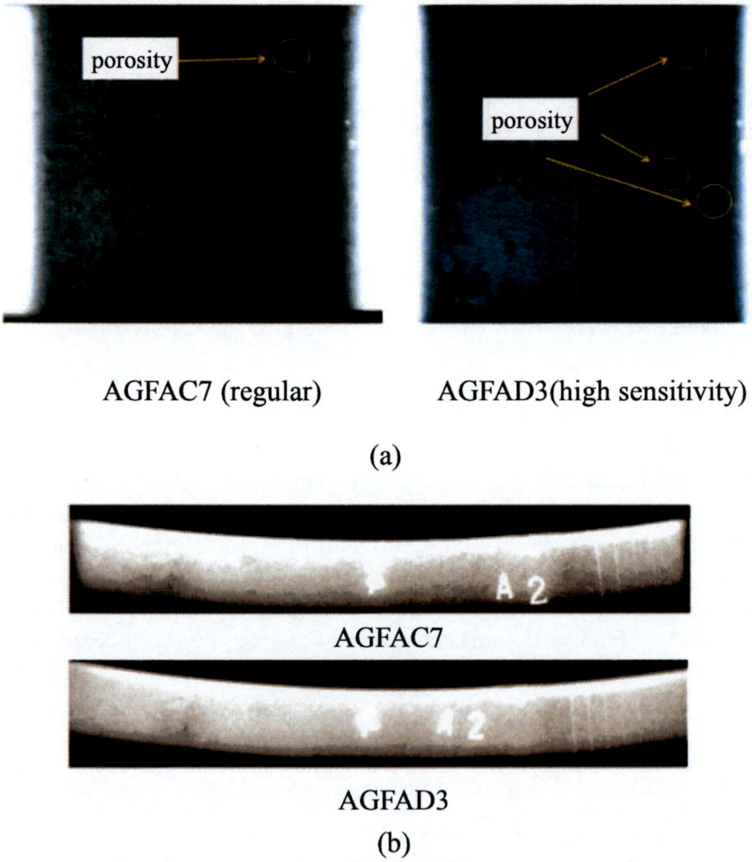

Fig. 6.1 Film for the nondestructive testing (AGFAC7 and AGFAD3)

parts to improve the detection accuracy, whereas the traditional film is used for thick parts to reduce the cost.

6.1.3 Optimization of X-ray Detection Technology for Large Castings

In order to overcome the shortcomings of the existing X-ray nondestructive testing technology for large castings, researchers developed an X-ray testing tooling and a testing method for round castings to reduce the omission ratio of casting defects and to improve the stability of testing process, as shown in Fig. 6.2. The method includes translation, turnover, rotation, and expansion devices. The method is to adjust the turning bracket to the horizontal state by adjusting the turning handle and locking the locating pin of the driven gear to fix the turning bracket. Placing the casting to be tested horizontally on the main and driven pulley, and adjusting the length of telescopic boom so that the main and driven pulley fix the casting to be tested from 4 directions to the flip bracket. Pulling the tooling base to the bottom of the X-ray detection probe, and locking the universal wheel through the universal wheel lock buckle. Opening the driven gear locating pin, shaking the turning handle to adjust the turning bracket to the predetermined detection angle, and locking the driven gear locating pin. Opening the lock of the active pulley, rotating the casting to be tested to the predetermined detection site by shaking the handle of the active

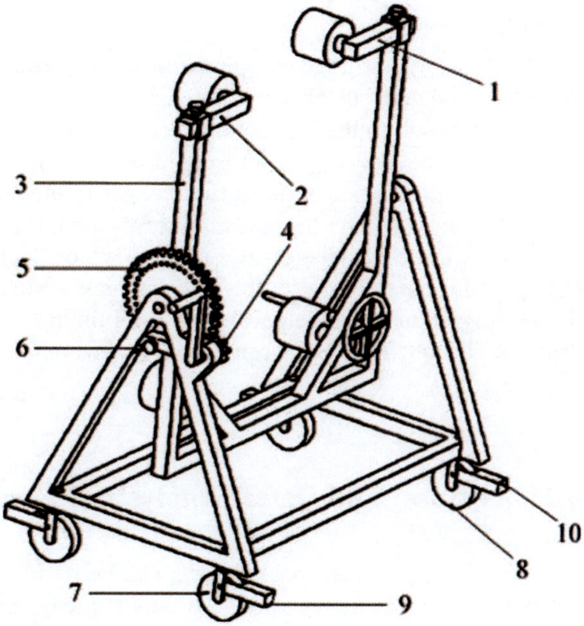

Fig. 6.2 X-ray testing tooling for round castings

Fig. 6.3 Omniscan MX UT ultrasonic testing system

pulley, and locking the lock of the active pulley. Opening the probe for imaging and film examination, and repeating the above operation to complete the 100% all-round X-ray film examination of the complex ring precision casting.

6.1.4 Exploration of New Nondestructive Testing Methods

For the thick bulk of large complex thin-walled castings, it is difficult to determine the depth of defects using the traditional X-ray detection. Ultrasonic phased array technology refers to the interaction between ultrasonic wave and casting, which is a technology of detecting macroscopic defects, measuring geometric characteristics, detecting and characterizing the changes of organization structure and mechanical properties based on the research of reflected, transmitted, and scattered waves. It has significant advantages in the detection of thick parts of large castings. For example, the phased array ultrasonic imaging system developed based on FPGA (Field Programmable Gate Away) has a minimum delay of 20 ms and a detection depth of 60 mm. The Omniscan MX UT ultrasonic testing equipment is shown in Fig. 6.3.

6.2 Anatomy and Defect Analysis of Large Castings

In engineering practice, loose defects can be divided into macroscopic shrinkage cavity and microscopic porosity according to their size [1]. The former is a fine pore shrinkage that can be seen with naked eyes or a magnifying glass, whereas the latter corresponds to the gathered or dispersed fine pores observed using a microscope [2, 3]. The traditional X-ray nondestructive testing technology can only evaluate in general the distribution of macroscopic shrinkage cavity and porosity defects, cracks, inclusions, and other defects in the casting. It is impossible to accurately conduct distribution of the micro-loose in the casting which is closely related to the mechanical properties. Micro-porosity defects reduce not only the compactness of the structure of large complex thin-walled castings, but also the mechanical properties of the castings under service conditions, in particular, the fatigue properties [4–7], resulting in the reduction in service life of components. Therefore, it is necessary to conduct ontological anatomy of the castings to get the information of internal defects.

Taking a certain type of rear casing casting as an example, the basic steps and defect analysis methods of anatomy of large castings are illustrated in the following. Firstly, the anatomical location is determined according to the structural characteristics of the casting, including the main bearing parts of the casting (lifting lugs, internal and external flanges, small boss et al.) and thin-wall parts with large area (the outer circumferential torus and support plate). Under a metallographic microscope with 100 magnification, the area percentage of micro-porosity defects in the area of 1 mm * 1 mm at the most severe position of micro-porosity defects is expressed as the micro-porosity index (MPI), and the metallographic pictures of different defect levels are shown in Fig. 6.4.

Due to the difference of solidification conditions in each parts of casting, the micro-porosity index of large castings varies greatly. The micro-porosity distribution of sampling at the lifting lug is shown in Fig. 6.5. It can be seen that the micro-porosity index at the final solidification site of the lifting lug reaches 5.93 due to insufficient riser feeding at the large lifting lug with a thickness of 23 mm.

A total of 107 samples were prepared for rear casing casting, 33 of which had a micro-porosity index higher than 1. Nine samples were prepared from the thick section of the 33 samples, accounting for 27.3% of the total. There were 12 samples from the section of variable section, accounting for 36.4% of the total. There were 12 samples in the thin-wall parts with thickness less than 3 mm, accounting for 36.4% of the total. It can be seen that the overall micro-porosity index of the thick parts of turbine rear frame (TRF) castings is relatively small. However, many micro-porosity defects existing in the variable-area parts and thin-walled parts with large area are one of the bottleneck problems that restricts the manufacturing of large and complex thin-walled TRF castings.

6.3 Structural Effects of Casting Defects

The structure of large castings is different and its local cooling conditions are also different, which leads to a great difference in microstructure of castings. The analysis of internal defects of a TRF casting shows that there is no apparent relationship

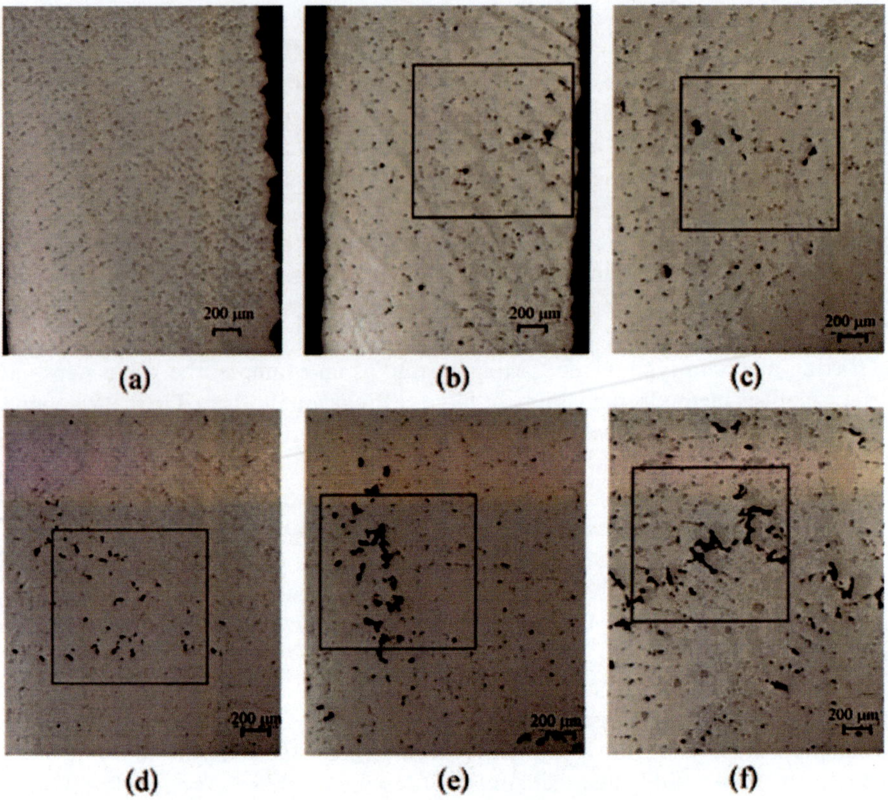

Fig. 6.4 Optical image of the micro-porosity and corresponding micro-porosity index (MPI). **a** MPI = 0, **b** MPI = 0.5, **c** MPI = 1, **d** MPI = 2, **e** MPI = 4, **f** MPI = 5

between the micro-porosity index and the structural thickness of the casting, as shown in Fig. 6.6.

However, there is an obvious relationship between the micro-porosity index and the section thickness ratio (the ratio of sectional area to thickness) of the local structure of the casting, as shown in Fig. 6.7. When the ratio of the section and thickness is less than 1000:1, the micro-porosity index is larger, indicating that the micro-porosity defects are easy to form in the thick part of the casting. With the increase of this ratio, the micro-porosity index decreases gradually. As a result, this section-to-thickness ratio is inversely proportional to the micro-porosity index where it is ranging from 1000:1 to 6000:1. However, when this ratio increases continuously to 10,000:1, a large number of micro-porosity defects appear again in large complex thin-wall castings.

Figure 6.8 shows the relationship between the micro-porosity index and the thermal-modulus of large complex thin-walled castings. The thermal-modulus of castings is mainly between 0.3 and 0.9 mm and the micro-porosity index decreases with the increase of the thermal modulus.

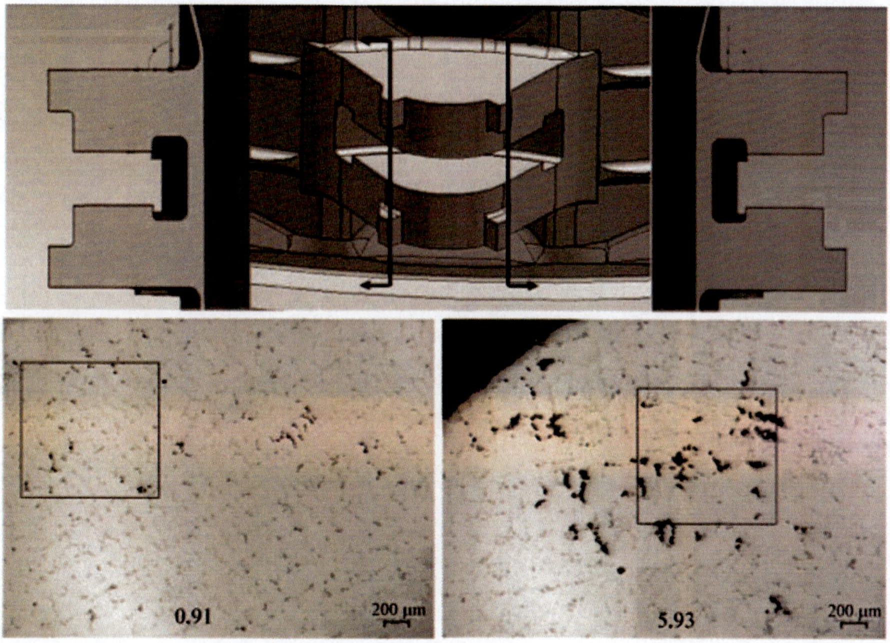

Fig. 6.5 Micro-porosity inside the lifting lug of the frame

Fig. 6.6 Relationship between the micro-porosity index and the thickness

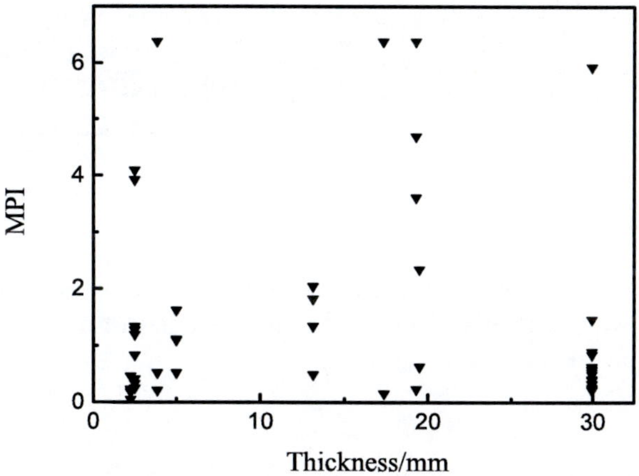

6.4 Micro-porosity Defects in Large Castings

With the increasing demand of the high-quality advanced equipment, it is necessary to reduce the internal defects of castings, especially the micro-porosity defects.

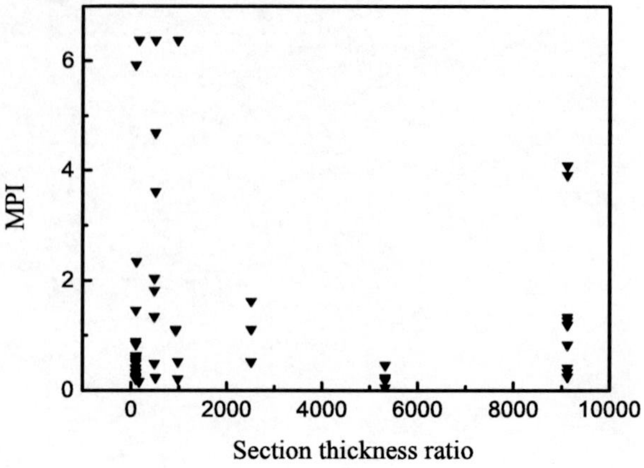

Fig. 6.7 Relationship between the micro-porosity index and the section thickness ratio

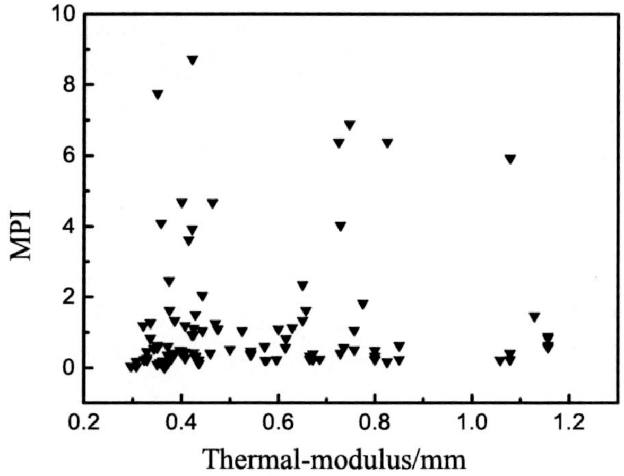

Fig. 6.8 Relationship between the micro-porosity index and the thermal-modulus

With the development of solidification theory and computer simulation capability [8], mathematical modeling and quantitative prediction of micro-porosity become possible.

6.4.1 Quantitative Prediction of Micro-porosity

Carlson K. et al. proposed a dimensionless micro-porosity criterion model based on the classical Niyama criterion, which could predict the total micro-porosity volume fraction formed in the solidification process only by knowing the temperature curve of solid phase fraction and the total shrinkage rate of the alloy. The prediction results of WCB steel and AZ91D magnesium alloy were in good agreement with the experimental results [9]. Based on the dimensionless Niyama micro-porosity criterion and by considering the trace gas precipitation, quantitative prediction of micro-porosity of thin-wall parts of high-temperature alloy castings can be achieved.

6.4.1.1 Mathematical Model of Thin-Wall Micro-porosity Prediction

From the simulation of casting temperature field, it can be seen that the thin-wall part of casting has the characteristics of near-directional solidification. Therefore, a dimensionless micro-porosity prediction function can be derived using one-dimensional directional solidification model. Darcy's law of the system can be expressed as follows:

$$f_1 V_1 = -K dP/(\mu dx) \tag{6.1}$$

where, f_1 is the liquid volume fraction, V_1 is the liquid rate (shrinkage rate) in the solid/liquid two-phase region (paste region), μ is the dynamic viscosity of the alloy, P is melt pressure, and x is the spatial coordinate. Permeability K in the solid–liquid two-phase region is determined by the Kozeny-Carman relationship:

$$K = \lambda_2 f_1^3/(180(1 - f_1)) \tag{6.2}$$

where λ_2 is the secondary dendrite spacing for the alloy.

The liquid and solid densities of the alloy are assumed to remain constant throughout the solidification process. In this way, the total solidification shrinkage of the alloy can be defined using the density, $\beta = (\rho_s - \rho_1)/\rho_1$. Then, the one-dimensional mass conservation equation is simplified by using β. Through integral calculation, it can be known that the contraction rate of the solid–liquid two-phase region is constant, which can be expressed as $V_1 = -\beta L$, where L is the solidification rate. In addition, L can be expressed in terms of temperature gradient $G(dT/dx)$ and cooling rate $R(dT/dt)$ N, which allows the solidification contraction rate of the solid–liquid two-phase region to be expressed as:

$$V_1 = -\beta R/G \tag{6.3}$$

After substituting Eqs. (6.3) into (6.2), we get:

$$dP/dx = \mu\beta R f_1/(KG) \tag{6.4}$$

As seen from Fig. 6.9, with the increase of solid phase fraction during solidification, the melt pressure in the paste region will drop from liquid phase pressure P_{liq} to a critical pressure P_{cr}, under which a micro-porosity defect will be formed. Here, the critical pressure drop (value of liquid pressure minus critical pressure) is defined as $\Delta P_{cr} = P_{liq} - P_{cr}$, where the liquid pressure is the sum of the system

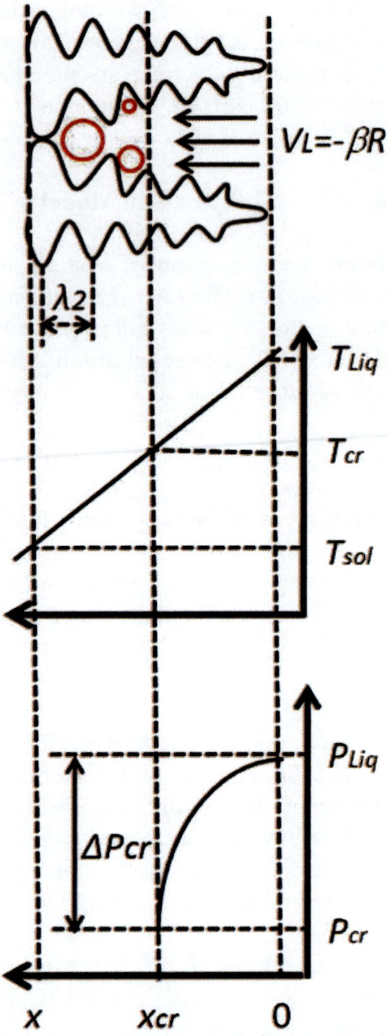

Fig. 6.9 Schematic view of the mushy zone during solidification

environmental pressure and the partial liquid pressure; while the critical pressure is related to the pressure balance when pores exist in a stable manner. The balance can be expressed by Young–Laplace equation, $P_{cr} = P_p - P_\sigma$, where P_p is the internal pressure and P_σ is capillary pressure. The capillary pressure can be calculated using $P_\sigma = 2\sigma/r_0$, where σ is the surface tension between micro-pores and surrounding liquid. r_0 is the initial radius of curvature when micro-porosity forms, which relates to the thermo-physical properties, size, and local geometrical features of the micro-porosity. For the pure solidification contraction without gas solubility in metal, the internal pressure (generated only by the vapor of alloy elements) can be neglected. Thus, the Young–Laplace equation can be simplified as

$$P_{cr} = -P_\sigma = -2\sigma/r_0 \tag{6.5}$$

From Eq. (6.5), it can be seen that the critical pressure is negative. During the formation of micro-porosity, the surface tension has to be overcome firstly. Assuming that the dynamic viscosity, temperature gradient, and cooling rate keep constant during the formation of micro-porosity, we then integrate Eq. (6.4) from the critical point to liquid zone, which leads to the following equation:

$$\Delta \mathrm{P}_{cr} = \int\limits_{P_{cr}}^{P_{liq}} dP = \int\limits_{x_{cr}}^{0} \frac{\mu\beta R f_1}{KG} dx = \int\limits_{T_{cr}}^{T_{liq}} \frac{\mu\beta R f_1}{KG}\frac{dx}{dT} dT = \frac{\mu\beta R}{G^2} \int\limits_{f_{1,cr}}^{1} \frac{f_1}{K}\frac{dR}{df_1} df_1 \tag{6.6}$$

where x_{cr},T_{cr}, and $f_{l,cr}$ correspond to the position, temperature, and liquid phase fraction, respectively, when the micro-porosity starts to form as the pressure decreases to P_{cr}.

In order to normalize the integral value obtained from Eq. (6.6), a dimensionless temperature θ is introduced as $\theta = (T - T_{\mathrm{sol}})/\Delta T_f$, where $\Delta T_f = T_{\mathrm{liq}} - T_{\mathrm{sol}}$ is the temperature range of solidification, T_{liq} is the liquidus temperature, and T_{sol} is the temperature at which the alloy solidifies completely. Meanwhile, substituting Eq. (6.2) into (6.1), we can obtain:

$$\Delta \mathrm{P}_{cr} = \frac{\mu\beta\Delta T_f}{\lambda_2^2}\frac{R}{G^2} I(f_{1,cr}) \tag{6.7}$$

where

$$I(f_{1,cr}) = \int\limits_{f_{1,cr}}^{1} 180\frac{(1-f_1)^2}{f_1^2}\frac{d\theta}{df_1} df_1 \tag{6.8}$$

The integral value in Eq. (6.8) can be calculated according to the solid phase-temperature curve obtained from the thermo-physical properties of the alloys. The square root of $I\ (f_{1,\mathrm{cr}})$ is defined as dimensionless *Niyama* criterion value (N_y^*). After rearranging Eq. (6.7), a new equation can be obtained:

$$N_y^* = \sqrt{I(f_{1,cr})} = \frac{G\lambda_2\sqrt{\Delta P_{cr}}}{\sqrt{\mu\beta\Delta T_f R}} \tag{6.9}$$

This dimensionless *Niyama* criterion considers not only the thermodynamic and kinetic conditions for solidification (crystallization temperature range ΔT_f, temperature gradient G and cooling rate R) but also the thermo-physical properties of alloys (the dynamic viscosity μ, overall solidification shrinkage ratio β) and the critical pressure decrease ΔP_{cr} at solid–liquid two-phase region. Assuming that during the formation of the micro-porosity, local liquid flow will be ceased. Then the final micro-porosity index f_p is related to the liquid phase fraction $f_{l,cr}$ under critical pressure, which can be expressed as:

$$f_p = \beta f_{l,cr} \tag{6.10}$$

The advantage of this dimensionless criterion lies in that the micro-porosity index can be quantified with the given solidification conditions and thermo-physical properties. According to Eqs. (6.7) to (6.10), the micro-porosity index can be expressed as a function of the dimensionless *Niyama* value. For specific alloys, the dimensionless *Niyama* value can be directly obtained from local solidification parameters and thermo-physical properties of the alloys:

$$N_y^* = \frac{G\lambda_2\sqrt{\Delta P_{cr}}}{\sqrt{\mu\beta\Delta T_f R}} \tag{6.11}$$

After the determination of the dimensionless *Niyama* value, the micro-porosity index can be obtained from the $N_y^* - f_p$ function; besides, the spacing of secondary dendrites λ_2 can be quantitatively calculated using Fisher and Kurz's theory:

$$\lambda_2 = C_\lambda R^{-1/3} \tag{6.12}$$

where C_λ is a constant that depends on the alloy system and can be acquired via experimental fitting results.

6.4.1.2 Quantitative Prediction Functions for Micro-porosity in K4169 Superalloy Castings

During the prediction for dimensionless micro-porosity, thermo-physical calculation software package JMatPro is employed to calculate the thermo-physical property parameters, including melt density-temperature curve, solid phase fraction-temperature curve, melt dynamic viscosity, and crystallization temperature intervals, etc. Then the functional relationship between the dimensionless *Niyama* value and the micro-porosity index is deduced according to equations from (6.7) to (6.10).

The relationship between the temperature and solid phase fraction in K4169 superalloy calculated using JMatPro software is presented in Fig. 6.10. In order to simplify the calculations, the relationship curve of dimensionless temperature and liquid phase fraction is shown in Fig. 6.11.

Polynomial fitting is performed using the data curves to obtain the gradient in Fig. 6.11. According to the structural features, the fitting is divided into 4 sections on the curve. The results are shown as dotted lines in Fig. 6.11 and listed in Table 6.1. According to the fitting equations, the expression for $d\theta/df_1$ can be obtained and is then substituted it into Eq. (6.8), which leads to a 4-segment integral formula. In

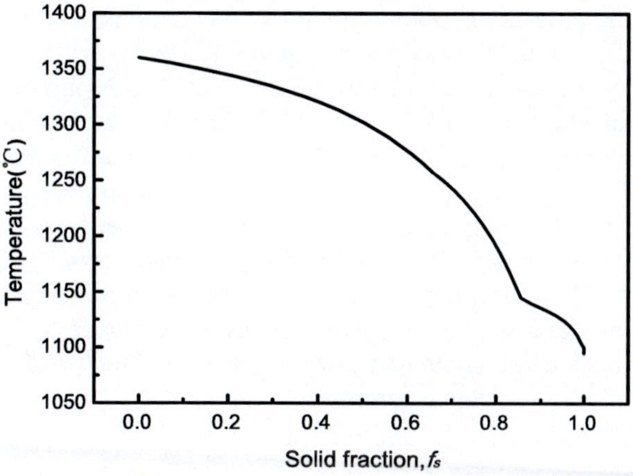

Fig. 6.10 Relationship between the solid fraction and temperature

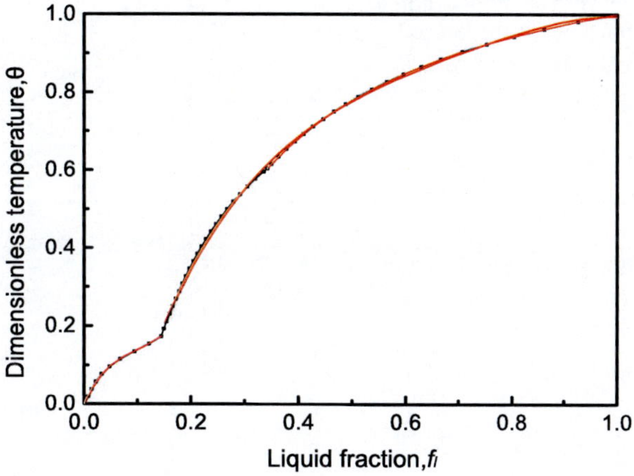

Fig. 6.11 Relationship between the liquid fraction and dimensionless temperature

Table 6.1 The fitting functions between liquid fraction (f_L) and dimensionless temperature (θ)

Liquid fraction (f_L)	Fitting functions
0.75248–1	$\theta = 0.23 + 1.65 f_L - 1.22 f_L^2 + 0.34 f_L^3$
0.51169–0.75248	$\theta = -0.115 + 3.1 f_L - 3.3 f_L^2 + 1.344 f_L^3$
0.14456–0.51169	$\theta = -0.4284 + 5.54 f_L - 10.04 f_L^2 + 9.32 f_L^3 - 3.27 f_L^4$
0–0.14456	$\theta = 3.25 f_L - 32.84 f_L^2 + 172.72 f_L^3 - 300.82 f_L^4$

the process of solving this integral equations, a corresponding $I\ (f_{\mathrm{l,cr}})$ value can be determined for any given f_{l} value. When the critical liquid phase fraction of K4169 superalloy is set to be in the range of 0–1 during the whole solidification process, an $f_{\mathrm{l,cr}}$ *versus* $I\ (f_{\mathrm{l,cr}})$ function curve can be established, as shown in Fig. 6.12. Combining Eqs. (6.4), (6.5) and the function relations in Fig. 6.12, the $f_{\mathrm{l,cr}}$ *versus* $I\ (f_{\mathrm{l,cr}})$ function can be converted into the dimensionless N_y^* *versus* f_{p} relation. In order to reflect the functional relationship between them more directly, we can take the log of N_y^* when plotting the graph, as shown in Fig. 6.13.

Therefore, for the predictions of micro-porosity in K4169 superalloy castings, provided that the dimensionless *Niyama* criterion N_y^* is calculated, the micro-porosity index can be obtained from Fig. 6.13 and the predictions for micro-porosity index of castings can be completed.

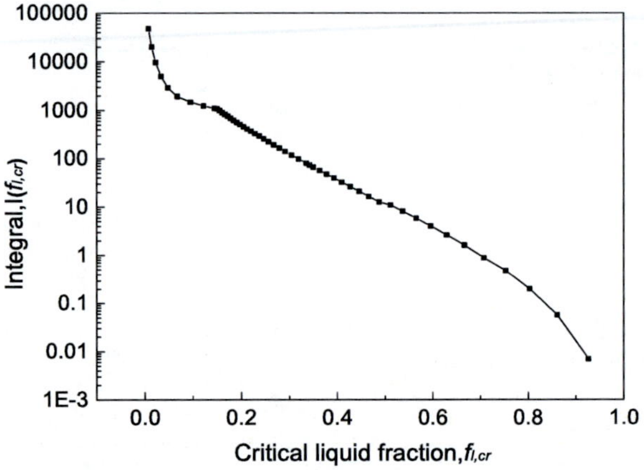

Fig. 6.12 Relationship between the critical liquid fraction and the integral function

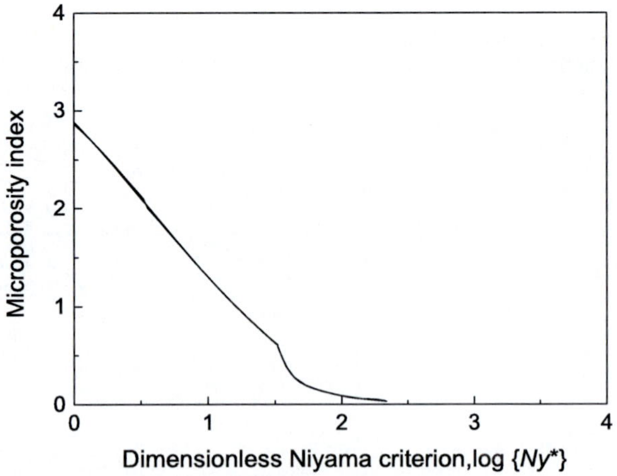

Fig. 6.13 Relationship between the dimensionless Niyama criterion and the microporosity index

6.4.2 Quantitative Prediction and Result Analysis of Thin Wall Micro-porosity

According to the dimensionless micro-porosity criterion Eq. (6.11), the calculation of dynamic viscosity, solidification shrinkage rate, temperature gradient, and secondary dendrite spacing, etc. are needed. The dynamic viscosity of K4169 superalloy is 8.9 mPa·s according to JMatPro thermodynamic calculations and the corresponding alloy density *versus* temperature curve is shown in Fig. 6.14.

From Fig. 6.14, it can be seen that the density of liquid phase is 7.47 g/cm^3 and solid phase density is 7.8 g/cm^3. The calculated total solid shrinkage is 0.0428. The solid–liquid phase temperature is determined using DSC, where the crystallization temperature interval is calculated to be 150 °C, as shown in Fig. 6.15.

Fig. 6.14 The density curve of K4169 superalloy at different temperatures

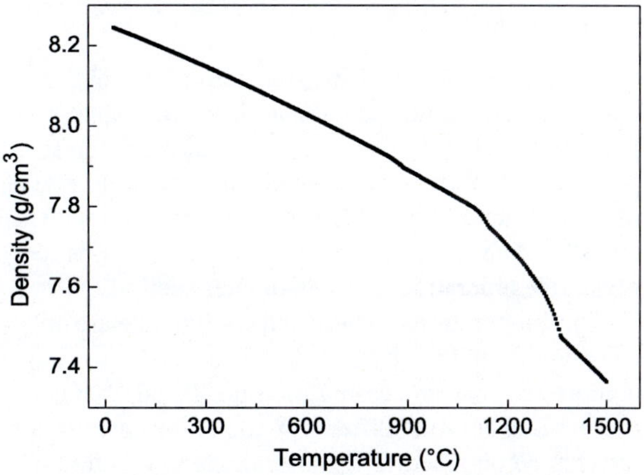

Fig. 6.15 The DSC curve of K4169 superalloy

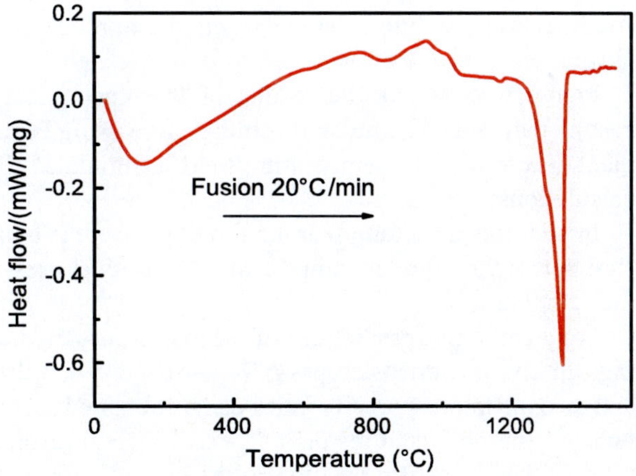

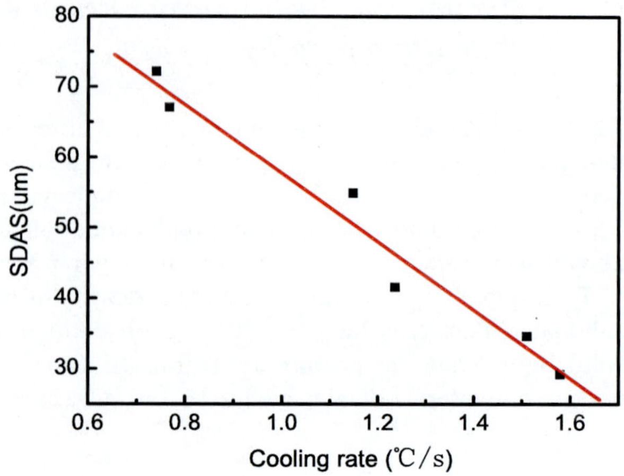

Fig. 6.16 Relationship between the cooling rate and SDAS

The secondary dendrite spacing can be obtained through fitting the empirical equations. As mentioned above, in the classical secondary dendrite spacing calculation Eq. (6.12), C_λ is a constant that is related to alloy compositions. Figure 6.16 presents the experimental results of the cooing rate and secondary dendrite spacing of K4169 superalloy castings, which are fitted with a slope C_λ value of 4.8×10^{-5} m (°C/s)$^{1/3}$. The result will be used for calculations for the secondary dendrite spacing of K4169 superalloy castings in later sections.

The temperature gradient and cooling rate are obtained based on the finite element calculations. After the simulation of casting filling and solidification process are completed, post-processing module Visual Cast is run and the temperature gradient and cooling curve of different parts of the castings are determined. Then the cooling rate can be obtained. The mesh and temperature field simulated using ProCast software are shown in Fig. 6.17. From the simulation results of temperature field, it can be seen that the solidification condition at thin-wall areas are close to the directional solidification condition, which meets the application conditions of dimensionless *Niyama*.

Figure 6.18 shows the results of thermocouple test at a certain point in the casting part and the simulated cooling curve using ProCast, which are in good agreement. Therefore, the temperature field is simulated in the subsequent cooling rate calculations.

In addition, assuming that the liquid pressure is balanced with atmospheric pressure below the liquidus temperature, the critical pressure drop is set to 0.1 MPa in the calculation.

For quantitative predictions of the micro-porosity index in K4169 superalloy castings, firstly, the dimensionless N_y^* value need to be calculated according to Eq. (6.11) and then the micro-porosity index can be obtained according to Fig. 6.10. Figure 6.19 presents the predicted results and experimental results of the micro-porosity index for each characteristic position in the large complex thin-walled K4169 superalloy castings, showing good agreement with each other. For example, Fig. 6.20 shows

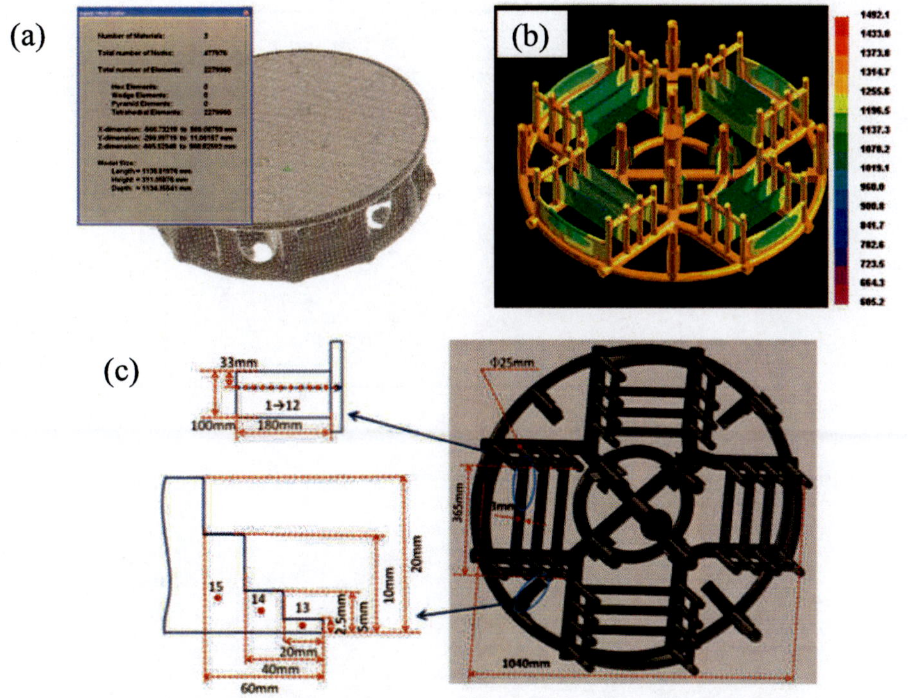

Fig. 6.17 The finite element mesh and temperature field for the casting: **a** mesh, **b** temperature field, **c** geometry

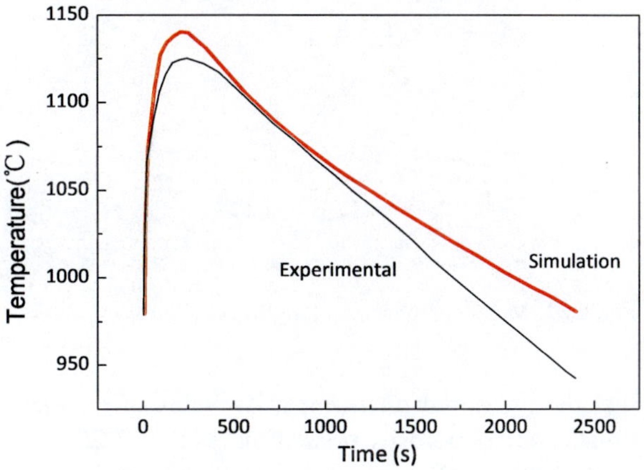

Fig. 6.18 Comparison between the experimental and simulated cooling curve

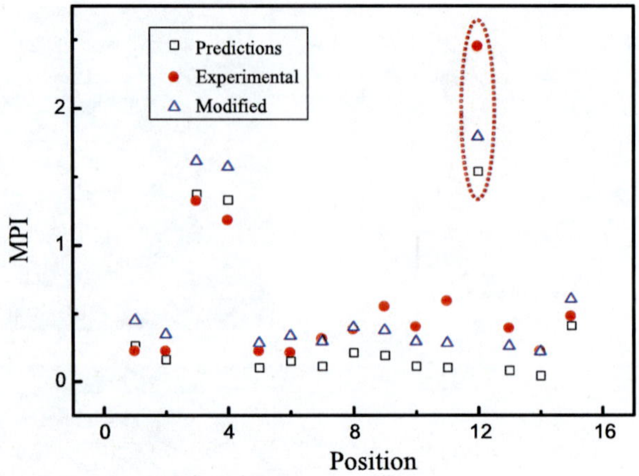

Fig. 6.19 Experimental results and predictions at different positions of the characteristic casting

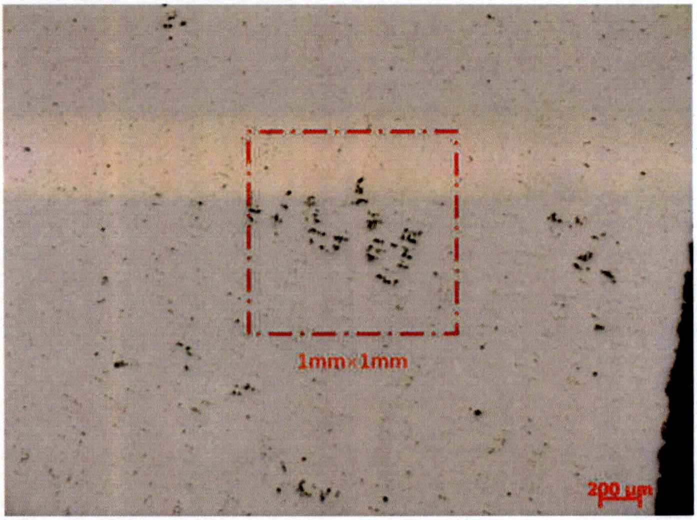

Fig. 6.20 Metallographic structure for MPI = 1.26

that the measured micro-porosity index of characteristic point 3 is 1.26, and the dimensionless porosity prediction result is 1.25.

According to Campbell et al. [10], the melt is inevitably involved with oxide layers during pouring process in the industrial vacuum furnace, forming double-layered oxide films in castings, on which the micro-porosity is finally formed. However, this dimensionless criterion function cannot predict the amount of bubbles involved due to unreasonable pouring system design or the gas generated via reactions between shell and the melt. Although the melt has been refined, the micro-porosity index

Table 6.2 The gas content in the master alloy and casting

Gas element	Oxygen (O)	Nitrogen (N)	Hydrogen (H)
Master alloy (%)	0.0009	0.0040	0.00012
Casting (%)	0.0045	0.0030	0.00007

increases due to shell reactions or the aggregation and nucleation of trace vapor-forming elements in the melt. Referring to the content of vapor-forming elements in master alloys and superalloy castings, as shown in Table 6.2, it can be found that the total content of vapor-forming elements in castings increases by 51% compared with that in the master alloy.

At present, although it is difficult to quantify the proportion of trace vapor-forming elements in the micro-porosity of large complex thin-walled castings, the influence of trace vapor-forming elements listed in Table 6.2 on the predicted results of dimensionless micro-porosity based on solidification shrinkage can be investigated using data fitting. Equation (6.13) shows the fitted relationship between predicted and experimental results of micro-porosity in large complex thin-walled castings in Fig. 6.16:

$$y_{experiment} = n_1 \cdot y_{prediction} + n_2 \tag{6.13}$$

where n_1 and n_2 are the coefficients related to the vapor phase elements in the alloys, which are 1.05 and 0.18, respectively, in this study. Substituting the dimensionless micro-porosity prediction results modified by Eq. (6.13) into Fig. 6.19, it is found that the overall difference between the modified prediction and experimental results is reduced.

Furthermore, there is a big difference between the experimental and predicted results at characteristic point 12. This point is used to verify the applicability of dimensionless micro-porosity criterion. The thickness at characteristic point 12 in the casting is much higher than that of other points, which no longer meets the directional solidification conditions. In fact, the macro-closed thermal section has been formed at this point, indicating again that the dimensionless criterion model is not suitable for forming thick parts of isolated molten pools.

6.4.3 *Three-Dimensional X-ray Imaging of Micropores*

The micropores in the casting usually have complex three-dimensional shapes. The traditional metallographic method, which can only observe the section morphology of micropores, is difficult to fully characterize the size, morphology, and distribution of the micropores. However, the shape, size, and distribution of the micropores greatly affect the fatigue properties of castings [11]. With the advancement of detection methods in the field of materials research and the development of computer technology, the characterization of micropores has received extensive attention. The current characterization techniques can not only reveal the formation mechanism of micropores in the solidification process of nickel-base superalloys, but also provide solid data support for the study of the relationship between micropores and mechanical properties. In particular, it provides a database for the fatigue life prediction of the rear casing casting under the alternating load.

6.4.3.1 Spatial Distribution Characteristics of Micropores

It is an important trend to apply new technologies and methods to study micropores in the field of materials science and engineering. Computed tomography imaging technology, which has the advantages of precise and three-dimensional visualization, has been widely used in non-destructive testing of key castings for aviation. Industrial CT uses the rays from different angles of the object to reconstruct the cross-sectional image of the object. The tomographic image corresponds to the material, geometry, composition, and density characteristics of the casting. By three-dimensional continuous scanning of the casting, information such as the position, shape, and size of the micropores can be obtained. In contrast to the traditional negative film technology, industrial CT does not create structural overlap [12], which is helpful to study the formation mechanism of micropores in castings.

According to the X-ray test results, the position of micropores on one hollow strut of the casting is determined, then the wire cutting sampling is performed. In industrial CT imaging, the sample is fixed to the translational and rotating stage of the stepper motor for 3D scanning and reconstruction. The sampling position and reconstruction result of the TRF casting are shown in Fig. 6.21.

In order to have a more intuitive study on the internal micropores morphology of the casting, the grayscale threshold is set to segment the reconstructed three-dimensional image, and the spatial distribution of micropores is shown in Fig. 6.22.

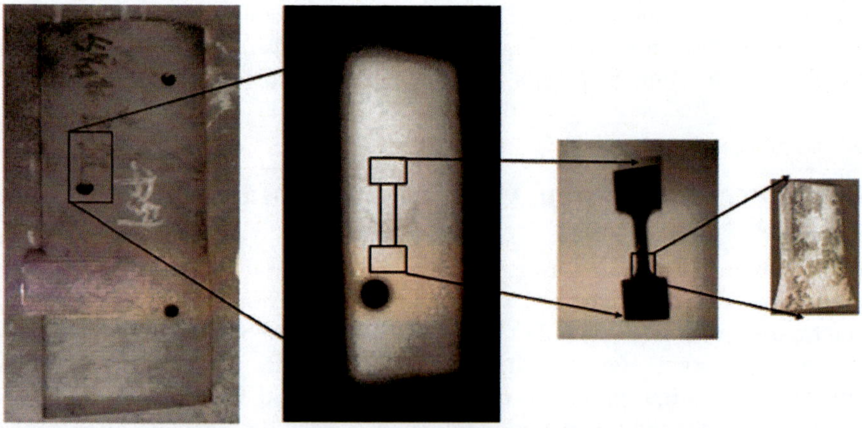

Fig. 6.21 Schematic diagram of sampling of one hollow strut from the TRF casting and 3D scanning reconstruction

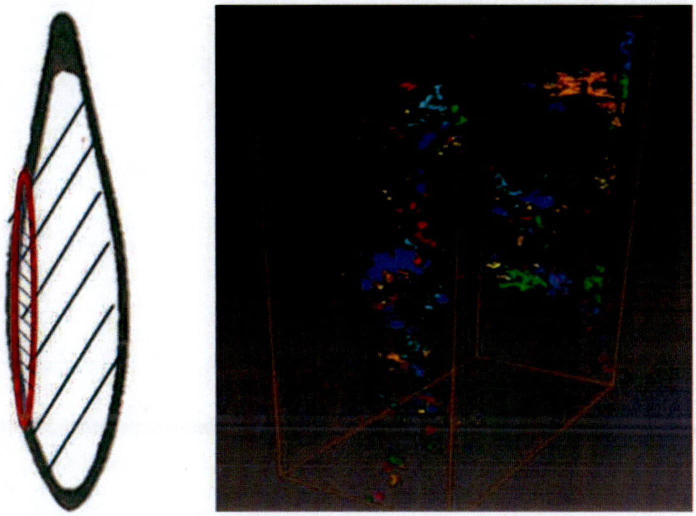

Fig. 6.22 Spatial distribution of micropores (Left: the section of the hollow strut and the sampling position, Right: the reconstructed morphology of micropores)

The micro-porosity presents a "sponge" shape as a whole, with agglomerated micropores and scattered micropores, which are mainly distributed on the inside of the strut (near the sand core side).

In addition, the size distribution of the internal micropores of the TRF casting is quantitatively counted. As shown in Fig. 6.23, it is found that with the increase of the micro-porosity equivalent diameter, their number in the casting is reduced and the semi-normal distribution characteristics are exhibited.

Fig. 6.23 Size distribution of micropores

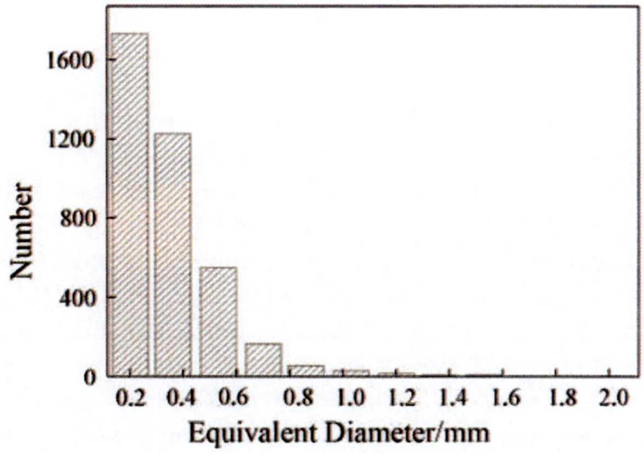

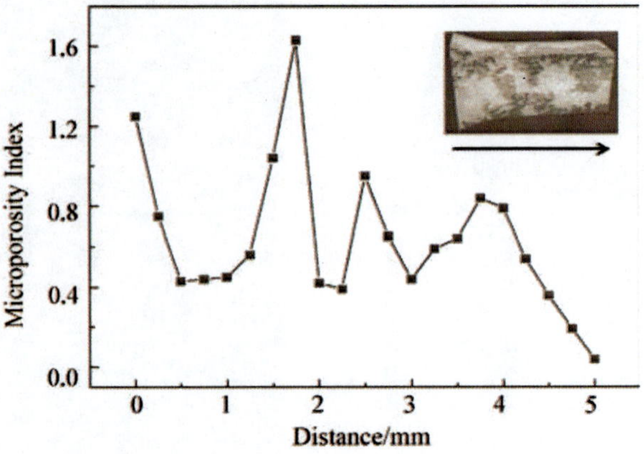

Fig. 6.24 Micropores index on cross section

Figure 6.24 is the micropores exponential distribution curve from the sample chuck to the middle of the sample. It is found that the micropores index on the cross-section shows irregular waveform characteristics, which directly reflects the complex shape of the micropores inside the casting appearance.

It can be seen from the above industrial CT test results that the micropores of the thin-walled support plate of the rear casing casting is mainly distributed on the inner side of the support plate near the inner sand core. The heat dissipation is slower at this part, which is equivalent to the existence of the structural heat knot. And this part belongs to the final solidification part, which eventually produces a lot of micropores. This kind of micropores defect is different from the typical thin-walled micropores defect. When designing the pouring system, the influence of the sand core should be fully considered, and the design margin of the riser should be appropriately increased, which can effectively suppress the occurrence of such microscopic loose defects.

6.4.3.2 Three-Dimensional Topography of Micropores

The high-energy and high-resolution synchrotron X-ray imaging technique makes it possible to further characterize the finer structural features of the micropores inside the casting. In order to characterize the fine structure of the micropores morphology of large complex thin-walled castings, the high-resolution CT scan of the static samples of K4169 superalloy castings is carried out using the synchrotron radiation tomography of the Shanghai Light Source BL13W line station. Then the advanced image processing technology is used to reconstruct the three-dimensional shape of the micropores between the dendrites in the thick/thin-walled part of the TRF casting, as shown in Fig. 6.25. In order to quantitatively describe the micropores morphology, the micropores characteristic factor-length to diameter ratio is introduced and the micropores of the thick-walled part of the casting is found to be "elliptical" with an aspect ratio of about 1.53. The thin-walled portion of the casting has a flattened

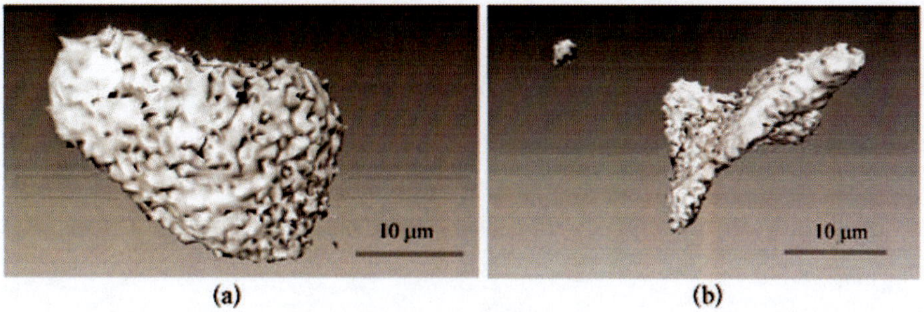

Fig. 6.25 Three-dimensional morphology of micropores. **a** Thick-walled part **b** Thin-walled part

'saddle shape' with a length to diameter ratio of about 1.57. The quantitative characterization of micropores, distribution, and morphology provides a basis for the study of the formation of micropores defects in K4169 superalloy large-scale complex thin-walled castings and its quantitative prediction.

6.4.4 Relationship Between Micropores and Mechanical Properties

Superalloy castings usually have two usage states, standard heat treatment state and hot isostatic pressure plus standard heat treatment state. The relationship between casting defects and mechanical properties in the two states is constructed. In particular, the relationship between microscopic porosity defects and mechanical properties has a great theoretical significance and application value.

6.4.4.1 Effect of Micropores on Tensile Properties

Figure 6.26 shows the relationship between the micropores indices and tensile properties of standard heat treatment and hot isostatic pressing + standard heat treatment. Although the tensile strengths of the standard heat treated and hot isostatic pressing + standard heat treatment samples both decrease linearly with the increase of micropores index, the tensile strength and micropores of the samples are different for the two states. The degrees of correlation are different. The micropores index of the standard heat treatment samples was reduced by 0.05, and the tensile strength was increased by about 21 MPa. The micropores index of hot isostatic pressing + standard heat treatment sample was reduced by 0.05, and the tensile strength was increased by more than 130 MPa. The least squares method was used to fit the tensile strength against the micropores index of the standard heat treatment samples. The result is given as

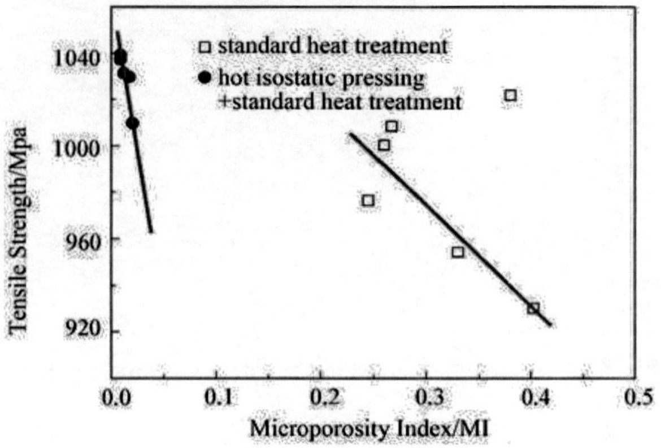

Fig. 6.26 Relationship between micropores and tensile strength

$$UTS_{ST} = -433.3MI + 1104.4 \tag{6.14}$$

$$R_{UTS,ST} = -0.8741$$

In the equation above, UTS_{ST} is the tensile strength of the standard heat treatment sample, MI is the micropores index, and $R_{UTS,ST}$ is the correlation coefficient.

The least squares method was used to fit the tensile strength and micropores index of the hot isostatic pressing + standard heat treatment samples. The result is

$$UTS_{HIP+ST} = -2694.7MI + 1064.5 \tag{6.15}$$

$$R_{UTS,HIP+ST} = -0.9795$$

In this equation, UTS_{HIP+ST} is the tensile strength of the hot isostatic pressing + standard heat treatment samples and $R_{UTS,HIP+ST}$ is the correlation coefficient.

Comparing the linear fitting results of the standard heat treatment with the hot isostatic pressing + standard heat treatment samples, the absolute value of the correlation coefficient of K4169 superalloy standard heat-treated samples is significantly smaller than that of hot isostatic pressing + standard heat treatment samples. After the standard heat treatment, there are still many Laves phase and large carbides remaining in the castings, which not only reduce the tensile strength of the casting but also reduce the impact of the micropores on the tensile strength. After the hot isostatic pressing + standard heat treatment of the casting, the Laves phase and the bulk carbide in the casting are effectively eliminated, with the degree of homogeneity of the structure being improved. The absolute value of the correlation coefficient of the hot isostatic pressing + standard heat treatment sample is close to 1, which shows that the linear correlation between the tensile strength of the alloy and the micropores index is higher in this state.

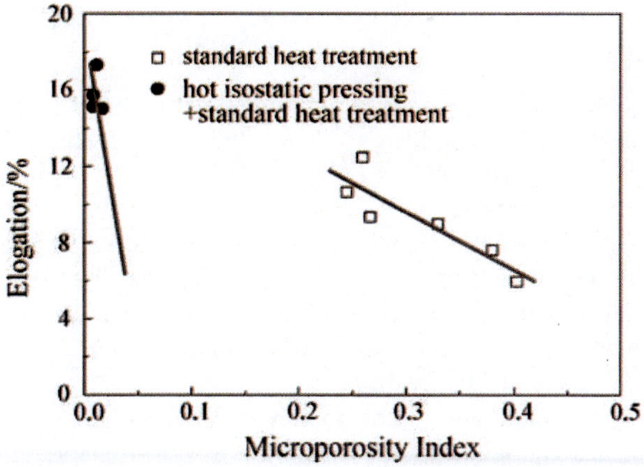

Fig. 6.27 Relationship between micropores and elongation after fracture

Regardless of the influence of the post-treatment state of the casting, the correlation coefficient R_{UTS} between the tensile strength and the micropores index of all the samples in Fig. 6.26 is equal to −0.6085, which is much smaller than 1. This indirectly shows that although the micropores is an important factor affecting the tensile strength, it is not the only factor.

Figure 6.27 shows the relationship between the micropores index and the elongation after fracture. It can be seen that with the increase of the micropores index, the elongation after fracture of the standard heat treatment and hot isostatic pressing + standard heat treatment samples both decrease linearly. But there is a significant difference in terms of the correlation between elongation and the micropores index after fracture. The micropores index of the standard heat treatment sample is reduced by 0.1, but the elongation after fracture is increased by about 3%. In addition, the micropores of the hot isostatic pressure + standard heat treatment sample is reduced by 0.01, but the elongation after fracture is increased by 3%. The least squares method is then used to fit the elongation after fracture against the micropores index of the standard heat treatment sample. The result is

$$EL_{ST} = -30.5MI + 18.8 \tag{6.16}$$

$$R_{EL,ST} = -0.8929$$

where EL_{ST} is elongation after fracture of standard heat treatment castings and $R_{EL,\,ST}$ is the correlation coefficient.

The least squares method is also used to fit the elongation after fracture against the micropores index of the hot isostatic pressure plus standard heat treatment samples. The result is

$$EL_{HIP+ST} = -344.2MI + 19.4 \tag{6.17}$$

$$R_{EL,ST} = -0.9141$$

where EL_{HIP+ST} is the elongation after fracture of hot isostatic pressing + standard heat treatment and $R_{EL,ST}$ is the correlation coefficient.

Comparing with the linear fitting results of standard heat treatment and hot isostatic pressing plus standard heat treatment samples, the absolute values of the correlation coefficient between the elongation after fracture and the micropores index of the two groups are close to or more than 0.9, with a relatively small difference. Even in the presence of a large amount of Laves phase and bulk carbides in the standard heat treatment castings, the micropores index can strongly influence the elongation of the casting.

The elongation after fracture and micropores of the samples treated by standard heat treatment and hot isostatic pressing + standard heat treatment shows a good linear correlation, in that the correlation coefficient of the elongation after fracture and micropores index of all the sample $R_{EL} = -0.8810$ in Fig. 6.27. This further shows that the elongation after fracture of the K4169 superalloy casting is closely related to the content of micropores. The hot isostatic pressing, which can effectively eliminate the micropores defects in the casting, greatly improves the elongation after fracture of the alloy. The resulting microstructural changes have a relatively weak effect on the elongation after fracture of the casting.

6.4.4.2 Effect of Micropores on Stress Rupture Property

The existence of micropores in the casting changes the internal stress distribution of the casting under the action of permanent load and forms a crack source around the smooth micropores of the surface, which then aggregates and grows to cause fracture failure of the casting. Figures 6.28 and 6.29 respectively show the relation-

Fig. 6.28 Relationship between high temperature rupture life and micropores index

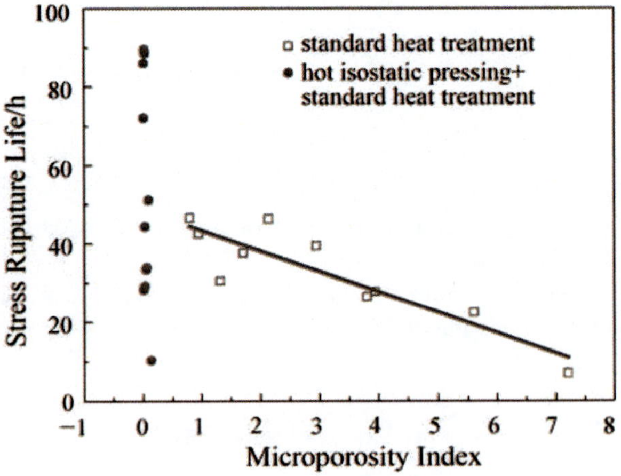

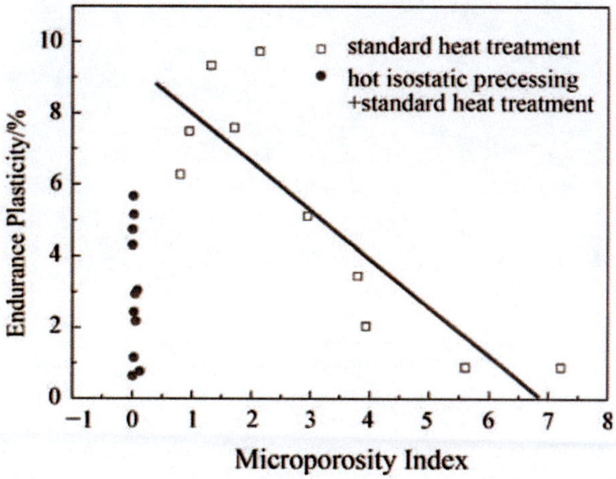

Fig. 6.29 Relationship between high temperature rupture ductility and micropores

ship between the stress rupture life, the endurance plasticity, and the micropores under the standard heat treatment and hot isostatic pressing + standard heat treatment at 650 °C/620 MPa. It can be seen that the standard heat treatment state and the endurance plasticity decrease with the increase of the micropores index, with the following linear fitting relationship

$$SR = 48.56 - 5.21MI \tag{6.18}$$

$$EL = 9.42 - 1.35MI \tag{6.19}$$

where SR is the stress rupture life and EL is the endurance plasticity. However, after hot isostatic pressing + standard heat treatment, the micropores index is generally small, but the stress rupture life and endurance plasticity change within a large range, which is close to the hot isostatic pressing treatment but does not completely heal the micropores.

6.4.4.3 Effect of Micropores on Fatigue Performance

The test sample taken from the TRF casting is processed as required and the room temperature axial tension and compression control stress mode fatigue test is performed. The test environment is a static air medium and the loading waveform is a triangular wave, together with an axial stress of 380 MPa in the control mode. The stress ratio is −1 and the test frequency is 0.3 Hz. The number of fatigue cycles (fatigue strength) before the sample fails is then recorded. Figure 6.30 shows the metallographic morphology near the fracture of the fatigue specimen, in which it is found that there is a fracture and a large number of micro-porosity defects

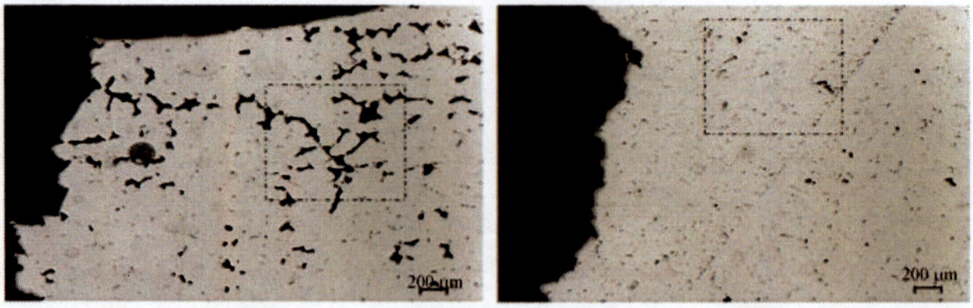

Fig. 6.30 Micropores morphology near the fracture (1 mm × 1 mm area). **a** Sample I **b** Sample II

Table 6.3 Fatigue life of the two samples as shown in Fig. 6.30

Sample No.	I	II
Fatigue life /cycles	36 405	>130 000
Micro-porosity index	12.27	0.47

near the sample I, but the micropores near the fracture of the sample II is less. The corresponding quantitative micropores index and fatigue life are shown in Table 6.3.

6.5 Formation and Prediction of Laves Phase

The K4169 alloy has a severe segregation tendency during solidification due to its high alloying degree and wide freezing temperature interval. Among the alloying elements, niobium is the main segregation element, with 4.0%–6.4% mass concentration in the matrix. During solidification, niobium may be enriched in the melt and form harmful Laves phase via the $L \rightarrow \gamma + $ Laves eutectic reaction at the interdendritic region, or even primary Laves phase. Since a large amount of Nb and other strengthening elements are captured in the Laves phase, the solid solution Nb element in the matrix is reduced. As a result, the solid solution strengthening effect is reduced. The formation of Laves phase also simultaneously reduces the amounts of γ' and γ'' phase, weakening the effect of precipitation strengthening. Moreover, the low-melting Laves phase is enriched between dendrites and on grain boundaries. On one hand, the Laves phase is a favorable nucleation site of cracking, and a large amount of interdendritic cracks are formed on the tensile fracture. On the other hand, the Laves phase is prone to be a source of welding cracks, affecting the quality of repair welding of castings.

It can be seen that the chemical segregation and the Laves phase formed during the solidification process of K4169 alloy casts are two forms of the main defects, which deteriorate the mechanical properties. Guo et al. studied K4169 alloy and found that when the alloy contains high amounts of Laves phase, the room temperature tensile strength and ductility of the alloy are decreased by 20% and 60%, respectively, and the endurance life under 650 °C and 620 MPa is reduced by 60%. Therefore, the elimination of segregation and the suppression of the formation of harmful Laves brittle phase are the key technical issues for the development and manufacturing of large-scale complex thin-walled sloping turbines with K4169 nickel-based superalloy.

The formation of chemical segregation during solidification is the result of multi-field and multi-scale coupling of macroscopic temperature field, solutal field, and micro-dendritic growth process. These factors include solid back diffusion, liquid finite diffusion, dendrite arm coarsening, dendritic tip supercooling, eutectic interface supercooling, and so on. Although researchers have established analytical and numerical models of micro-segregation with increasing solid fraction during solidification based on the influence of solute redistribution, it is impossible, only based on the study of solute distribution with single solidification conditions, to predict the degree of segregation of TRF casings that have large scale and complex thin-walled casting structures. Therefore, criteria and prediction methods for micro-segregation degree of solidification process of different structured K4169 superalloy complex structural castings are studied and established, which provides a basis for optimizing and adjusting casting process design and is significantly important for guiding the casting manufacturing of large scale complex structural components.

6.5.1 MSI Criterion of Micro-segregation Degree

According to the formation elements of Laves phase, the Laves phase is formed owing to the high segregation of Nb and Mo elements during solidification of superalloys. Therefore, the content and distribution of Laves phase produced by $L \rightarrow \gamma +$ Laves eutectic reaction at the final stage of solidification is related to the segregation degree of Nb and Mo elements, i.e., the solute concentration of Nb and Mo in mushy zone of liquid phase. The evolution laws of Nb and Mo solute segregation depend on their diffusion time and distance in liquid and solid phase.

According to the classical solidification theory, the microscale solute redistribution behavior and the resulting micro-segregation not only depend on the physical properties of alloys (e.g. liquid phase diffusion coefficient D_l, solid phase diffusion coefficient D_s, solute distribution coefficient k) and external process conditions such as solidification rate v (or local solidification time t_f) and temperature gradient G, but also are closely related to the dendrite growth mode, complex changes in solidification dendritic morphology, and end effect of the solidification process the curvature of dendrite tip. In recent years, researchers have studied the solidification characteristics of composition changes of multi-component alloys, solid back diffusion, liquid finite diffusion, dendrite arm coarsening, dendritic tip supercooling, eutectic interface supercooling, solidification volume change. Based on these studies, some analytical models and numerical models of micro-segregation in which the solute concentration increases with solid fraction during solidification are established. However, the existing micro-segregation models for solute distribution are based on the change of solute concentration with the increase of solid fraction during solidification process, which can only indicate the interfacial concentration C_1^* as a function of the solid fraction f_s, but cannot be used to determine the composition distribution of final solid phase. Therefore, these existing models can only study the solute distribution under a certain solidification condition; in addition, while the dynamic parameter of solid fraction must be considered, the segregation degree under various factors of different parts of a large-scale complex component cannot be determined.

The so-called segregation is due to the spatial concentration difference exerted by different solubility and the mobility of solutes in the liquid and solid phase, which is related to the law of solute atom motion on both sides of solid–liquid interface. When the liquid phase flows are not considered, the solute segregation during dendrite growth is mainly determined by solute diffusion flux on both sides of the liquid–solid interface. If the solubility of the solute in solid phase is lower than its solubility in liquid phase (k < 1), the solute will be discharged as the interface advances. As the interface advances and the solute is continuously discharged, the concentration in liquid phase at solid–liquid interface will decay away from the maximum at solid–liquid interface in the light of the exponential law.

According to the definition of supersaturation in the solute diffusion processes, the supersaturation Ω is the ratio of the amount of solute change ΔC at the tip to the equilibrium concentration difference ΔC^*, which is given as

$$\Omega = \Delta C/\Delta C^* = \left(C_1^* - C_0\right)/C_1^*(1 - k) \tag{6.20}$$

Supersaturation Ω represents the driving force for solute diffusion at the front of solidification interface. According to the mathematical solution of parabolic diffusion problem derived by Ivantsov in 1947, for the hemispherical crown dendrites, the solution of diffusion equation indicates that the supersaturation Ω equals to the ratio of the tip radius R of the dendrite to the characteristic diffusion distance δ_C (the characteristic diffusion length indicates that the contained total solute within the solute diffusion length equals to the total of solute contained in the infinite boundary layer), i.e.

$$\Omega = R/\delta_C \tag{6.21}$$

where, $\delta_C = 2D_l/v$. This dimensionless ratio is the Peclet, i.e., $P_{e_l} = \Omega$, therefore,

$$P_{e_l} = \frac{vR}{2D_l} \tag{6.22}$$

It can be seen that the Peclet number characterizes the driving force for diffusion, which is determined by the supersaturation at the front of solid–liquid interface. If only the radial diffusion is considered, the thickness of boundary layer around the growing equiaxed grain equals to the radius of sphere, i.e. the solute diffusion length at the end of dendrite approximately equals to the radius of sphere R. Therefore, Eq. (6.22) can be expressed as,

$$P_{e_l} = \frac{vL}{2D_l} \tag{6.23}$$

Since the eutectic usually occurs during solidification process, the concentration of liquid phase cannot increase continuously. In order to truly describe the chemical distribution at the end of solidification, it is necessary to consider the back diffusion in solid phase. According to the principle of mass conservation, the degree of back diffusion of solute atoms in the solid phase depends on the dimensionless parameter α, which is described as F_{O_s} and represents the proportionality between the thickness of diffusion boundary layer in the solid phase $\delta_s(\delta_s = 2D_s/v)$ and the system size L, namely,

$$F_{O_s} = \alpha = \frac{D_s t_f}{L^2} \tag{6.24}$$

To dimensionally analyze the solute distribution during solidification, the widely used Buckingham π dimensionless analysis theory in mathematical physics field is applied. Based on the study of diffusion fluxes of liquid and solid solute atoms on both sides of solid–liquid interface, and according to solute mass conservation, the segregation index characterizing the degree of solute segregation during solidification can be expressed as [13]

$$\text{MSI} = \frac{P_{e_l}}{kF_{O_s}} \tag{6.25}$$

where $P_{e_l} = \frac{vL}{2D_l}$, $F_{O_s} = \frac{D_s t_f}{L^2}$. Substituting them in the Eq. (6.25), we can get

$$\text{MSI} = \frac{P_{e_l}}{kF_{O_s}} = \frac{1}{2kD_lD_s}\frac{vL^3}{t_f} = A \cdot \frac{vL^3}{t_f} \tag{6.26}$$

where $A = \frac{1}{2kD_lD_s}$ is a physical parameter that depends on the material. The coefficient A can be considered as constant for different structural parts on the same component.

It can be seen from the Eq. (6.2) that the solidification conditions that affect the segregation index MSI include solidification rate, characteristic diffusion distance, and local solidification time. The segregation index increases with solidification rate v and characteristic diffusion distance L, and decreases with local solidification time t_f. According to the dendrite growth theory, in general, the rate at which the solid–liquid interface exits the solute is positively proportional to the dendrite growth rate, and the discharged solute must diffuse out along the concentration gradient of interface. Therefore, the larger growth rate makes the concentration gradient and solute segregation index larger.

For the equiaxed dendrite grains in the supercooling melt, the average growth rate v during the dendrite growth process is substituted for the dendrite instantaneous velocity v, so

$$v = \frac{L}{t_f} \tag{6.27}$$

where L is the characteristic diffusion distance. Consequently, the segregation index MSI can be furtherly expressed as

$$MSI = \frac{P_{e_l}}{kF_{O_s}} = \frac{1}{2kD_lD_s}\frac{L^4}{t_f^2} = A \cdot \frac{L^4}{t_f^2} \tag{6.28}$$

Local solidification time t_f is the time required for a fixed point in the system to start from solidification start to the end, which is the time from the dendrite tip to the dendrite root. It can be seen that the local solidification time characterizes the time for dendrite arms being in contact with liquid phase. According to Kattamis and Flemings in 1965 and Feurer and Wunderdlin in 1977, the secondary dendrite spacing λ_2 was approximately changed with the cubic root of local solidification time.

$$\lambda_2 = 5.5(Mt_f)^{1/3} \tag{6.29}$$

where $M = 166 \cdot \frac{\Gamma D_l \ln\left(\frac{C_{eut}}{C_0}\right)}{m(1-k)(C_0 - C_{eut})}$, Γ is Gibbs–Thomson coefficient, and C_{eut} is eutectic composition.

It can be seen from the Eq. (6.29) that the longer the local solidification time is, the larger the secondary dendrite spacing λ_2 and the longer the dendrite arm is in contact with liquid phase will be, and the more the solid phase solute atoms will be back diffused, lowering the concentration gradient in solid phase and the degree of segregation of solute and resulting in a decrease of the segregation index.

The secondary dendrite spacing λ_2 after solidification of the metal is largely determined by the solute diffusion field during dendritic growth. When back diffusion occurs in the mushy zone during dendrite arms coarsening, it mainly occurs between the secondary dendrite arms. For the equiaxed dendritic growth mode in the supercooling melt, the secondary dendrites arm spacing λ_2 is the characteristic length of solute diffusion, which can largely reflect the local solidification time. Therefore, the chemical segregation and the distribution of second phase are directly affected by the secondary dendrite arm spacing, thereby affecting the microstructure and mechanical properties of the casts. In general, the secondary dendrite arm spacing is the diffusion length, i.e. $L = \frac{\lambda_2}{2}$.

Substitute into Eq. (6.28),

$$MSI = \frac{P_{e_l}}{kF_{O_s}} = \frac{1}{32kD_lD_s}\frac{\lambda_2^4}{t_f^2} = A' \cdot \frac{\lambda_2^4}{t_f^2},\ A' = \frac{1}{32kD_lD_s} \tag{6.30}$$

For column grains with directional solidification, the secondary dendrites are limited, and the solute elements in mushy zone are mainly concentrated in the primary dendrites, and their characteristic diffusion distance corresponds to half of the primary dendrite arm spacing, i.e. $L = \lambda_1/2$.

Substitute $L = \lambda_1/2$ into Eq. (6.28),

$$MSI = \frac{P_{e_l}}{kFo_s} = \frac{1}{32kD_lD_s}\frac{\lambda_1^4}{t_f^2} = A' \cdot \frac{\lambda_1^4}{t_f^2},\ A' = \frac{1}{32kD_lD_s} \quad (6.31)$$

6.5.2 Prediction of Micro-segregation at the Characteristic Positions of Large Casting Using the MSI Criterion Function

In recent years, researchers have studied the influence of solute distribution on solidification properties of binary and multi-alloys, back diffusion of solid phase during solidification process, limited diffusion of liquid, coarsening of dendrite arms and supercooling of dendrite tips. Some analytical and numerical models of micro-segregation with increasing solid fraction during solidification were established. However, they are based on the study of solute distribution under single solidification condition, which cannot be applied to predicting and determining the degree of segregation under different couplings of different large-scale complex thin-walled casts. The rear casing casts contain large-area hollow support plates, varying cross-sections, lifting lugs, bosses and ring collars. The use of the MSI criterion function to predict the degree of segregation at each characteristic structure is presented in the following.

Firstly, the ProCast software is used to simulate the filling and solidification of the casts, and the related temperature field and microstructure are obtained. The material of large castings is nickel-base superalloy K4169 with a pouring temperature of 1500 °C and a preheating temperature of mold shells of 950 °C. Many parameters are required when predicting segregation index, which can be divided into two categories: one is the thermal physical parameters of the alloy, which belongs to the fixed value; the other one is solidification parameter, which changes with the structure and process conditions of the casts.

The thermal property parameters involved in the micro-segregation index MSI include equilibrium partition coefficient, liquid phase diffusion coefficient D_l, and solid phase diffusion coefficient D_s, which is assumed as a constant in solidification interval. Passing the alloy composition of the cast into JMatPro thermodynamic calculation software, it can be calculated by the thermophysical parameters in JMatPro. The specific calculation procedures include:

1. By calculating the phase diagram of K4169 alloy by JMatpro software and combine with relevant literatures, the thermophysical parameters used in the calculation in this example can be obtained. The parameters are listed here: $\Gamma = 3.65 \times 10^{-7}$ km, $k_{Nb} = 0.48$, $D_{l,Nb} = 3 \times 10^{-9}\ m^2\ s^{-1}$, $D_{s,Nb} = 2.82 \times 10^{-13}\ m^2\ s^{-1}$, $C_{0,Nb} = 4.3$ wt.%, $C_{eut,Nb} = 19.1$ wt.%, $m_l = -10.9$, $T_l = 1360$ °C, $T_s = 1180$ °C.

2. The ProCast finite element software is applied to simulate the casting filling and solidification process of castings, and the local solidification time t_f of each part of the casting was obtained.
3. The λ_2 value of each structural part of the casts is calculated by using the secondary dendrite spacing calculation module provided by ProCast software.
4. The local solidification time t_f, secondary dendrite arm spacing λ_2, and the calculated thermophysical parameters by JMatPro are substituted into Eq. (6.30), and the micro-segregation index MSI of structural parts of the complex structured casts is calculated by ViewCast module in ProCast. According to the micro-segregation index MSI of each structural part, the micro-segregation degree of each part of casts are determined.
5. The prediction results are compared with the anatomical results. As shown in Fig. 6.31, the quantitative metallographic statistic of the 12 parts of the casts are compared with the predicted results. The comparison results are shown in Fig. 6.32, where it can be seen that the prediction results are consistent with the anatomical results. It shows that the larger micro-segregation index results in the severer micro-segregation and the content of Laves phase formed at the end of solidification would be higher.

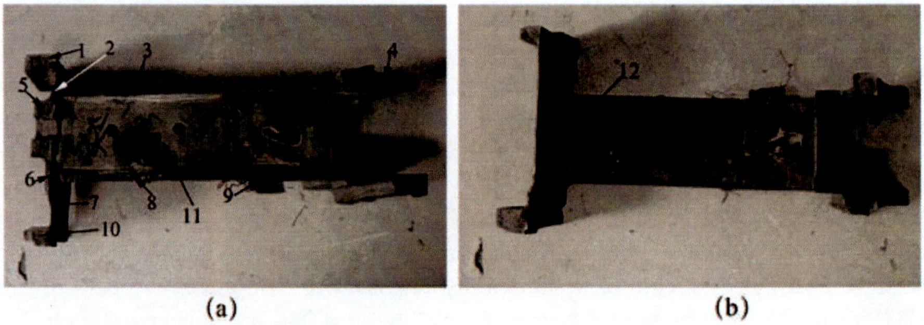

Fig. 6.31 Castings used in the experiment

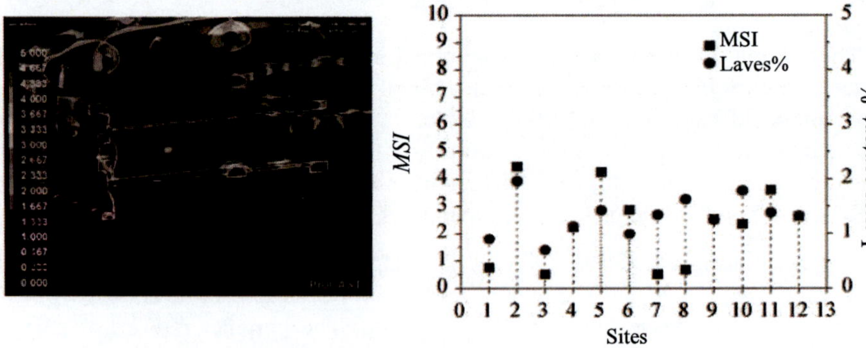

Fig. 6.32 Comparison of the predicted MSI with the experimental results

6.6 Repair Welding Technology of Large Castings

6.6.1 Necessity of Repair Welding of Large Castings

Although precision casting technology for large structure is always in the progress and development, there are still inevitably defects such as loose, porosity, inclusions and cracks in castings. These defects can bring serious damages to the performance and life of castings under high temperature service conditions and may even become a major hidden trouble. They must be repaired by welding. In addition, there are often some technological holes reserved in some thin-wall parts due to the needs of forming technology in the process of precision forming of large complex thin-wall castings. These technological holes also need to be repaired by welding to ensure the structural and functional integrity of components.

6.6.2 Several Key Problems to Repair Welding of Defects

6.6.2.1 Welding Deformation Control of Thin-Walled Parts

The defect repair welding of thin-wall casting will undergo local welding heat process. A large amount of residual stress and strain will be accumulated in the weld as well as its surrounding area. Due to the weakness of rigid constraint in the thickness direction of thin plate, if there is no external rigid constraint to the workpiece in the local repair welding, the stress will be partially released and the deformation will appear. The main form is wave deformation, which is the result of the combined actions among the casting material characteristics, welding process parameters, groove forms, and constraint conditions. If there is a rigid constraint, the stress and strain will accumulate in the weld and its heat-affected zone (HAZ). When the joint cannot withstand the large stress and strain, it will induce the generation of crack defects. The key to suppressing welding deformation is to control the stress and strain, and to prevent the welding deformation by using restraint technology, as well as to ensure the joint not to crack due to the residual stress. Welding residual deformation is the deformation that remains in the structure after welding. It includes longitudinal contraction deformation, transverse contraction deformation, bending deformation, angular deformation, and so on, among which longitudinal and transverse contraction deformation is the main cause for other deformations. The main factors that affect the longitudinal contraction deformation are cross-section area of welds, welding seam length, welding energy input, and groove form, etc. The main factors affecting transverse contraction deformation are welding energy input, weld thickness, and groove form, etc.

The welding deformation of nickel-base superalloy sheet can be measured by triangular method with laser. When an IN718 plate with thickness of 2 mm is welded in a straight line along the center of the plate without constrain, and its stress is fully released after cooling, the welding deformation is measured as shown in Fig. 6.33.

It can be seen from the Fig. 6.33 that, without constraint, the residual welding deformation of the nickel-base superalloy plate is large, which generally presents the deformation law that the middle part is concave along the weld width but is bulged along the weld length. In addition, it is also accompanied by local random wave deformation. Using thermal elastoplastic 3D-FEM method, the total welding deformation of nickel-base superalloy sheet can be simulated and predicted. Figure 6.34 shows the finite element simulation grids for the welding straight along the centerline of a nickel-base superalloy plate with thickness of 2.0 mm (For simplification, only half of them are calculated according to the symmetry principle). The result of simulation for the welding deformation is shown in Fig. 6.34b. It can be seen that as an overall deformation in welding process, the welded plate begins to be concave in

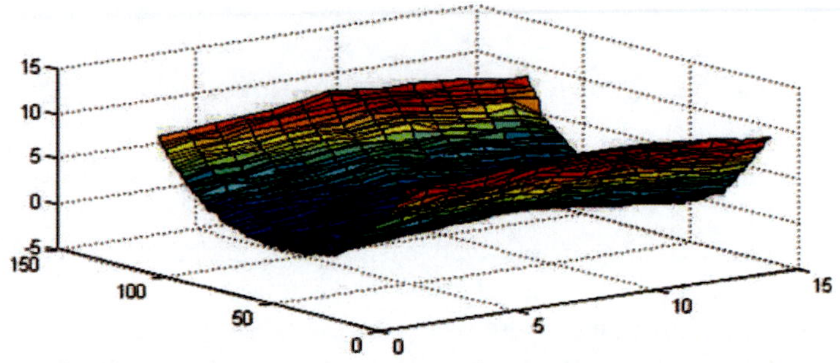

Fig. 6.33 Welding deformation measuring results of a plate (2 mm)

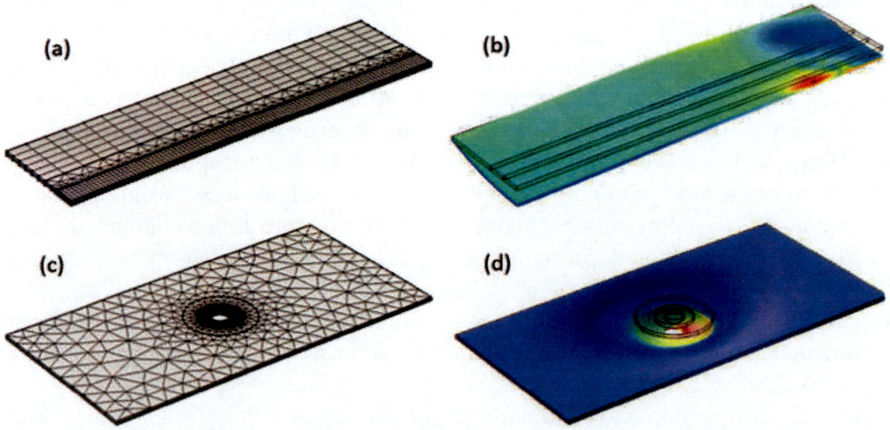

Fig. 6.34 Thermal elastoplastic 3D-FEM welding process simulation. **a** straight weld simulation grids of a plate; **b** straight weld deformation and thermal stress distribution on a plate; **c** simulation grids for a circular weld of a hole; **d** weld deformation and thermal stress distribution of a hole on a plate

the middle part of the straight weld joint along the weld width, and to be convex in the middle part along the weld length. Figure 6.34c shows the calculation grids for the simulation of welding a hole with diameter of 10 mm on a plate with thickness of 2.0 mm (for simulating the repair welding of technological holes on thin-walled castings). The simulated results of welding deformation are shown in Fig. 6.34d.

In the simulation above, the welding temperature field can be simulated according to the varied welding parameters and process conditions. Thus, the influence on welding deformation can be predicted for varied welding parameters (such as welding current, welding speed), welding sequence, and the constraint. The corresponding measures can be adopted to control the welding deformation in allowed range as much as possible.

6.6.2.2 Control of Secondary Defects in Repair Welding

The so-called secondary defect refers to the welding defects such as pores, inclusions, lack of penetration, undercut, and cracks found in the repaired weld zone and its HAZ after the repair welding of a large thin-walled casting. In other words, the secondary defects are found in the same place after the repair welding for the primary defects, i.e., casting defects. The occurrence of secondary defects is related to many factors, among which the key factors include casting material, structure, and welding process.

For large complex thin-walled IN718 alloy castings, many secondary defects may occur during the repair welding process. Practical experience shows that most of them are welding hot cracks, which is related to the high sensitivity of cracking for nickel-base superalloy, as well as their structure characteristics. Therefore, how to control the secondary defects in repair welding for nickel-base superalloy will be mainly discussed in terms of the welding hot crack.

To study the sensitivity of welding hot crack of IN718 alloy sheet, fluorescence detection and X-ray examination are carried after welding. It has been found that there are mainly five types of welding hot cracks. (1) Longitudinal and transverse cracks at the centerline of the weld; (2) HAZ cracks; (3) Mixed cracks composed of transverse cracks and HAZ cracks in adjacent weld toe; (4) Arc crater cracks; (5) Mixed cracks of the former weld bead arc crater cracks and the latter weld bead solidification cracks in multi-layer welding. The position diagram of all kinds of welding cracks are shown in Fig. 6.35.

According to the analysis of the cross-section of the weld, crack surface morphology and composition analysis based on EDS, the longitudinal, transverse and arc crater cracks are the solidification cracks, whereas the HAZ cracks are the liquefied cracks. The solidification crack is also called the crystallization crack. The nickel-base superalloys have different degrees of crystallization crack sensitivity. The superalloys strengthened by solution have little sensitivity to the crystallization crack. However, the precipitation strengthened superalloys with low Al and Ti content (<4%) have medium sensitivity to the crystallization crack, while the precipitation strengthened superalloy with high Al and Ti content as well as the casting superalloy have greater sensitivity to the crystallization crack. In addition, the state of

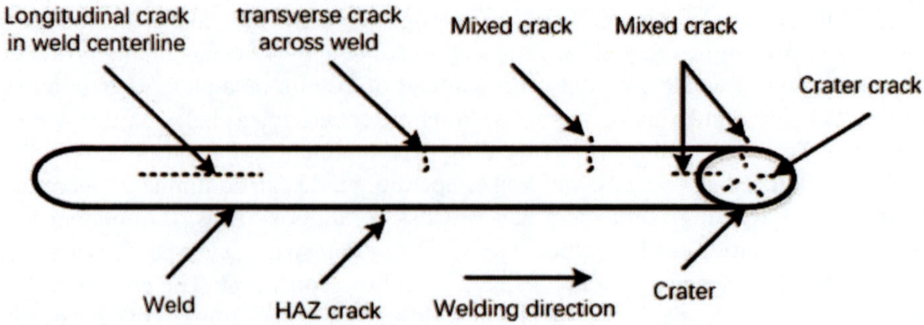

Fig. 6.35 Schematic diagram of welding hot crack

superalloy also affects its crack sensitivity. The alloy treated via homogenization has less crack sensitivity than the alloy after aging. The alloy strengthened by solution has less crack sensitivity than the alloy in flat state and cold rolling state. Because after flat pressing and cold rolling, the strength and hardness of the alloy increase, while the plasticity decreases. The same is true for aging treatment, which increases the restraint degree and crack sensitivity of welding parts.

The solidification crack sensitivity for the precipitation enhanced nickel-base superalloy can be presented with sensitivity index P_{SC} as below:

$$P_{SC} = 69.21w(Ti) + 27.3w(Nb) + 9.0w(Mo) + 300w(Si) - 55.3 \quad (6.32)$$

The larger the P_{SC} value is, the more likely it is to produce solidification cracks. Liquation cracks in the HAZ near the weld may also be combined with solidification cracks in the weld zone to form mixed cracks, as shown in Fig. 6.36. The phase with

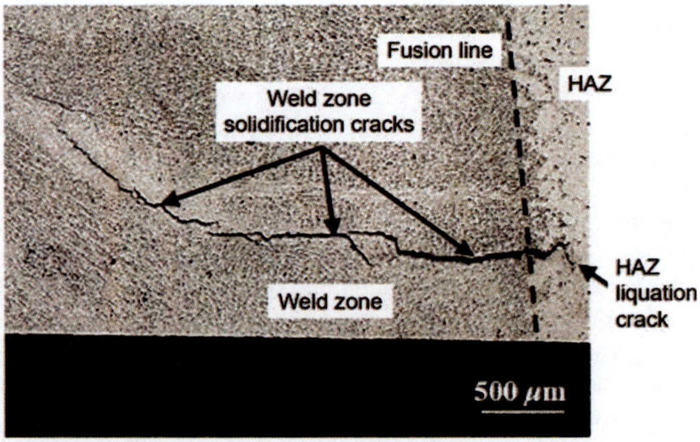

Fig. 6.36 HAZ liquation crack and solidification crack in weld zone form mixed cracks

low melting point in the HAZ near the weld liquefies in welding heating process, and then separates under the action of contraction stress and thermal stress in subsequent cooling process. The liquid film is torn, and the crack is generated, and it then propagates to the fusion line. According to the welding temperature field, the drop of temperature in the melting zone near the fusion line lags behind that in the HAZ. When the liquation crack in the HAZ extends to the partially melted zone (PMZ), the crack front will induce cracking in the segregation region of the PMZ under the combined action of additional stress, contraction stress and thermal stress. The structure of PMZ is also dendritic and there is a chain Laves eutectic phase between the dendrites, which partially melts in the welding process to form a liquid phase with low melting point and high viscosity between the dendrites and the dendrite axis. When the liquid film forms, it will separate under the action of contraction stress, thermal stress, and crack tip stress, resulting in the expansion of the crack. There are stratified segregation regions in the weld zone, as well as low melting point liquid phase in these segregation regions during solidification. When the crack tip approaches this solid–liquid coexisting region in the weld zone, the crack front stress field, the contraction stress, and the thermal stress work together to induce cracking in the weld zone and extend to the center of the weld. According to the initiation and propagation mechanism of solidification crack, the propagation of solidification crack in the weld zone is multi-dimensional, leading to a complex morphology of the formed crack.

Another case is that the HAZ liquation crack initiates in the near weld zone and expands to the PMZ and specimen surface. Meanwhile, the solidification crack occurs in the weld zone near the fusion line, which mainly extends to the weld zone and extends very closely to the fusion line. In this case, there are two cracks on both sides of the fusion line. When they are close to each other, the tip strain zones of the cracks are connected, resulting in a mixed crack. If their tip strain zones are not connected, the two cracks keep separate in a certain distance, despite that the tips of the two cracks can be relatively close.

The occurrence of HAZ liquation crack is closely related to the micro-segregation of the base metal and the low melting point phase and its size, quantity, and distribution. After homogenization heat treatment (1090 °C, 1 h) of the casting, the amount and length of liquation cracks in the HAZ decrease significantly, as shown in Fig. 6.37a, c. The content of the Nb-rich low-melting point Laves phases decrease around the crack, as shown in Fig. 6.37b, d. The reason is that after homogenization heat treatment, the segregation degree of the base metal decreases and part of the low-melting point phases dissolve into austenite, resulting in a decrease in their size and quantity. Since it is not easy to form continuous liquid film in the welding heat process, the crack sensitivity thus becomes small under the same heat input condition.

Arc crater cracks often appear in the primary and secondary axes of columnar dendrites, as shown in Fig. 6.38. Towards the finish of the welding process, the welding torch moves to crater, arc stops to move along the welding direction, but only moves in the vertical direction. The arc length is elongated and the heat transmitted by arc to the molten pool decreases to zero. This will affect the flow behavior of the liquid metal and the heat conduction direction of the molten pool. It is obvious that

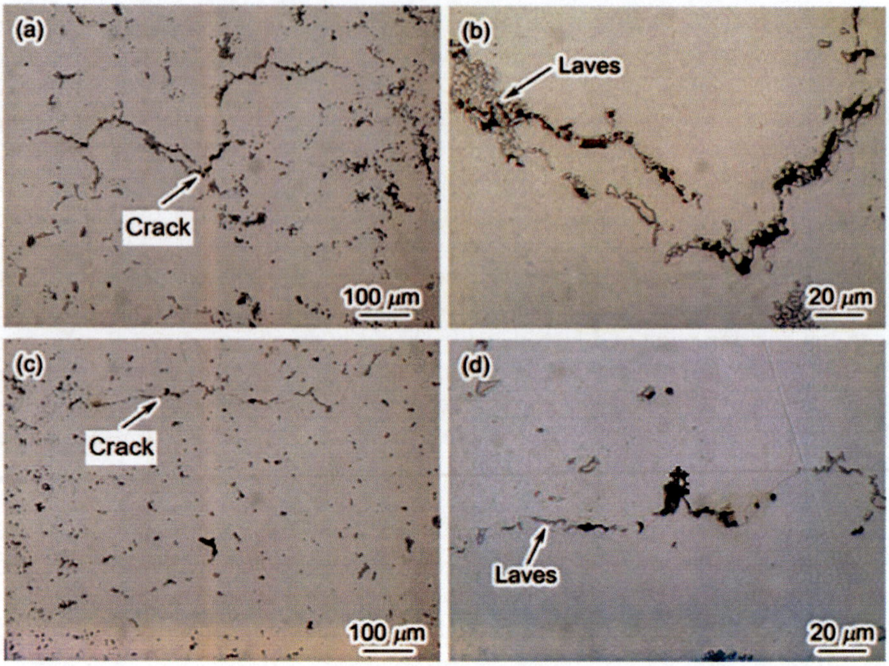

Fig. 6.37 HAZ liquation crack in as-cast and homogenization treated (1090 °C, 1 h) IN718 alloy **a** HAZ liquation crack in as-casted base material; **b** crack amplified for Fig. (**a**); **c** HAZ liquation crack in homogenization treated base material; **d** crack amplified for Fig. (**c**)

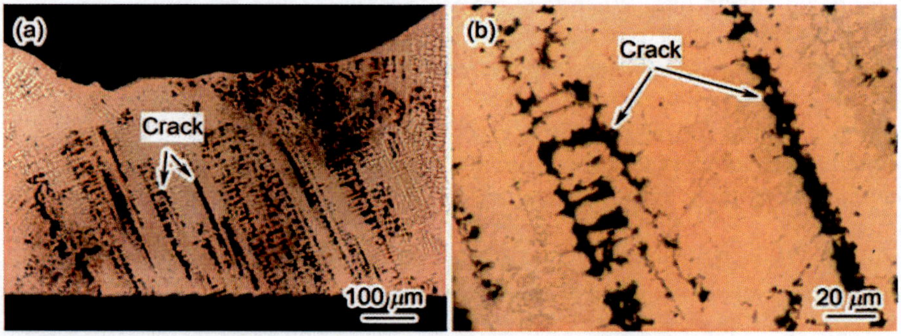

Fig. 6.38 **a** Metallography of arc crater crack; **b** crack amplified for Fig. (**a**)

the columnar dendritic crystal grows from the crater edge to the center. During the solidification of arc crater molten pool, dendrites in the mushy zone collide with each other, forming a relatively independent micro-scale closed liquid zone, in which the liquid viscosity is relatively large. In addition, the dynamic strain rate in arc crater is relatively fast, thus producing large stress and strain. The effective cross-section area at the arc crater is smaller than other areas of the weld bead, leading to larger solidification contraction stress and thermal stress. Under the action of these stress forces, Nb-rich and C-rich liquid film between dendrites is split to form arc crater crack.

In multilayer and multi-pass welding, next welding pass covers last welding pass, and some metals in the arc crater of last welding pass enter the weld bead of next welding pass. It can be seen from the welding temperature field and the flow behavior of the weld pool that the interlayer metal cannot fully and uniformly be mixed into the weld bead of last welding pass. Some micro-sized areas are affected by the transfer effect of molten metal composition in arc crater, and more micro-scale segregation zones can be generated. When they are solidified, Nb-rich and C-rich liquid phase are relatively more in the mushy region, which improves the sensitivity of solidification crack in the next welding pass.

When the molten metal at the fusion line of last welding pass is adjacent to the crater metal of next welding pass, under the action of cooling solidification and contraction stress, the low-melting-point liquid film is split to form a new hot crack. The strain field at the crack tip contacts with the strain field at the original crater crack tip. These two cracks then join together to form a mixed crack. Another case is that the Nb-rich low-melting-point eutectic at the arc crater of last welding pass is liquefied under the heating action of next welding pass. When it is solidified, the liquid film is torn to form a crack. This crack and the arc crater crack of last welding pass expand to the weld bead of next welding pass under the action of thermal stress and contraction stress, which induces a solidification crack to be initiated and propagate and eventually form a mixed crack.

According to the results of hot crack sensitivity test of 718 alloy, the following rules can be concluded:

(1) With the increase of welding line energy, the sensitivity of hot crack of 718 nickel-base superalloy increases.
(2) The degree of micro segregation in the alloy has a great influence on the sensitivity of hot crack. The micro segregation for as-casted base metal is most serious, thus the hot crack is most sensitive to occur. The micro segregation for homogenization treated base metal is relatively small, and the sensitivity of hot crack is lower. For the base metal in rolling state, there is almost no micro segregation and the hot crack sensitivity is the minimum.
(3) Under the same original condition of base metal and welding line energy, the hot crack sensitivity of multi-pass welding is greater than that of single-pass welding, the hot crack sensitivity of component is greater than that of plate, and the hot crack sensitivity of closed circular weld is greater than that of butt joint.
(4) In multi-pass welding, the arc crater of last weld bead is easy to induce the crack in the next weld bead, which is the sensitive area of welding hot crack. The casting defects will also remarkably increase the hot crack sensitivity.

According to the above rules, the hot crack in repair welding process can be controlled using the corresponding technological measures, such as controlling the welding line energy, homogenizing the casting before the repair welding, using single pass welding as far as possible, or filling the arc crater at the end of welding. Hence, the occurrence of welding crack can be effectively avoided.

6.6.2.3 Structure and Performance Control of Repair Welded Joint

Welding temperature field distribution and heating and cooling speeds are not controllable to some certain extent. The lamellar structure is likely to form in the weld metal due to the micro segregation. When the segregation becomes serious, the eutectic structure will be formed between the dendritic crystals. In welding HAZ, the grain boundary is partially melted and the grains grow up, thus the microstructure and mechanical properties of the weld and its HAZ are inevitably inhomogeneous. The key to repair welding process is how to minimize the inhomogeneity of the structure and properties of the welded joints.

Both the comparative test and practical experience show that the as-cast structure of the base metal with composition segregation has a great influence on the microstructure and properties of the weld. The serious one will lead to the generation of cracks and other defects. Therefore, the homogenization treatment of a thin-wall 718-alloy casting can reduce its micro segregation, improve the weldability of base material, and reduce the sensitivity of welding hot crack. Figure 6.39a–d comparatively show the structures of the welding joint and the PMZ for the as-cast base metal and homogenization treated base metal, respectively. It can be seen that the weld zone of joint is full of columnar dendrites and the dendrite arms are relatively small, growing from the fusion line to the weld center with a strong direction. The structural characteristics of base metal has a small effect on the weld area. The segregation zone in the welding HAZ disappears and the precipitation phase between dendrites decreases. They change from the dendrite structure to an equiaxial crystal structure. However, there are more residual dendrite precipitation in the HAZ of the as-cast base metal, with more dendrite structure characteristics being retained. The PMZ of the joint is kept to be the dendrites structure. It is more obvious for the PMZ of a welded joint for as-cast base metal. Comparing Fig. 6.39b and d, there are more precipitates in the PMZ of a joint for as-cast base metal, especially more chain-typed Laves with strong continuity. However, there are fewer precipitates in the PMZ of a joint for homogenization treated base metal, with relatively weak continuity.

Figure 6.40 shows the relationship between the PMZ width of the welded joint and the welding line energy. It can be seen that the PMZ width of joint increases with the increase of welding line energy. The reason is that the distance between isothermal surfaces of welding temperature field increases with the increase of welding line energy. When the welding energy keeps the same, the PMZ of joints for as-cast base metal is wider, whereas the PMZ of joints for homogenized base metal is narrower. For two different microstructures of base metals, the temperature fields of the welding joints are consistent when the line energy and the plate thickness are the same. The width of PMZ depends on the micro distribution of alloy composition, as well as on the melting point temperature difference between high-melting point phase and

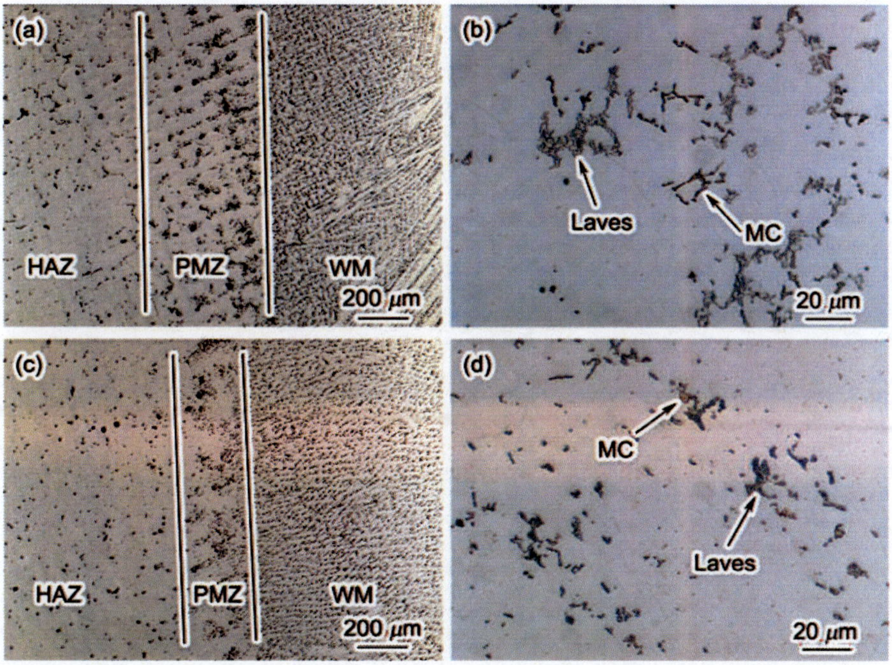

Fig. 6.39 Microstructure of weld joint (2.4 kJ/cm) for as-casted and homogenization treated IN718 alloy. **a** Structure in joint of as-casted base metal; **b** Structure of PMZ in joint of as-casted base metal; **c** Structure in joint of homogenization treated base metal; **d** Structure of PMZ in joint of homogenization treated base metal (HAZ—welding heat affected zone; PMZ—partially melted zone; WM—weld metal)

Fig. 6.40 Width of melted zone in weld joint of as-casted and homogenization treated 2.8 mm thick IN718 alloy plate welded with varied line energies (2.4, 3.6, 4.8 kJ/cm)

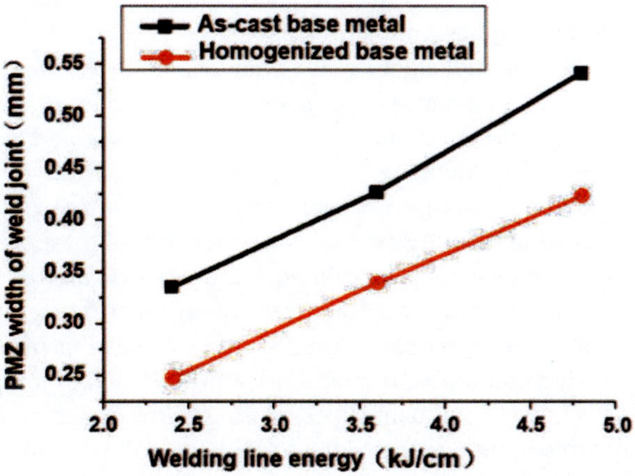

low-melting point phase, such as the liquidus temperature of dendrite austenite and the solid-phase temperature of low-melting point phase between dendrites

Since there are segregation zone and Laves phase in both the as-cast structure of base metal and homogenized structure of base metal, the melting point of Nb-rich interdendritic Laves phase is lower than that of the dendrite austenite. During the welding process, the interdendrite segregation zone and the Nb-rich second phase in PMZ are melted, while the dendrite austenite keeps as the solid state. The liquid film forms between the dendrites, which is in the temperature range of solid–liquid coexistence. The content of Nb in the low-melting point phase of the as-cast base metal is higher than that of the homogenized base metal, and its melting point is lower, so it liquefies at a lower temperature. The temperature range of solid–liquid coexistence of the as-cast base metal is larger, which leads to a wider PMZ of the weld joint than that of the homogenization treated weld joint after casting. In the welding process, the low-melting point phase in PMZ forms less liquid film between dendrites. The liquid film distribute in a small range of welding joint, and they are not easy to be connected with each other. Thus, the sensitivity of welding hot crack is low.

From the main difference between the hot crack characteristics of the welded joints for as-cast and homogenized structures of base metal, it can be seen that the improvement of PMZ microstructure in weld joint is due to the homogenization treatment of the base metal. Therefore, homogenization of base metal after casting is an important way to improve its weldability.

In addition to the influence of the microstructure of base metal on the microstructure and properties of the weld joints, the welding thermal process itself is also an important factor that causes the inhomogeneous microstructure and properties of the joints. It is mainly shown that under the action of welding temperature field, the inhomogeneous composition and structure of welded joint appear, which lead to the inhomogeneous joint performance.

The inhomogeneity of the composition in the weld is mainly manifested through the micro segregation and macro segregation in the weld. The presence of composition segregation in the weld can be simply verified using the metallographic test of a weld. The metallographic sample is cut parallel to the weld bead and its macroscopic morphology is characterized, as shown in Fig. 6.41. It can be seen that the isothermal solidification line on welding pool presents a periodic composition segregation along the welding direction. The reason for this feature is that when the molten pool is solidified at a certain speed along the welding direction, there is solute element segregation at the isothermal solidification front of the welding temperature field, which is continuously pushed forward. When the elements are segregated to a certain extent, the constituent super-cooling forms under the fluctuation of molten pool and its temperature field, and the solidification thus occurs. As the alternative segregation and solidification continue to repeat, the periodic macroscopic lamellar segregation forms in the weld. Among them, the amount of second phase precipitation at A-1 adjacent to the weld fusion line is less than that of A-2. In addition, the amount of second phase precipitation at B-1 in the weld center is less than that of B-2. The weld

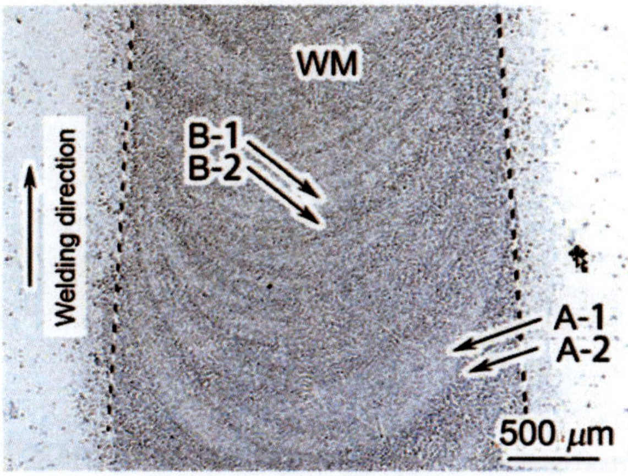

Fig. 6.41 Lamellar segregation in weld metal (2.4 kJ/cm)

microstructure has the characteristics that the layered structure appears alternately, namely, the lamellar segregation along the welding direction.

When the weld crystallizes, the columnar dendrites near the fusion line grow first and move towards the center of the welding pool. The solute concentration in the center of the welding pool gradually increases and finally the columnar crystals form on both sides of the weld contact and solidify at the weld center, resulting in macroscopic regional segregation, that is, the total amount of the second phase at B is greater than that at A. In addition, due to the higher cooling rate at the weld center, the dendrite microscopic segregation is small, that is, the second phase particles at B are small, so is its continuity. In the crater area, the macroscopic and microscopic segregation differences in the central and marginal areas are more serious. Further quantitative segregation detection can be carried out by SEM combined with EDS.

On one hand, the inhomogeneity of weld structure is related to the structure state of base metal and composition segregation of weld structure. On the other hand, it is related to the gradient distribution of local temperature field generated by welding thermal cycle. Figure 6.42 shows a typical weld joint for 1.8 mm thick 718-alloy sheet in TIG welding. It can be seen that, under the action of welding temperature field, different gradient structures are formed from the high temperature area to the low temperature area in the weld joint. The weld center area is mixed with melted base metal and filler metal, which is called the weld zone. Since 718-alloy is a precipitate-hardened superalloy and since the weld metal underwent the melting and solidification process, the original γ' or γ'' phase, carbide phase, and boride phase were all dissolved into the matrix equally, forming a single γ phase (i.e., Ni solution). Due to the high cooling rate of the weld metal, the transverse dendrites are short, whereas the dendrites along the spindle are very long. There is large composition segregation between the dendrites and spindle, resulting in the production of eutectic composition structure in the weld.

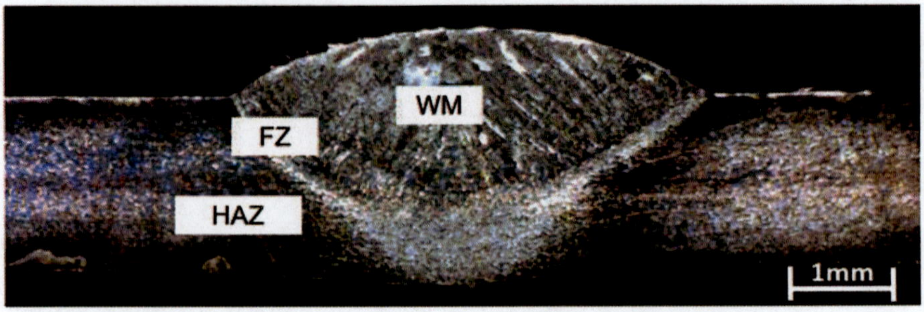

Fig. 6.42 Typical TIG weld joints of 718 alloy with a thickness of 1.8 mm

The fusion zone, or semi-melting zone, is the area where the liquid metal in the molten pool begins to crystallize and grow on the semi-melted grain. The orientation of original grain on the side of base metal is different from the growth direction of the grain in liquid metal. Meanwhile the composition of the liquid metal is also different from that of the base metal. These lead to sudden changes of composition, structure, and performance in this area, which is a weak area in the welding joint. Although the base metal near the fusion zone is not melted, it is affected by the thermal field of welding temperature, which is thus called the heat-affected zone. In the thermal cycle region, the periodic heating temperature gradient is large, resulting in the dissolution of enhanced γ' and γ'' phase and transformation of the carbide phase, which makes the HAZ structure very complex.

To sum up, under the action of welding temperature field, the weld joints of superalloy are characterized by inhomogeneity of composition, structure, and performance in the weld zone, fusion zone, and HAZ. All kinds of factors must be taken into account to make the joint meets the requirements of casting.

6.6.3 Repair Welding Method and Process

6.6.3.1 Repair Welding Method

At present, the main methods to repair defects of nickel-base superalloy castings in the world include fusion welding, laser cladding, self-spreading welding, brazing, and transition liquid phase diffusion welding, etc. These repair-welding methods have their own features. The most suitable method can be selected based on the material properties, working temperature, and structural characteristics of superalloy components. Among all the repair welding methods mentioned above, except the fusion welding, most of them are mainly characterized by filling gaps. Therefore, they are mainly applicable to repairing welding cracks of nickel-base superalloy castings, especially for the repair welding cracks of superalloy blade. Laser cladding can also repair welding the defects other than the cracks to a certain extent, but it needs multilayer cladding and has low efficiency. Only fusion welding can be applied to repair welding a variety of defect other than the crack defects. Because in nickel-base superalloy castings, there are many types of casting defects, including loosen, porosity, inclusions, cracks, etc. Once this kind of defects are found, it generally needs to adopt the mechanical grinding method to remove these defects, even if it is a crack defect, it also needs to clean the harmful impurities and oxidation film in the crack by grinding method, in order to prevent the second defect. Therefore, where the defect has been polished, it is usually necessary to fill in more filler metals, which must be close to or consistent with the base metal in composition, structure and performance, and form a solid metallurgical bond with the base metal. The filler wire metal, matching the composition, structure, and performance of the base metal, can be melted into the welding pool by fusion welding, so that the weld metal and the base metal become a unity. If the brazing or the diffusion method is adopted, there are differences in terms of structure and performance between the filler metal and base metal after all. When the filler metal in the gaps is less, the joint performance can be improved close to that of the base metal by diffusion process, however, when the filler metal is very large in quantity, it is not possible to improve the joint performance by diffusion with the base metal. In view of this, it is reasonable to choose fusion welding to repair the defects of large thin-wall nickel-based superalloy castings. Among the fusion welding methods, TIG welding is an economic and practical mainstream repair welding method with the most stable welding quality, the most flexible operation, and the easiest to achieve the equipment conditions. Therefore, the TIG welding process will be mainly focused in followed introduction of welding process in this chapter.

6.6.3.2 Repair Process with TIG Welding

In repair welding of large complex thin-wall castings, to ensure the welding quality, the welding deformation and the welding hot crack sensitivity should be reduced, thus the segregation of base metal and the welding heat input should be reduced. To get a higher quality of repair welding, the following measures can be used, such as homogenization heat treatment before welding, subsection welding with low line energy input, control of inter-layer temperature, and removing each segmental arc crater by polishing.

Due to the small thickness of thin-wall parts, the deformation is easy to be produced during TIG welding. The main method is to control the welding heat input, which is usually expressed as the line energy, namely, arc heat input per unit length of weld. The calculation formula is:

$$q = \frac{IU}{v} \tag{6.33}$$

where, I—welding current (A), U—arc voltage (V), v—welding speed (mm/s), q—line energy (J/mm).

According to the above equation, the line energy is related to the welding current, arc voltage, and welding speed. In TIG welding, the welding power supply is from a constant current source. If the arc length is constant, the arc voltage U is also constant. In automatic welding or stable manual TIG welding, arc length is unchanged. Therefore, the welding current and welding speed coordinate with each other to determine the line energy per unit length in the welding seam. The higher the welding current, the slower the welding speed, and the higher the line energy, and vice versa.

When the welding line energy is constant, the total heat input to welding joint increases with the increase of weld length. To minimize welding deformation, the overall heat input of welded joints should be minimized. Meanwhile, in order to reduce the sensitivity of welding hot crack, the welding speed should not be too high. As a result, it is necessary to reduce the welding current. However, when welding with a low current, the welding pool becomes smaller and the amount of filling metal deposition is limited, which lowers the welding efficiency.

Since the ambient temperature is generally lower than the welding temperature, the joint is in an exothermic state during the welding process. The overall heat input of the welded joint is shown as follows:

$$Q_W = \eta q s - Q_C \tag{6.34}$$

where, Q_W—overall heat input welding joint (J), η—welding arc heat efficiency, s—weld length (mm), Q_C—heat transfer from welding joints (J).

Generally, η is the thermal efficiency of the arc-plasma to the molten pool of TIG welding in the air, and its value is 0.7. However, in the arc welding with different shielding gases, the shape of arc will change. For example, the first ionization energy of helium atom is 2372.3 kJ/mol, and the argon atom is 1520.6 kJ/mol. According to the principle of the minimum voltage for arc, when the shielding gas is helium, the arc will contract, thus the arc radius is small, and the current density increases. It can effectively increase the melting depth and reduce the arc heating area under the same line energy, which is beneficial to the increase of deposition amount and the control of welding deformation.

Increasing the heat dissipation of a joint during welding is also an effective method to reduce the overall heat input. One of them is to increase the joint cooling rate, such as the use of high thermal conductivity of copper gaskets and copper cooling fixtures in the welding process, and water or other coolant can be used to increase the joint cooling rate further. However, in this way, the joint is cooled while arc heating, which will easily lead to the excessive gradient of joint temperature in space and time, increase the strain rate of the joint and the crack sensitivity, thus it is limited in use.

The second method is to use subsection welding and control the temperature between each subsection, i.e., after the completion of a subsection welding, cooling to a certain temperature and then welding the next subsection. The overall heat input of welded joints in subsection welding is shown as follows:

$$Q_W = \sum_{i=1}^{n} \left(\eta q s_i - q_C^i\right) \tag{6.35}$$

where, n—number of subsections of weld seam, s_i—weld length of each subsection (mm), $q_c{}^i$—heat dissipation in cooling of each subsection (J).

If the cooling time of welds of the same length increases, the overall welding heat input decreases, which is beneficial to the precise control of the cooling rate, as well as to inhibiting welding deformation and hot crack sensitivity. However, the influence of arc crater on the sensitivity of hot crack increases with the increase of the joining area of the front and rear segments. Consequently, it is necessary to control it by closing arc with slow-down current and polishing the arc crater after welding.

On this basis, pulse and waveform can be further used to control arc and to reduce heat input for same welding deposition quantity. For repair welding of holes larger than 10 mm diameter, homogeneous fillers of corresponding size and shape can be prepared in advance and then placed in the hole to weld the gap between the filler and the casting, which can help reduce the amount of welding deposition, the welding deformation, and the probability of hot crack generation.

The repair welding for defect of 718 nickel-base superalloy thin-wall casting is taken as an example to illustrate the repair process with TIG welding as bellow.

Due to the randomness and arbitrariness of the position and shape of the defect in repair welding of 718 nickel-base alloy thin-wall casting, it is difficult to adopt the automatic welding. Meanwhile, according to the characteristics of 718 nickel-base alloy thin-wall casting components, it is necessary to adopt the welding with a relatively low line energy and a stable process. Therefore, the most suitable welding method is the manual TIG welding. It can also realize one-side welding with back formation, which is an ideal method for defect repair welding of nickel-based alloy thin-wall components.

(1) **Repair welding equipment**

TIG welders with good performance and good arc stability under low current conditions are recommended, such as DA300P from OTC of Japan, INVERTIG PRO digital 350/AC/DC from REHM of Germany, MagicWave2200 from FRONIUS of Denmark, etc. These welders have excellent performance, pulsed output current, and good maneuverability. They are suitable for the manual TIG welding.

(2) **Supplementary materials for repair welding**

a. IN718 wire with diameter of 1.2 mm can be selected as filler material;
b. Pure argon (purity $\geq$ 99.99%) can be used as shielding gas (2 bottles are needed, one for TIG welding, one for filling in thin-wall casting cavity).

(3) **Welding tools**

a. Head type protective mask with automatic color changing is recommended;
b. Protective clothing, shoes and gloves;
c. Diagonal cutting pliers for cutting filler wires.

d. Stainless steel wire brush for mechanical cleaning of the local surface before welding;
e. Hammer (with a sharp head and a flat head) for peening the welding surface, to reduce the surface stress and crack sensibility.
f. Pneumatic or electric milling cutter, grinding wheel, electric stainless steel wire brush, used for grinding and cleaning groove before welding;
g. Acetone, cleaning cloth without gauze head, used for cleaning groove before welding;
h. Electric heat blower for fast drying of water vapor and impurities in welding area;
i. Strong light flashlight, used for observing the back of welding seam inside cavity;
j. High temperature resistant aluminum foil tape, used for sealing other holes on the cavity during repair welding, so as to fill argon into the cavity to protect the back of welding seam;
k. Argon flow meters (2 sets), one set is used for TIG welding machine to control the flow rate of shielding gas, and the other set is used for filling the shielding gas into thin-wall casting cavity.

(4) **Defect cleaning**

a. Before welding, the defects (such as pores, inclusions, cracks, loose, etc.) that need to be repaired should be grinded and removed by electric or pneumatic milling cutter, grinding wheel and other mechanical processing methods, until the defects are not visible to the naked eye;
b. For the parts where secondary defects are caused by the original repair welding, the HAZ of the original welding bead should also be grinded off while the secondary defects are removed.
c. After the defect is cleaned, the dye penetrant inspection method should be adopted to ensure that the defect is thoroughly cleaned.

(5) **Groove grinding**

a. In order to ensure the one-side welding with back formation, the groove angle of the welding site should be grinded to about 45° after the defect is cleaned, and the groove direction should be a smooth transition without any abrupt changes.
b. For the repair of technological holes larger than 10 mm in diameter, it is recommended to add the same thick fillers in the center to reduce the amount of filling and welding deformation. In addition, the bevel angle of fillers and grooves should be grinded to more than 45°, and the gap between the two should be controlled at about 2 mm, to ensure that the backside can be thoroughly welded with single-side welding.
c. The casting surface within 20 mm of the welding zone from the groove shall be cleanly grinded to avoid adhesion of molding sand, impurities, rust spots, etc. Where possible, sand and impurities within 10 mm of the backside of the welding zone should also be cleanly grinded.

(6) **Groove cleaning**

a. The welding groove and its adjacent area after mechanical grinding shall be further cleaned by manual or electric stainless steels wire brush to remove all oxide impurities and sand particles dropped by grinding;
b. Use acetone and cleaning cloth without gauze to wipe the groove and surrounding area of the welding area and the filler wire (the filler wire should be kept dry, and its surface should be free of oil, rust and other impurities).
c. When acetone is used to wipe the welding parts, the electric heat blower can be used to blow the welding parts for a short time, to ensure that the welding parts and the surrounding areas are free of any moisture absorbing oxides, grease, water vapor, fingermarks, and other impurities that affect the welding quality.
d. At present, there is still no effective methods to clean the back surface of welding position of a small closed cavity, which may lead to impurities on the back entering into the weld pool during welding, affecting the welding quality.
e. Welding should be performed as soon as possible after cleaning in the above way, in order to avoid absorbing moisture in the air on the cleaned surface.
f. The welding surroundings shall be kept dry and equipped with appropriate ventilation and smoke exhaust devices to ensure the health and safety of welders. However, during welding the shielding effect of argon gas on the welding pool shall be prevented from being directly affected by the ventilation.

(7) **Repair welding process**

a. In the welding process, high quality, nickel-base alloy filler wire should be selected to mix with the casting, to improve the weld structure and performance.
b. While welding, the back surface shall be protected by argon gas, of which the main purpose is to prevent oxidation on the back of the joint and reduce the impact of defects on the molten metal flow in the welding pool. For cavity type thin-wall casting, high-temperature resistant aluminum foil tape should be used to seal other technological holes being not repaired, and then the pure argon should be inflated into the cavity to achieve back protection.
c. TIG welding with direct current electrode negative (DCEN) and small line energy is adopted. The welding current should be between 40 and 80 A, and arc should not swing during welding. Compared with large line energy, the small line energy can reduce cracking tendency, but too small heat input may lead to too large molten pool viscosity, affecting weld formation.
d. With the aid of the viscosity of molten metal in the welding pool, one-side welding with back formation can be completed without gasket. It can be welded from the edge of the groove to the central area or from one side of the groove to the other.

e. Subsection welding can fully release the stress in the welding process, control the temperature between layers, and reduce the tendency of hot cracking. The welding time of each weld should not be too long, and it is advisable not to exceed 5 min to prevent overheating of the weld.
f. After each segment is welded, the surface of the weld shall be gently peened with a hammer immediately. Through vibration and peening, compressive stress shall be applied on the surface of the weld to alleviate and reduce the tensile stress in the solidification and cooling process of the weld, and reduce the tendency of weld cracking. When the temperature drops to 300 °C, the hammering can be stopped.
g. Interlayer temperature should be controlled during welding. After finishing welding of every weld bead, one should stop for a moment, to avoid overheating in local area and to reduce deformation, or move to other parts to be welded. After the weld cooling, one should go back to continue welding.
h. Arc ignition and arc extinguish points should be staggered with each other. The arc should avoid extinguish on base material. Because the cracking tendency at arc crater is large, the cracking sensitivity can be reduced by avoiding arc extinguish in the defective structure and impurity of the base metal.
i. Once a defect is found in the repair welding process, the welding shall be stopped immediately. It can only be continued after the defect is polished.

6.7 Inhibition Effect of Hot Isostatic Pressure Technology (HIP) on Defects of Large Casting

The structure of K4169 superalloy TRF castings are very complex, and it is difficult to control during casting. Metallurgical defects are often found in the casting, e.g., microporosity, segregation and Laves phase, which are difficult to eliminate. The existence of these defects deteriorates the properties of castings seriously, especially the high-temperature rupture and fatigue properties, as well as reduces the reliability of castings. Hot isostatic pressing (HIP) can not only eliminate pore defects such as microporosity and macroporosity in castings effectively, but also improve the microstructures and eliminate segregation and brittle phases of alloys.

6.7.1 Elimination of Microscopic Porosity by HIP

The HIP treatment process of K4169 superalloy is 1170 °C/140 MPa/4 h. After HIP treatment, the porosity of the castings decreases significantly, and the microporosity of the castings is improved considerably.

Table 6.4 shows the porosity of the castings before and after HIP, which indicates

Table 6.4 Porosity of K4169 alloy under different heat treatment conditions

Material status	As-cast	HIP state
Microporosity/%	1.4	0.12

that HIP can eliminate the void defects of the castings effectively and make the castings well compact. The morphological changes of microporosity before and after HIP are shown in Fig. 6.43.

The number of microporosity is expressed using the microporosity index, which is equal to the percentage of the total area of micro-voids in the field of view to the total area of observation. According to the national standard, the distribution of microporosity before and after HIP treatment of castings is evaluated, as shown in Table 6.5.

The results show that there are significant differences in terms of the size and quantity of microporosity before and after HIP treatment. The as-cast specimens contains a lot of microporosity, which are different in size, irregular in shape, and uniformly dispersed on the surface of polished specimen. There is no obvious tendency of aggregation. This indicates that the filling of liquid metal during casting process is complete, but it is difficult to replenish at the later stage of solidification, as shown in Fig. 6.43a. After HIP treatment, the matrix structure of the castings is compact and uniform, and there is almost no visible microporosity under the optical microscope,

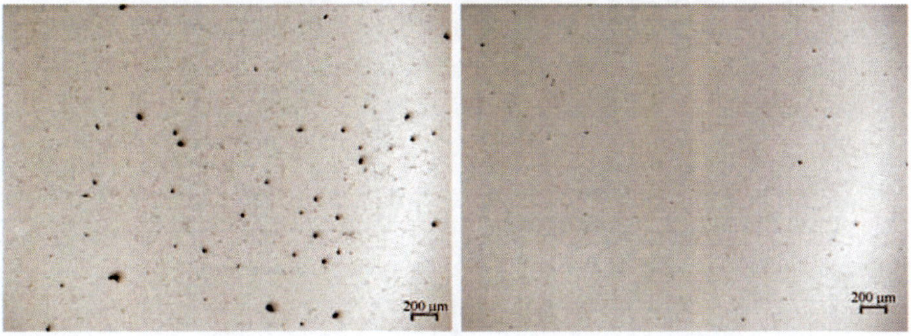

(a) (b)

Fig. 6.43 Changing of microporosity before and after HIP. **a** As-cast **b** HIP state

Table 6.5 Microporosity Index of castings in different post-treatment states

Groups and post-processing status		As-cast	HIP state
Microporosity index	Distribution interval	0.20–1.10	0–0.12
	Average value	0.63	0.03

as shown in Fig. 6.43b. Comparing the microporosity index of the two kinds of specimens, it can be concluded that the microporosity of the specimens treated by HIP can be reduced by more than 90% compared with that of the as-cast specimens. In order to further study the improvement of casting defects of superalloy by HIP, the size distribution of microporosity in as-cast and HIP states was analyzed by image processing software. The results are shown in Fig. 6.44.

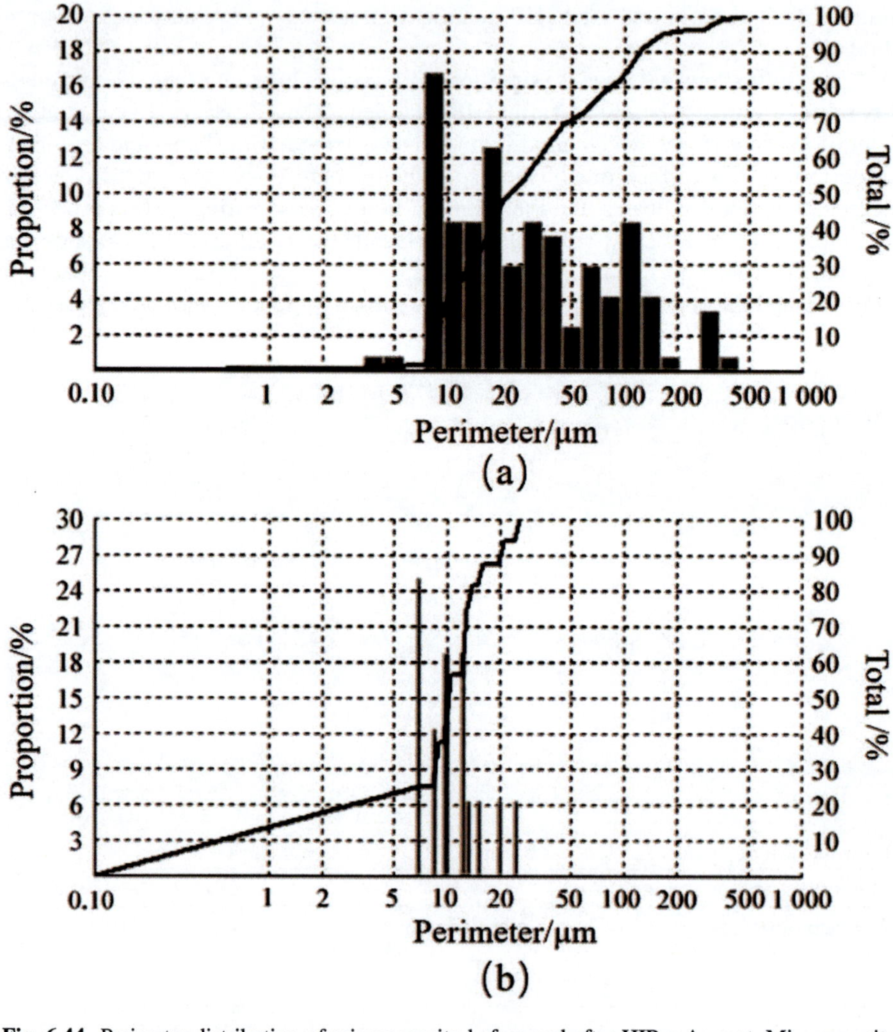

Fig. 6.44 Perimeter distribution of microporosity before and after HIP **a** As-cast, Microporosity size content distribution/μm, **b** HIP, microporosity size distribution/μm

The outer perimeter of the projection of microporosity in the field of view is used to represent the size of microporosity. In as-cast specimens, the perimeter range of microporosity is widely distributed in the range of 3–200 μm, which may be even up to nearly 500 μm. Among them, the microporosities below 10 μm are very few. The microporosities in the range of 10–50 μm are the most portion, accounting for 74%. With the increase of perimeter, the number of microporosities decreases. The number of microporosities between 50 and 150 μm increases, accounting for about 20% of the total number, and only 4% of microporosities above 200 μm as shown in Fig. 6.44a.

After HIP treatment, the number and total amount of microporosities in the matrix of castings are significantly reduced. The circumference distribution of microporosities shifts to the left and narrows obviously, mainly in the region of 6–20 μm, and the number of microporosities decreases with the increase of circumference. Microporosities with circumference of less than 10 μm accounted for more than 50% of the totality, while that with circumference of more than 20 μm accounted for only about 6%. Large-scale microporosities with circumference of more than 50 μm healed completely as shown in Fig. 6.44b.

6.7.2 Elimination of Segregation and Laves Phase by HIP

As-cast structure of large complex superalloy castings has larger segregation zone and higher Laves phase content as shown in Fig. 6.45.

Microstructure inhomogeneity is another unavoidable defect of as-cast microstructures besides microporosity. The inhomogeneity of microstructures leads to the decrease of mechanical properties. The standard heat treatment system prescribed by the aviation standard is not enough to eliminate Laves phase. Usually, HIP is used to eliminate Laves phase and reduce component segregation before homogenization at 1095 °C.

As-cast microstructures of K4169 alloy are typical dendritic structures, the dendritic trunk area is white and the interdendritic area is black as shown in Fig. 6.45a. At higher magnification, a clear black "island" segregation zone can be seen, especially in the middle of the "island" structure, which is Nb-rich white block Laves phase. Laves phase and segregation zone together constitute the interdendritic region, and the skeleton particles are MC carbide phase.

Compared with the as-cast microstructures, standard heat treatment can not eliminate Laves phase, and there are still obvious segregation zones as shown in Fig. 6.45b. The volume fraction of Laves phase shows little decrease, and a large number of needle-like delta phases are produced in the Nb-rich black segregation region around the Laves phase, which is mainly due to the exclusion of Nb and Mo to the interdendritic zone during solidification. Therefore, the Nb content in the interdendritic region is relatively high with about 10 wt.%. Laves phase was formed by $L \rightarrow \gamma +$

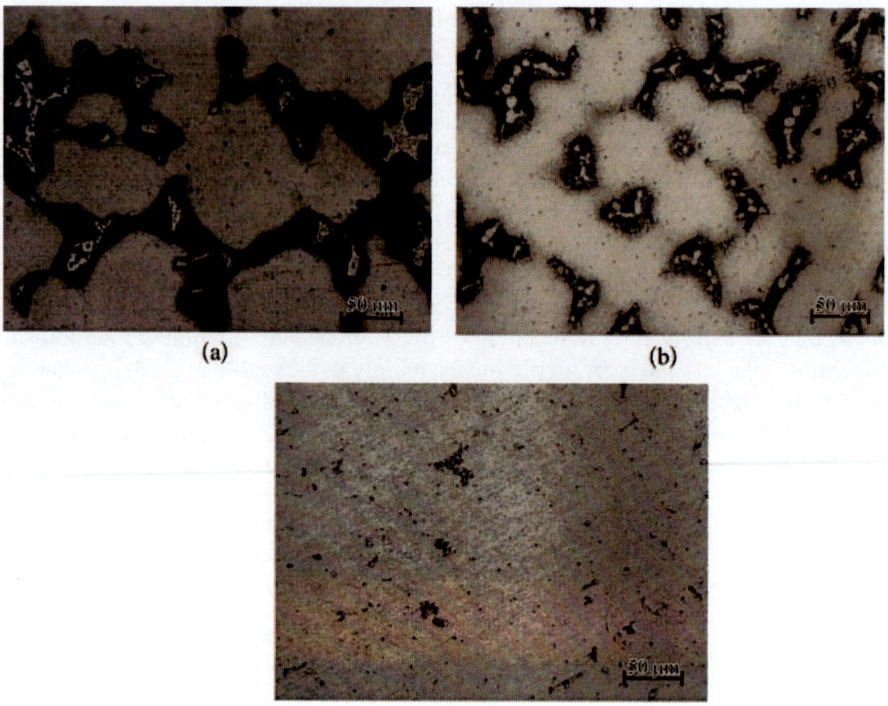

(a) (b) (c)

Fig. 6.45 Microstructure of K4169 Alloy under different heat treatment conditions. **a** As-cast, **b** Aviation standard heat treatment, **c** HIP+ Aviation standard heat treatment

Laves eutectic reaction at the end of solidification. The Laves phase was enriched by 33 wt.% Nb, which was about 6 times of the original alloy composition. However, the Nb element in the segregation area is still at a high level although it has been reduced. The enrichment of Ni, Ti, and Nb in these segregation areas is conducive to the nucleation and growth of the δ-Ni_3Nb phase. A large number of acicular δ-Ni_3Nb phases were formed under standard heat treatment. However, due to the high Nb concentration in Laves phase of K4169 alloy and due to the high activation energy of Nb diffusion in K4169 alloy, the difficulty for homogenization treatment is obviously increased. Therefore, it is difficult to eliminate segregation of Nb elements by standard homogenization treatment. The Laves phase is completely eliminated after heat treatment + HIP and the eutectic characteristics tend to be invisible as shown in Fig. 6.45c. Due to the different segregation degree of different structure parts in large complex thin-walled castings, the elimination degree of segregation is somewhat different. For the thin-walled part [see Fig. 6.46a], the Laves phase and its associated segregation zone are largely eliminated and no acicular δ-Ni_3Nb phase is formed. However, with the increase of thickness, although the Laves phase is completely dissolved in the matrix, the surrounding segregation zone still exists. After HIP+ standard heat treatment, acicular δ-Ni_3Nb phase is precipitated again in the segregation region as shown in Fig. 6.46b.

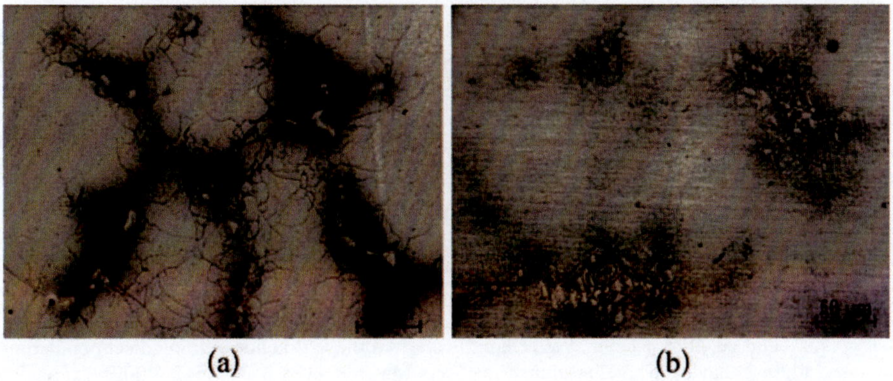

Fig. 6.46 Elimination of segregation after HIP in thick and large parts. **a** HIP, **b** HIP plus standard heat treatment

6.8 Future Development Trend of Casting Defect Prediction and Control

With the development of computer, intelligent identification system, and non-destructive testing technology, the inspection of large castings will develop towards to automation, three-dimensional, intelligent, and refinement. Automatic recognition and evaluation of casting defect types, grades, and spatial distribution will be realized, and more and more attention will be paid to the detection and characterization of micro-defects.

Computer-aided casting will proceed from macro-temperature field and shrinkage prediction to microporosity and harmful phase prediction, eventually towards digitalizing the whole casting process. With the development of artificial intelligence technology such as big data technology and machine learning, the prediction and control of casting defects will be transformed. The technology of repairing and restraining casting defects will be further developed, which is expected to break through the difficult problems of repairing large castings.

References

1. P. Lee, A. Chirazi, D. See, Modeling microporosity in aluminum-silicon alloys: a review. J. Light Met. **1**(1), 15–30 (2001)
2. H. Zheng, W. Tian, J. Sun et al., Overview on porosity forming mechanism of spheroidal graphite iron. Foundry **54**(11), 1063–1065 (Chinese) (2005)
3. National Standards of the People's Republic of China, Method for determination of grain size, primary dendrite spacing and microporosity of Superalloy Castings. Vol. GB/T 14999.7-2010 (Chinese) (2010), pp. 1–14
4. C.D. Lee, Defect susceptibility of tensile strength to microporosity variation in as-cast magnesium alloys with different grain sizes. Met. Mater. Int. **16**(4), 543–551 (2010)

5. D.G.L. Prakash, D. Regener, W.J.J. Vorster, Effect of position on the tensile properties in high-pressure die cast Mg alloy. J. Alloy. Compd. **470**(1–2), 111–116 (2009)
6. C. Do Lee, Variability in the tensile properties of squeeze-cast Al–Si–Cu–Mg alloy. Mater. Sci. Eng. Struct. Mater. Prop. Microstruct. Process. **488**(1–2), 296–302 (2008)
7. C. Do Lee, Constitutive prediction on the variability of tensile properties due to microporosity variation. Met. Mater. Int. **14**(1), 15–20 (2008)
8. D.M. Stefanescu, Computer simulation of shrinkage related defects in metal castings—a review. Int. J. Cast Met. Res. **18**(3), 129–143 (2005)
9. K.D. Carlson, C. Beckermann, Prediction of shrinkage pore volume fraction using a dimensionless Niyama criterion. Metall. Mater. Trans. A **40**(1), 163–175 (2009)
10. J. Campbell, M. Tiryakioğlu, Bifilm defects in Ni-based alloy castings. Metall Mater Trans B **43B**, 1–13 (2012)
11. P. Li, P.D. Lee, D.M. Maijer et al., Quantification of the interaction within defect populations on fatigue behavior in an aluminum alloy. Acta Mater. **57**(12), 3539–3548 (2009)
12. Y. Cao, Y. Xiong, Application of industrial computed tomography to testing complicated castings. NDT, **24**(8), 357–359 (Chinese) (2002)
13. L. Nastac, D.M. Stefanescu, Macrotransport-solidification kinetics modeling of Equiaxed dendritic growth. Ii. Computation problems and validation on Inconel 718 superalloy castings. Metall. Mater. Trans. A **27**(12), 4075–4083 (1996)

Chapter 7
Dimensional Precision Control of Large Castings

Changhui Liu, Sun Jin, and Xinmin Lai

A turbine rear frame (TRF) is an important load-bearing component with a complex shape in the aero-engine, which requires not only a high metallurgical quality but also an extremely strict dimensional accuracy. For example, the accuracy of key linear dimensions is CT4~CT6 [1] according to the tolerance requirements of the ISO standard, which is very difficult to manufacture. However, the final casting shape produced by investment casting is close to the final shape of the part, which can be used directly or with little additional machining [2, 3]. Therefore, investment casting has become the main method of aero-engine superalloy complex structural components.

In recent years, with the continuous improvement of thrust-to-weight ratio, the structure of key hot-end castings of aero-engine is becoming more and more integrated, thin-walled, and complicated with variable cross-section (the maximum outline size is 1.8 m, the minimum wall thickness is less than 2 mm, and the maximum casting weight is 1000 kg [1]). It is also developing towards the integration of structural bearing capacity and gas diversion structure and function. This not only puts forward the requirement of high bearing capacity for aviation castings but also poses a new challenge to the dimensional accuracy of aviation castings. The deviation due to deformation and the size mismatch lead to the poor qualification rate of the finished products of large and complex thin-wall castings, the low service performance, and the difficulty of ensuring the batch manufacturing ability. This is the bottleneck issue that must be overcome in order to achieve the independent development of the key parts of aero-engine. Therefore, the dimensional accuracy and dimensional stability of complex thin-wall castings have become a common but difficult problem in investment casting of aero-engine superalloy complex structures [2, 4, 5].

In this chapter, by studying the influence of complex processing parameters on the overall accuracy of castings, the technical system of deviation detection, evaluation, diagnosis, and control for large castings is constructed. The new theory of stream of variation simulation of large castings is established, and new technologies and methods such as quantitative analysis, process optimization, accurate detection, and rapid diagnosis of large castings are formed. Therefore, the quality control of large

B. Sun et al., *Precision Forming Technology of Large Superalloy Castings for Aircraft Engines*, https://doi.org/10.1007/978-981-33-6220-8_7

castings can be developed from qualitative analysis to quantitative control. It provides new technologies and methods for casting precision control.

7.1 Main Variation Sources Affecting the Dimensional Accuracy of Investment Castings

7.1.1 Analysis of Main Variation Sources in Wax Part Production

7.1.1.1 The Main Variation Source Which Affects the Deformation of the Wax Part

Through the deformation mechanism and performance test of wax materials, it is found that the technological parameters of wax pressing stage can affect the properties of wax materials, which leads to the deformation of wax materials in the filling process. In the process of wax pressing, there are many factors causing the dimensional deviation of the wax die. First of all, the difference between the die of wax pressing and the drawing of die design introduces the initial size error, which depends on the manufacturing accuracy of die construction. In general, the die can be used repeatedly for a long time after manufacturing, which allows the size error of the die to be regarded as a systematic error. In the process of wax injection, the exhaust property of the die will affect the filling of wax material, but the exhaust property of the die is determined by the design. The influence caused by the exhaust property of the die can be eliminated by modifying the die. The thermal conductivity of the die will affect the cooling rate of the wax material and the thermal conductivity of the die will be determined by the die material. When the mold material is determined, the thermal conductivity of the die will be determined, such that the thermal conductivity of the die will not introduce random error.

In the process of wax injection, we can set the parameters of wax pressing through the wax pressing program, including holding pressure, holding time, melt temperature, wax injection speed, die temperature, die locking force and so on [6–8]. Through the analysis of the deformation mechanism of wax materials in the process of wax pressing, it can be seen that these parameters directly affect the properties of wax materials and thus affect the deformation. In addition, these parameters will fluctuate within a small range around the set value due to the instability of the wax compressor system. Among them, as long as the die locking force is set at a reasonable level to ensure that the upper and lower die of the die can be closely fitted, its fluctuation will not cause the dimensional change of the wax die. The fluctuation of pressure holding pressure, holding time, melt temperature, die temperature and so on can directly affect the physical properties of wax in the filling process, which has an effect on the solidification and cooling shrinkage of wax and finally introducing the dimensional variation. Therefore, these processing parameters bring random error and finally reflect the dimensional variation of the wax part.

7.1.1.2 Main Variation Sources Affecting the Deformation of the Wax Part in the Storage Process

According to the analysis of the deformation mechanism of the wax part in the storage stage, it can be seen that the storage temperature and storage time of the wax part directly affect the deformation of the wax part. The deviation of storage temperature and storage time also directly lead to the dimensional deviation caused by the final wax part used in the preparation of shell. In the actual storage process, although the wax part is stored in the environment of constant temperature and humidity, the storage temperature will fluctuate slightly in different ranges according to the different production conditions near the set temperature level. On the other hand, the stored wax part will be used to manufacture the shell, and due to the production progress, there will be some deviations in the storage time. Therefore, in the storage stage of the wax part, storage time and storage temperature are the main random factors in the storage process, which are the main variation sources affecting the fluctuation of the wax part dimension.

In summary, the main variation sources of wax part dimension in the process of mold preparation include dimensional deviation of die, holding pressure, holding time, melt temperature, wax injection speed, storage temperature, storage time and so on. After the dimensional deviation of the die is introduced, it will not change in the process of wax die pressing in the future, which is the system error. In the process of batch preparation of the wax part, the actual process parameters of each wax part will be different, and the dimensional deviation of the final wax part will be random due to the variation of process parameters.

7.1.2 Analysis of Main Variation Sources in Shell Production

The preparation process of the ceramic shell is the most time-consuming process in the whole investment casting process. The preparation process of the whole ceramic shell is shown in Fig. 7.1. Before removed, the wax pattern was first covered completely by the slurry using the immersion method. In order to promote an even coating, the pattern was rotated over a slurry dipping pool so that the redundant slurry can drain from different points. Then using the rainfall method, a uniform distribution of stucco was formed. The shell coating is categorized into three types: the surface layer, the back layer, and the slurry sealing layer. After an adequate drying for the shell, the patterns were then melted in a furnace so that only the unfired or green shells are remained. The processing parameters such as the dewaxing temperature, the dewaxing time, the dewaxing pressure, and the heating direction are set in the dewaxing process. Before casting, the shell should be put into a furnace again. The

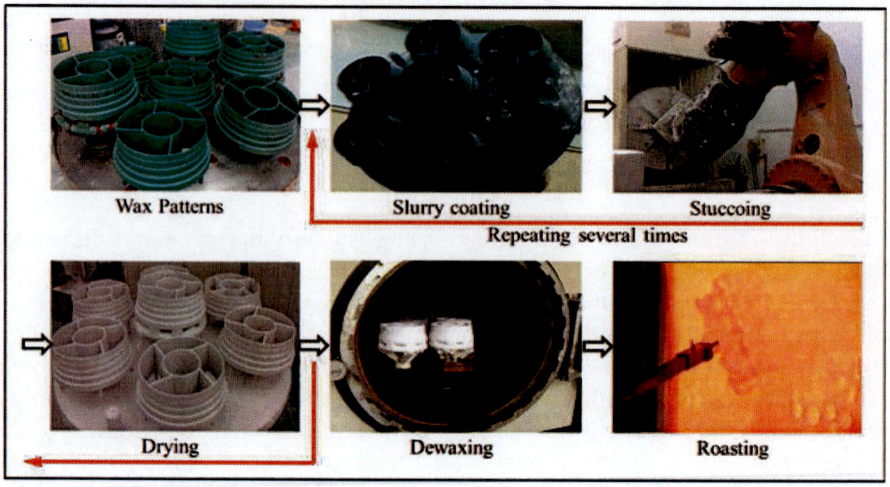

Fig. 7.1 The production process of ceramic shell

shell should be roasted by setting different heating rate, roasting temperature, and holding time in order to increase the strength of the shell and avoid the fracture of the shell introduced by the pressure of liquid metal in the pouring process.

Since the preparation process of the shell is very complex, many factors can affect the deformation of the shell. Through the study of the properties and deformation mechanism of the shell, combined with the test of the properties of the partial shell, it can be seen that the physical properties of the slurry are basically determined by the composition of the slurry. The composition and particle size determine the composition of the shell after drying, thus affecting the microstructure and thermal properties of the shell. However, the thermal properties determined by the composition of the material are basically stable. The immersion time, slurry coating time, stuccoing time, stuccoing angle, drying time, drying temperature, environmental humidity, dewaxing time, dewaxing pressure, dewaxing temperature, and heating rate are all uncontrollable and likely to fluctuate during the whole preparation process of the shell. These processing parameters are random. The random variations caused by slurry coating time, stuccoing time and stuccoing angle are reflected in the uniformity of the thickness and thickness of the shell. The thickness uniformity caused by the geometric structure and riser of the shell and the position of the dewaxing port is systematic, while the thickness uniformity caused by the processing parameters is random. Therefore, the variation sources in the preparation stage of the shell can be summarized as follows: the different thickness of the shell, the dewaxing time, the dewaxing pressure, the dewaxing temperature, the roasting temperature, the holding time, the heating rate and so on.

7.1.3 *Analysis of Main Variation Sources in the Casting Process*

In the casting process, when the shell is started in the baking furnace, the alloy is put into the crucible of the vacuum furnace, and the starting equipment is started to melt the alloy. When the temperature of the melt of the alloy reaches the casting temperature, the door of the roasting furnace is quickly opened. The shell and the pouring cup are quickly transferred to the ingot buggy by a forklift. Then the outer chamber is entered. The molten metal is injected into the shell through the runner. After the casting is finished, the air is ventilated. When the door is opened, the ingot buggy is withdrawn and the magnesium powder is sprayed. Finally, the casting is transferred to the designated area by a forklift and the alloy is cooled and solidified in the shell.

In the process of casting and solidification of the alloy, the alloy and the shell have non-uniform heat exchange and complex mechanical action. Therefore, the thermal properties of the shell have an important influence on the shrinkage of the alloy. In addition, casting temperature, casting speed, and other processing parameters are also the main factors affecting the shrinkage of the alloy, of which the fluctuations will introduce random deviations.

7.2 Method for the Precision Control of Large-Scale Castings

7.2.1 *The Accurate Measurement for the Dimensional Deviations of the Investment Casting Process*

7.2.1.1 Method for Accurate Dimensional Deviation Detection

1. Measurement of key dimensions

The key dimensions, such as the inner and outer flange diameter, are required to be measured effectively with high accuracy. According to the designed measurement specification and the arrangement of functional measurement points, the three-coordinate measurement machine is used (as shown in Fig. 7.2). The accuracy of the measurement is 4.5 μm, which meets the measurement requirements.

2. Measurement of 3D dimensions

For complex castings, curved surfaces are often designed. There are some limitations to evaluate its shape deviation by three-coordinate machines. Not only the measurement is time-consuming but also the measurement results are not intuitive. For this

Fig. 7.2 Three-coordinate measurement machine

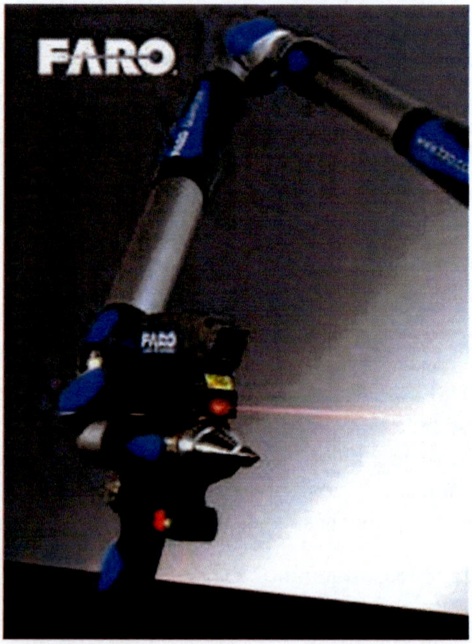

Fig. 7.3 FARO laser scanner

reason, we use a FARO laser scanner, as shown in Fig. 7.3, as a full-size measurement platform. Its measurement accuracy is 80 μm, which meets the requirements of measurement.

3. Measurement of the inner cavity dimensions of the shell

In order to monitor the dimension change of the whole process, the inner cavity dimensions of the shell must be measured. It is a great challenge to measure the inner cavity dimensions of the shell. This problem can be solved using the industrial computer tomography (CT) method [9]. The measurement accuracy is 25 μm. The measurement process and the post-processing of measurement results for a sample shell are shown in Figs. 7.4 and 7.5, respectively.

In order to analyze the dimension change of the whole process, the laser scanner and industrial computer tomography are used in the investment casting process for a sample. Its dimension deviations to the design dimensions of the die are shown in Fig. 7.6.

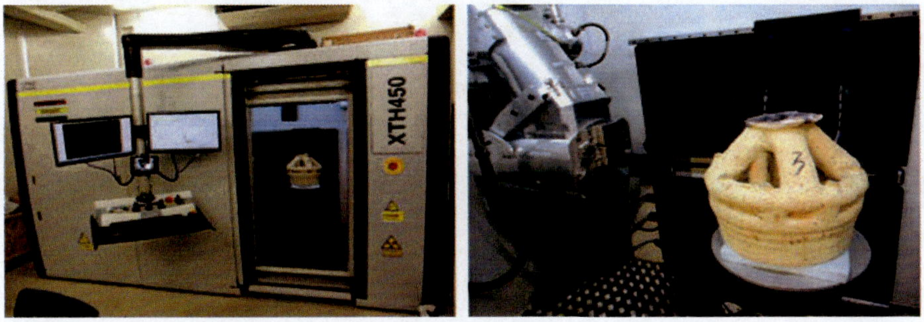

Fig. 7.4 The measurement process for a sample shell

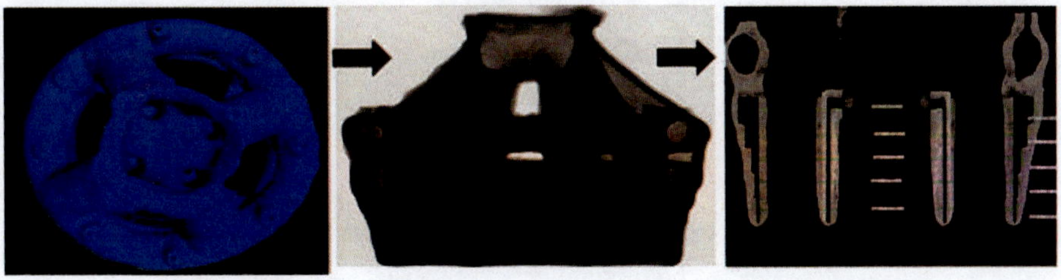

Fig. 7.5 The post-processing of measurement results for a sample shell

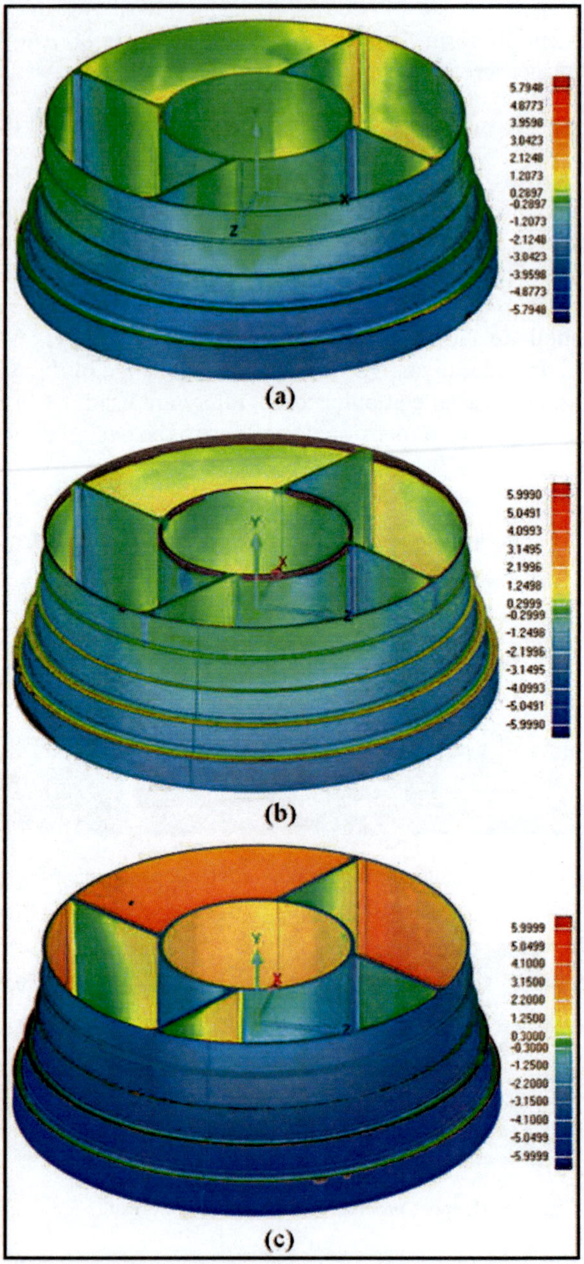

Fig. 7.6 Dimension deviations (unit: mm) of the measured point cloud model relative to the CAD model. **a** wax, **b** ceramic shell and **c** casting part [10]

7.2.1.2 Optimization and Design of Dimensional Deviation Detection System for Large-Scale Casting

During the deviation detection of each stage of the casting, the positioning mode influences the observability, the inheritance, and deviation detection. A good positioning method should guarantee the quality of parts in the assembly process. Different positioning methods can have a great influence on the assembly quality of the same part. The positioning mode can be divided into full positioning, incomplete positioning, under positioning, and over positioning. The six degrees of freedom of the workpiece being completely restricted is called full positioning. When the workpiece has dimensional accuracy requirements in the three directions of x, y, and z, the positioning method is generally adopted. When the processing requirement or the actual function of the workpiece does not need to limit the six degrees of freedom of the workpiece, the positioning method adopted at this time is incompletely positioned. According to the processing requirement and the actual function of the workpiece, the restricted degree of freedom is not limited, and the positioning mode is under-positioning. Two or more positioning elements on the clamp limit the positioning mode of the same degree of freedom of the workpiece as the over-positioning. When the parts are over positioned, it may lead to the workpiece cannot be installed, resulting in the deformation of the workpiece or positioning elements. However, the reasonable use of over positioning can improve the position accuracy between the workpiece positioning benchmarks and between the working surfaces of positioning elements.

For a specific positioning method, the quality of its positioning is usually determined based on engineering experience. At present, a quantitative representation method is lacking. As shown in Fig. 7.7, there are three feasible positioning methods for the same part. Empirically, the first one is the best and the third one is the worst. However, there are some more complex positioning methods, the traditional experience cannot directly say the quality of the positioning method, so for the characterization of the positioning method, we need a quantitative representation method, using the calculation data to explain the quality of the positioning method.

For the ring parts in this paper, because the ring parts have the characteristics of rotational symmetry, the general positioning method is axial and circumferential positioning (the second or third positioning method as shown in Fig. 7.7). The traditional positioning deviation research is missing for this type of positioning. This chapter is to solve the positioning problem of the ring.

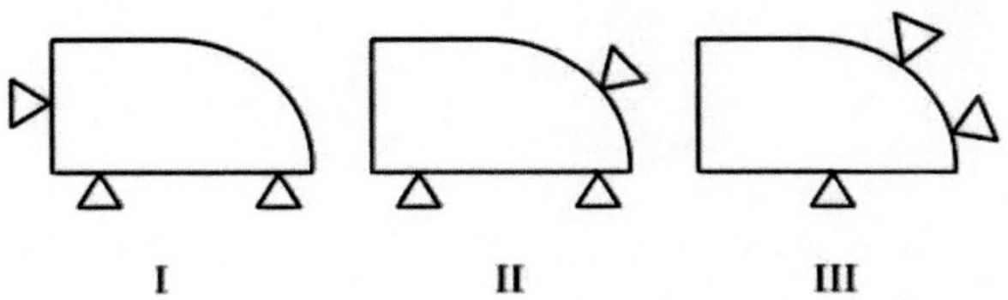

Fig. 7.7 Three different positioning methods of the same part

1. Positioning stability analysis

(1) Stability index of positioning Jacobian matrix

Suppose a part is located by n locating points and the contact between the locating point and the part can be regarded as point friction contact. The relationship between the contact force F of the fixture and the workpiece and the torque F_e of the workpiece is given as follows:

$$JF = F_e \tag{7.1}$$

where J is the positioning Jacobian matrix. in the deterministic positioning, the Jacobian matrix has a given form. It's a transfer matrix. In the rigid body kinematics, it mainly plays the role of transferring the positioning point deviation to the part positioning deviation in the positioning process.

In the two-dimensional case:

$$J_i = \left[-n_{ix}, -n_{iy}, n_{iy}x_i - n_{ix}y_i\right] \tag{7.2}$$

In the three-dimensional case:

$$J_i = \left[-n_{ix}, -n_{iy}, -n_{iz}, 2(n_{iz}y_i - n_{iy}z_i), 2(n_{ix}z_i - n_{iz}x_i), 2(n_{iy}x_i - n_{ix}y_i)\right] \tag{7.3}$$

In these formulas, $\overrightarrow{n_i} = (n_{ix}, n_{iy}, n_{iz})^T$ is the normal vector of location point i. The general solution of Eq. (7.1) is as follows:

$$F = (I - J^+J)\lambda \tag{7.4}$$

where $J^+ = J^T(JJ^T)^{-1}$ is the generalized inverse matrix of matrix J, I is the unit matrix, and λ is an arbitrary vector.

In Eq. (7.4), $F = J^+F_e$ represents the cooperative force of the fixture used to resist the external force and external torque, $F = (I - J^+J)\lambda$ is the internal force set of zero space J. Internal force is released at the locating point to correct F, avoiding the locating point moving at the contact point and causing no residual force and moment on the part surface. According to Formula 7.4, a Gaussian sphere can be established and its expression is $\|F\|^2 = 1$. The Gaussian sphere can be mapped to a hyperellipsoid in the workspace, which is called force hyperellipsoid. The equation for this sphere is defined as follows:

$$F^T(JJ^T)^{-1}F = 1 \tag{7.5}$$

The principal axis of the ellipsoid is $e_1\sigma_1, e_2\sigma_2, \cdots, e_n\sigma_n$, in which e_i is the i-th eigenvector of JJ^T and σ_i is the i-th singular value of matrix J, that is, the length of the i-th principal axis of the hyperellipsoid. Here, the product of all singular values is defined as the Jacobian matrix stability parameter w, that is:

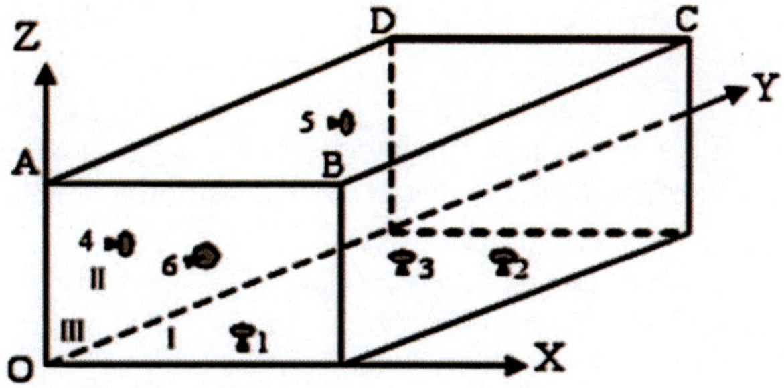

Fig. 7.8 Deterministic positioning scheme of parts

$$w = \sigma_1\sigma_2 \cdots \sigma_n \tag{7.6}$$

It can be seen from the formula that this parameter is related to the position of the contact point of the clamped part. Because $\det(JJ^T) = (\sigma_1\sigma_2 \cdots \sigma_n)^2$ the Jacobian stability index of the positioning can be expressed as:

$$w = \sqrt{\det(JJ^T)} \tag{7.7}$$

(2) Stability and its influencing factors under deterministic positioning

Suppose that the part adopts the deterministic positioning scheme of 3-6-1 mode, as shown in Fig. 7.8, the coordinates of main positioning points 1, 2, and 3 are respectively: $(x_1, y_1, 0)$, $(x_2, y_2, 0)$, and $(x_3, y_3, 0)$; the coordinates of secondary positioning points 4 and 5 are $(0, y_4, z_4)$, $(0, y_5, z_5)$; the coordinates of positioning point 6 are $(x_6, 0, z_6)$.

According to Eq. (7.3):

$$J = \begin{bmatrix} 0 & 0 & -1 & 2y_1 & -2x_1 & 0 \\ 0 & 0 & -1 & 2y_2 & -2x_2 & 0 \\ 0 & 0 & -1 & 2y_3 & -2x_3 & 0 \\ -1 & 0 & 0 & 0 & 2z_4 & -2y_4 \\ -1 & 0 & 0 & 0 & 2z_5 & -2y_5 \\ 0 & -1 & 0 & -2z_6 & 0 & 2x_6 \end{bmatrix} \tag{7.8}$$

Equation 7.8 can be written as follows:

$$\begin{bmatrix} 0 & 0 & 0 & 2y_1 & -2x_1 & -1 \\ 0 & 0 & 0 & 2y_2 & -2x_2 & -1 \\ 0 & 0 & 0 & 2y_3 & -2x_3 & -1 \\ -1 & 0 & -2y_4 & 0 & 2z_4 & 0 \\ -1 & 0 & -2y_5 & 0 & 2z_5 & 0 \\ 0 & -1 & 2x_6 & -2z_6 & 0 & 0 \end{bmatrix} = \begin{bmatrix} 0 & A \\ B & C \end{bmatrix} \tag{7.9}$$

where $A = \begin{bmatrix} 2y_1 & -2x_1 & -1 \\ 2y_2 & -2x_2 & -1 \\ 2y_3 & -2x_3 & -1 \end{bmatrix}$, $B = \begin{bmatrix} -1 & 0 & -2y_4 \\ -1 & 0 & -2y_5 \\ 0 & -1 & 2x_6 \end{bmatrix}$, $C = \begin{bmatrix} 0 & 2z_4 & 0 \\ 0 & 2z_5 & 0 \\ -2z_6 & 0 & 0 \end{bmatrix}$

It is obvious that J is invertible. From Eq. (7.6), it can be seen that the stability index of the Jacobian matrix in this positioning method is:

$$w = \|A\|\|B\| = 8|(x_2y_3 + x_3y_1 + x_1y_2 - x_2y_1 - x_1y_3 - x_3y_2)(y_5 - y_4)| \quad (7.10)$$

From geometry knowledge, we can get:

$$\begin{cases} S = \frac{1}{2}(x_2y_3 + x_3y_1 + x_1y_2 - x_2y_1 - x_1y_3 - x_3y_2) \\ L = y_5 - y_4 \end{cases} \quad (7.11)$$

where S is the area of a triangle composed of locating points 1, 2 and 3, and L is the y-direction distance between locating points 4 and 5.

Therefore, the stability index of positioning Jacobian matrix can be simplified as:

$$w = 16SL \quad (7.12)$$

From Eq. (7.12), it can be seen that the more dispersed the main positioning points are, the larger the positioning triangle they form. The larger w is, the more stable the positioning method is. The larger the distance of the second positioning point in the y-direction is, the more stable the positioning method is.

(3) Stability comparison of different positioning modes for different structures

Since the established expression of the positioning stability coefficient has been determined, the stability of different positioning methods under the same part is compared. The selection of locating points on the main locating surface generally follows that the locating points are scattered as far as possible to maximize the triangle area formed by the locating points. Therefore, we only need to study the two-dimensional positioning.

In the first case, as shown in Fig. 7.9, the coordinates of main positioning points 1 and 2 are (x_1, y_1), (x_2, y_2), respectively, and the coordinate of secondary positioning point 3 is (x_3, y_3), The curvature of the primary and secondary locating surfaces are 0 and parallel to the coordinate axis.

According to Eq. (7.2), the Jacobian matrix of the positioning method is:

$$J_1 = \begin{bmatrix} 0 & -1 & x_1 \\ 0 & -1 & x_2 \\ 1 & 0 & -y_3 \end{bmatrix} \quad (7.13)$$

According to Eq. (7.7), the Jacobian stability index of the positioning mode is as follows:

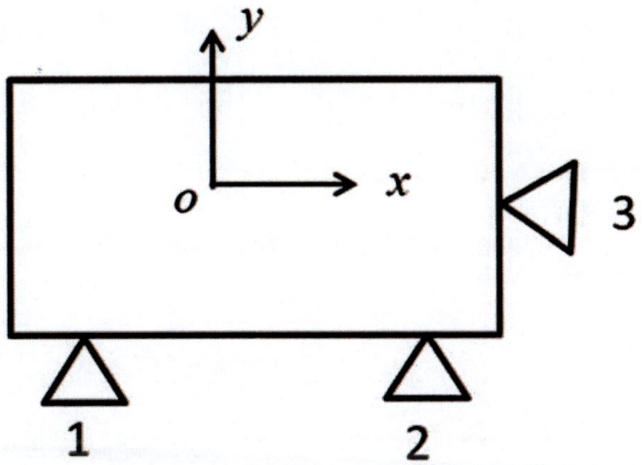

Fig. 7.9 Positioning method 1

$$w_1 = \sqrt{\det(J_1 J_1^T)} = x_1 - x_2 \tag{7.14}$$

Equation 7.14 shows that the larger the x-direction distance between the positioning points 1 and 2, the better the stability of the positioning method.

For the second case, as shown in Fig. 7.10, the coordinates of main positioning points 1 and 2 are (x_1, y_1), (x_2, y_2) ,respectively, the coordinates of secondary positioning point 3 are (x_3, y_3), the curvature of secondary positioning surface is not 0, and the curve equation is $y = ax^2 + bx + c$.

According to Eq. (7.2), the Jacobian matrix of the positioning method is:

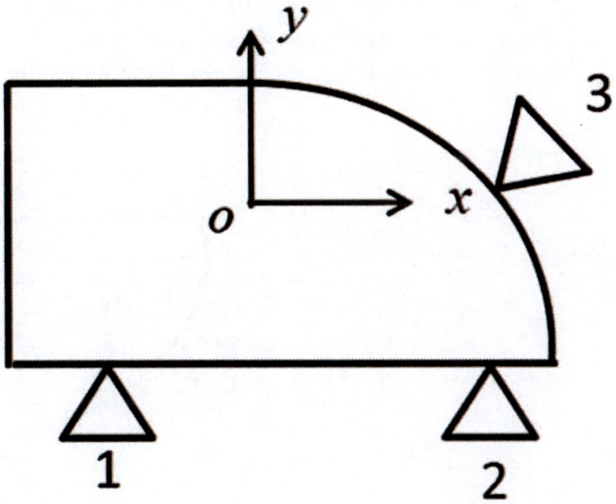

Fig. 7.10 Positioning method 2

$$J_2 = \begin{bmatrix} 0 & -1 & x_1 \\ 0 & -1 & x_2 \\ -(2ax_3+b)(\frac{\varphi}{|2a|})^{1/3} & (\frac{\varphi}{|2a|})^{1/3} & -(\frac{\varphi}{|2a|})^{1/3}[x_3 + y_3(2ax_3+b)] \end{bmatrix} \tag{7.15}$$

where φ is the curvature of the curve and the mathematical expression is $\varphi = \frac{|2a|}{[1+(2ax_3+b)^{3/2}]}$.

According to Eq. (7.7), the Jacobian stability index of the positioning mode is as follows:

$$w_2 = \sqrt{\det(J_2 J_2^T)} = (x_1 - x_2)\sqrt{1 - \left(\frac{\varphi}{2a}\right)^{2/3}} \tag{7.16}$$

It can be seen from the geometric relationship that the tangent slope of positioning point 3 is $k = 2ax_3 + b$, so Eq. (7.16) can be simplified as:

$$w_2 = (x_1 - x_2)\sqrt{\left(1 - \frac{1}{1+k^2}\right)} \tag{7.17}$$

From Eq. (7.17), it can be seen that, for the positioning method of positioning points on the arc surface, the larger the x-direction distance of positioning points 1 and 2, the better the stability of the positioning method. It also can be found that the greater the slope of the tangent line at the position of positioning point 3 on the arc surface, the better the stability.

For the third case, as shown in Fig. 7.11, the coordinates of main positioning points 1 and 2 are (x_1, y_1), (x_2, y_2), respectively, and the coordinates of secondary positioning point 3 are (x_3, y_3), the curvature of secondary positioning surface is not 0, and the curve equation is $y = ax^2 + bx + c$.

According to Eq. (7.2), the Jacobian matrix of the positioning method is:

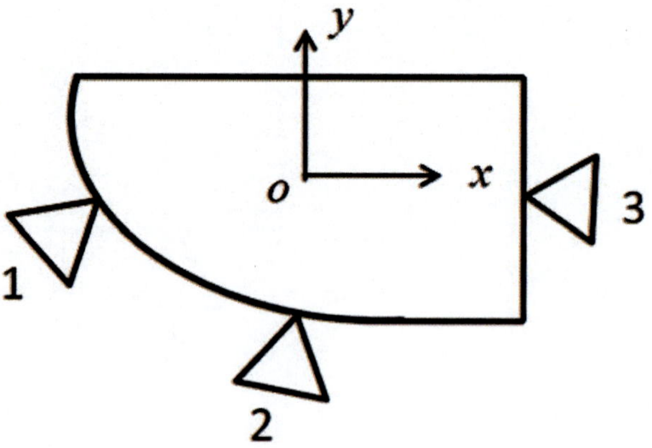

Fig. 7.11 Positioning method 3

$$J_3 = \begin{bmatrix} A & B & Bx_2 - Ay_2 \\ C & D & Dx_2 - Cy_2 \\ 1 & 0 & y_3 \end{bmatrix} \tag{7.18}$$

where, $A = 2ax_1 + b, B = 2ax_2 + b$ is the slope of the tangent of the locating points 1 and 2 on the locating surface;

$$C = \frac{2ax_1y_1 + y_1b - x_1}{\sqrt{A^2 + 1}}, D = \frac{2ax_2y_2 + y_2b - x_2}{\sqrt{B^2 + 1}} \tag{7.19}$$

According to Eq. (7.7), the Jacobian matrix of the positioning method is:

$$w_3 = \sqrt{\det(J_3 J_3^T)} = \sqrt{A(y_1 - y_3) - B(y_2 - y_3) + (x_1 - x_2)^2} \tag{7.20}$$

From Eq. (7.20), it can be seen that under the positioning method 3, the stability of the positioning mode is related to the x-direction distance of the positioning points 1 and 2 and the y-direction distance of the positioning points 1, 2 and 3, as well as the shape of the position of the positioning point on the arc positioning surface.

Comparing the stability of the above three positioning methods, under the same part structure, the positioning stability for different positioning methods are different. Obviously, the stability of positioning method 3 is affected by the most factors, which is also the most unstable. Therefore, when the main locating surface of the part is an arc surface, the selection of the positioning method has greater stability to the positioning of the part. When selecting the positioning method, it is suggested to avoid the selection of the arc surface.

2. Optimized selection of positioning mode for ring parts

From the previous section, it is known that the quality of the positioning method is manifested in the stability of positioning and the deviation transmission ability of positioning. Next, for specific parts, the positioning method can be optimized.

The part illustrated in Fig. 7.12 is a flange part diagram, where AB is the outer surface of the inner flange with the corresponding flange equation being $x^2+y^2 = a^2$. CD is the outer surface of the outer flange with the flange equation $x^2 + y^2 = b^2$. Due to the lack of positioning features on the part surface, the part adopts the axial positioning. The manufacture accuracy of the AB surface is forced to improve during the design process so that it can become the positioning surface which can limit the movement of the outer flange in the radial direction. At the same time, in order to limit the movement of the parts in the circumferential direction, a locating slot needs to be opened under the outer flange. To achieve the best positioning quality, the width of the positioning slot and the location of the positioning point on the AB surface are calculated.

The quality of the positioning method can be determined according to the stability of positioning and the error transmission ability of positioning. For this example, we also consider these two aspects. Since the AB surface is the datum plane and has

Fig. 7.12 Flange parts

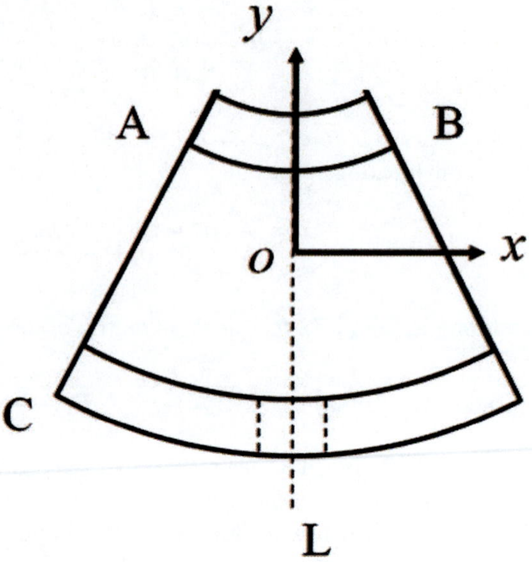

a large area, it is regarded as the main locating surface. Two locating points P1 and P2 are arranged on the *AB* surface. The whole part is symmetrical about the y-axis. Empirically, when P1 and P2 are symmetrical, the positioning is relatively stable. Thus, let the coordinates for P1 and P2 be (x, y) and $(-x, y)$, respectively. If the positioning point on the positioning slot is P3 (x_3, y_3), then the quality of the positioning method under different positioning point arrangements can be measured according to the theory of the first two sections.

According to Eq. (7.2), in this positioning mode, the positioning Jacobian matrix can be expressed as:

$$J = -\frac{1}{a}\begin{bmatrix} -x & y & -2xy \\ x & y & 2xy \\ a & 0 & ay_3 \end{bmatrix} \tag{7.21}$$

(1) Positioning stability

According to Eq. (7.7), under the positioning mode, the positioning stability parameters are:

$$w = \sqrt{\det(J^T J)} = a^2(y_3^2 + 3) + 8x^2y^2 \tag{7.22}$$

The larger the w is, the better the stability of the positioning mode is. According to Eq. (7.22), a larger y_3 will lead to a larger value of $8x^2y^2$ and hence a more stable positioning mode. Since the locating point is on the surface *AB*, the equation

$x^2 + y^2 = a^2$ is satisfied. As a result, when $x = -\frac{\sqrt{2}}{2}a$, $y = \frac{\sqrt{2}}{2}a$, $8x^2y^2$ takes the maximum value $2a^4$.

Thus, when the coordinate of P1 is $(-\frac{\sqrt{2}}{2}a, \frac{\sqrt{2}}{2}a)$, the coordinate of P2 is $(\frac{\sqrt{2}}{2}a, \frac{\sqrt{2}}{2}a)$, the larger the y_3 is, the better the stability of the positioning method is.

(2) Degree of space freedom restriction

In this way, the ill-conditioned index of the Jacobian matrix is:

$$k = \|J\| \cdot \left\|J^{-1}\right\| = 4\left[xy(y_3 - 2y)\right]^2 \tag{7.23}$$

It is thus suggested that a smaller ill-conditioned index of the positioning Jacobian matrix will lead to a smaller ill-conditioned degree of the positioning Jacobian matrix. In addition, less distortion of the positioning method in the process of error transmission will result in better positioning error transmission ability together with a higher accuracy.

When $y_3 = 2y$, the transmission ability of positioning error is the best when x and y are determined. That is, when $y_3 = \sqrt{2}a$, this positioning method is the best. Under this condition, $x_3 = \pm\sqrt{b^2 - 2a^2}$. Thus, when the width of the positioning slot is $L = 2\sqrt{b^2 - 2a^2}$, the positioning method is the best. The selected positioning method is shown in Fig. 7.13.

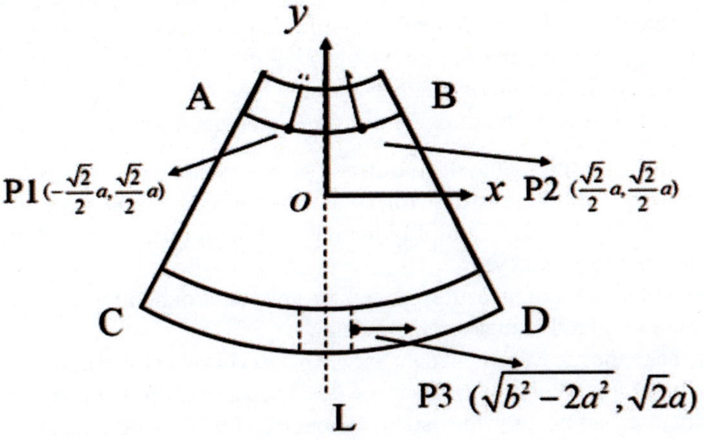

Fig. 7.13 Optimized positioning method of flange parts

7.2.2 *Specification for Measurement Points and Functional Dimensions of Complex System Casting*

In the quality inspection process for the complex casting, the concept of functional dimension (FD—Functional Dimension) was put forward and fully implemented by Volkswagen in the 1990s. The appearance of the functional dimension system brings great convenience for manufacturing dimension quality control and also provides a new idea for the optimal design of the dimension detection system. The casting function size is extracted from the general detection points of the casting, and it is an optimization of the general checkpoints.

According to the functional requirements of casting products, a series of dimensions that must be guaranteed and controlled can be put forward to ensure the quality of these dimensions and the quality of products can meet the requirements. The formation of a functional dimension is based on the information of measurement points, such that the distribution of measurement points and the distribution of functional dimensions are in pairs. The establishment of functional dimension can correspondingly reduce the number of measurement points, i.e., remove the measurement points with poor effect. In addition, it can reduce the measurement error caused by positioning error, which is completed by the mathematical calculation between measurement points. Finally, it can optimize the tolerance design of castings, organize the relevant measurement points according to their functions, and make the measurement purpose clear and the detection system clear and intuitive.

In the measurement of small samples such as complex castings, a large number of data cannot be obtained. Consequently, in order to detect and diagnose the measurement error, there must be inheritance between the measurement data. The functional dimension results in the inheritance between the measurement data. The benefits of measurement points and functional dimensioning are as follows:

i. Make the measurement standardized, at the same time, according to the list of functional dimensions, formulate the corresponding three-dimensional measurement procedure, simplify the measurement operation, and improve the measurement efficiency.
ii. The establishment of measurement points and functional dimensions can ensure the real and effective measurement data.
iii. It can make the measurement data between parts have an inheritance, and further make the implementation of accurate detection and fault diagnosis possible.
iv. Provide data support for the establishment of a failure mode database.

The functional dimension and measurement point are interrelated, and the selection of measurement points is determined according to the product function. In other words, the functional dimension can determine the layout of the measurement point. In addition, the functional dimension is expressed by measurement point, so the layout of the measurement point can optimize the corresponding functional dimension. The relationship between functional dimensions and measurement points is shown in Fig. 7.14.

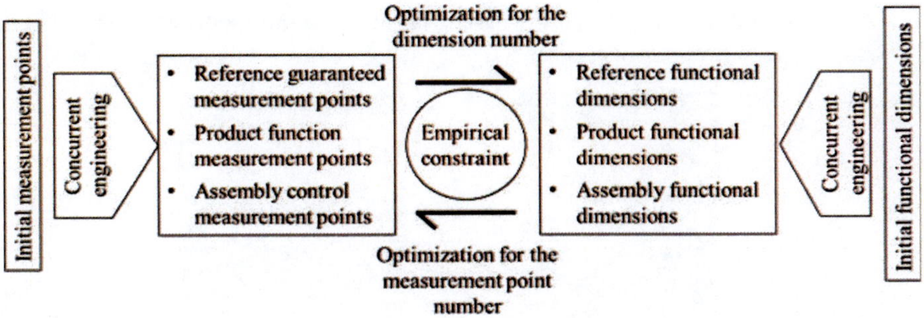

Fig. 7.14 Relationship between measurement points and functional dimensions

The targets of measurement points and functional dimensions of large and complex castings are as follows:

i. Formulate normative measurement points and functional dimension documents
ii. Unify the measurement method and formulate the corresponding CMM measurement procedure.

7.2.2.1 Selection of Measurement Information

The specific point selection method for the support plate is as follows.

Establish a feature surface with a specific radius and take points on the intersection line between the feature surface and the support plate surface, as shown in Fig. 7.15.

The specific point selection method of the flange is as follows.

Starting from the center of the circle, make the tangent plane at a fixed angle and take points on the intersection line of the tangent plane and the flange, as shown in Fig. 7.16.

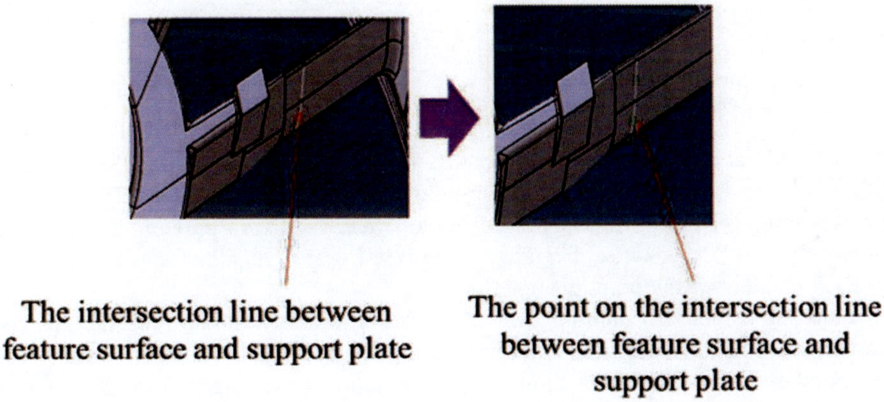

Fig. 7.15 Sampling method for measurement points of the large casting support plate

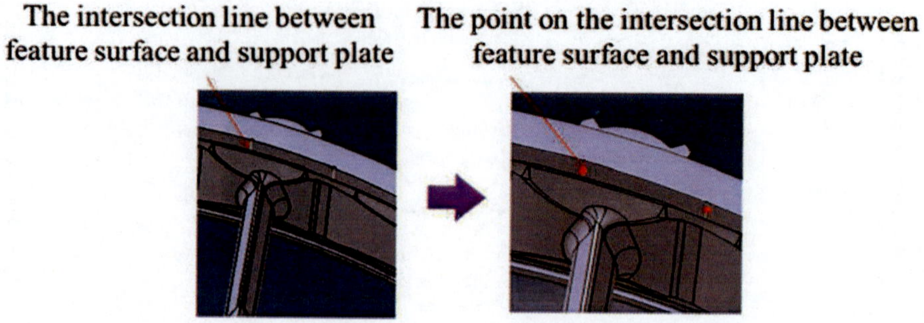

Fig. 7.16 Sampling method for flange measurement points of large castings

7.2.2.2 Naming of Measurement Points and Functional Dimensions

Each measurement point and functional dimension have corresponding names. The naming method of measurement point and functional dimension is as follows.

Nomenclature of measurement points: points in different areas of the part are represented by different numbers. The specific rules are as follows:

- Outer flange: A
- Inner flange: B
- Support plate: C
- Lifting lugs and others: D
- Outflow channel: E
- Inner runner: F.

After grouping, the parts are divided into 12 parts. Each part is replaced by capital letters A, B, …, L. Due to different features on the parts, the types of points are different. For different types of points, different letters are used. The rules are as follows:

- Hole center point: H
- The point on face: S
- Center point of threaded hole: W
- Edge point: E.

The number of points in each area starts from 001 and ends at 999. The points of lifting lugs and related feature holes are named directly without area. The order of point naming is as follows. The first lifting lug on the right is 1 part and the selection order of points is clockwise, from top to bottom. Figure 7.17 illustrates the meaning of each element of the point designation.

Function dimension naming method: Function dimension is represented by the capital letter F, different functional groups are represented by different letters, and the number of function dimensions of each function group starts from 001 to 999, with specific rules as follows.

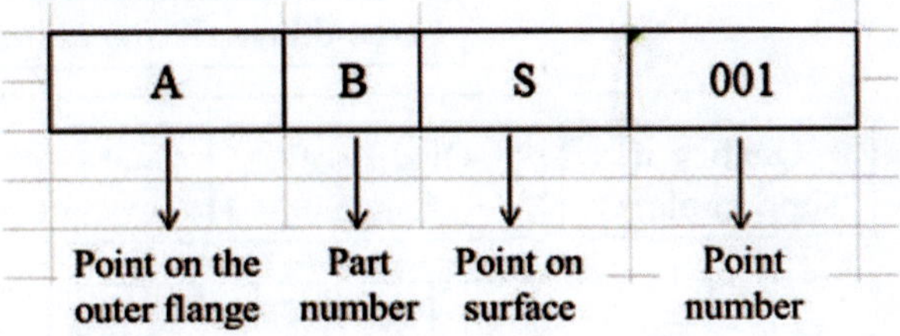

Fig. 7.17 Naming of measurement points

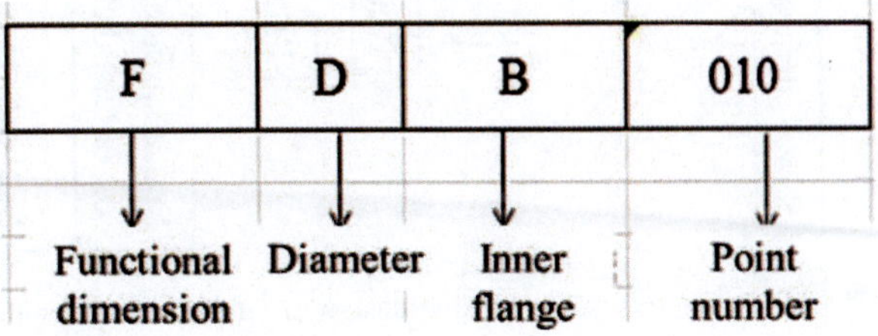

Fig. 7.18 Function dimension naming

- Diameter and related dimensions: D
- Roundness: R
- Flatness: F
- Surface profile: P
- Position: L.

For the areas where functional dimensions exist, different numbers are used to represent them. The specific rules are as follows:

- Outer flange: A
- Inner flange: B
- Support plate: C
- Lifting lugs and others: D
- Outflow channel: E
- Inner runner: F.

Figure 7.18 shows the meaning of each part of the functional dimension designation.

7.2.2.3 Design and Optimization of Measurement Points and Functional Dimensions of Large Thin-Walled Castings

According to the characteristics of large-scale, thin-walled and varying cross-section of large-scale thin-walled ring casting, combined with the functional requirements of the product and the requirements of the design drawings, the setting of the functional dimensions of the research object is shown in Fig. 7.19.

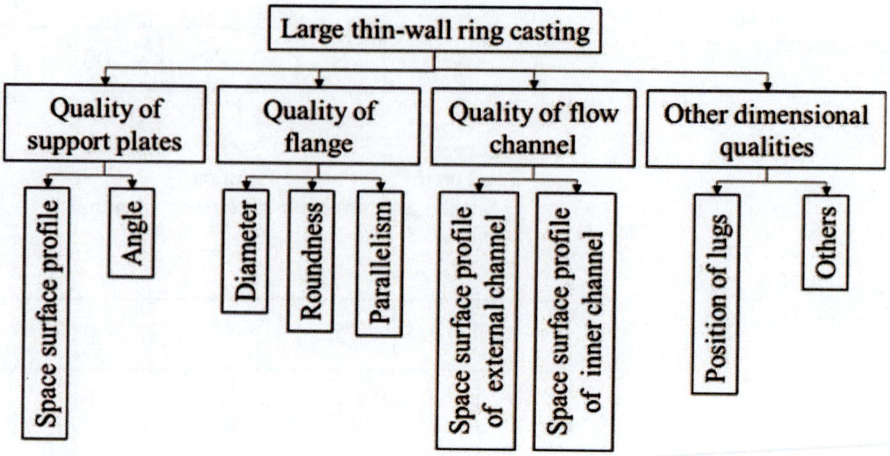

Fig. 7.19 Functional dimension design of large thin wall ring casting

According to the functional dimension design shown in the figure, combined with the actual measurement and verification of feasibility, a total of 238 sets of functional dimensions are set.

For each group of functional dimensions, there are corresponding measurement points to characterize them. Combined with the design functional dimensions of products and the design and functional requirements of products, there are 5000 initial measurement points in total. The design of the measurement points of the support plate is shown in Fig. 7.20.

Too many measurement points will lead to the reduction of measurement efficiency and the excessive loss of measurement machines. For the same type of measurement points, the number of measurement points should be reduced as much as possible in order to meet the requirements of product measurement quality. Once

Fig. 7.20 Measurement point design of the support plate

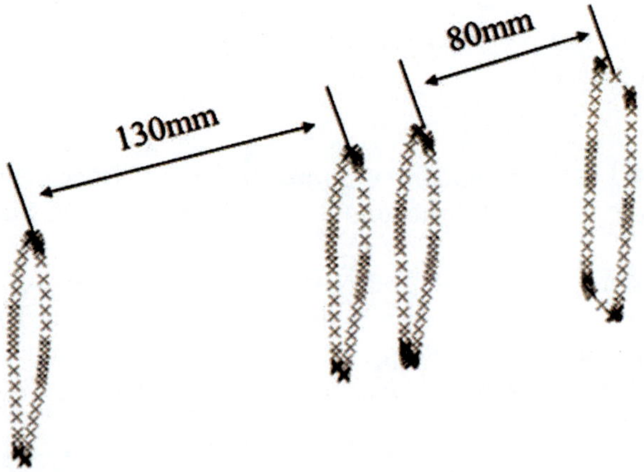

the number of measurement points is appropriately reduced, the measurement points need to be optimized. The optimization methods include correlation analysis, reliability analysis, function dimension accuracy analysis, etc. These methods have been described in detail in the literature and will not be described here. Figure 7.21 is an example of a correlation analysis of measurement points in the outer channel of the large casting support plate. The correlation of measurement points A, B, C and D is shown in Table 7.1.

It can be seen from the data in the table that the correlation between measurement point B and measurement point C is the highest, which indicates that the similarity between the two points is the highest of the four points. Because of this, either point B or point C should be deleted during the process of arranging measurement points.

Through a series of optimizations, after the initial measurement points are screened and reset, the optimization results are shown in Table 7.2. The quality information of the whole part can be learned by reading the information about the fixed measurement points.

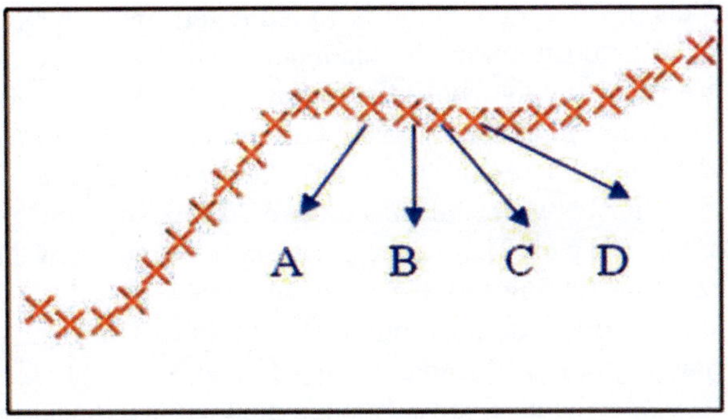

Fig. 7.21 Correlation analysis of measurement points of the outer channel

Table 7.1 Correlation analysis results of measurement points in the outer channel

	A	B	C	D
A	1			
B	0.9178	1		
C	0.9142	0.9798	1	
D	0.8500	0.9380	0.9485	1

Table 7.2 Optimization results of the number of measurement points

	Optimization of pre-measurement points	Optimized measurement point	Optimization rate (%)
Flange	156	120	23.07
Support plate	3600	2172	39.67
Runner	984	624	36.58

7.2.3 *Rapid Diagnosis of Error Sources in the Casting Process*

7.2.3.1 Rapid Diagnosis of Error Mode in the Casting Process

In the production and assembly process of ring parts, the parts are usually clamped and positioned according to the 3-2-1 principle. The traditional 3-2-1 positioning method usually follows the following principles:

i. Three locating points are selected in the normal direction of the largest plane of the part, and the distribution of the three points is scattered as far as possible;
ii. Select two locating points in the second plane of the part, and the distance between the two points should be as long as possible;
iii. Select a locating point in the minimum plane of the part.

For a workpiece, when a set of positioning scheme is determined, all the positioning points will not move and rotate when they are in contact with the fixture. This positioning method is called deterministic positioning. On the basis of deterministic positioning, the Jacobian matrix of the positioning method can be obtained by analyzing the constraint equation of the positioning point and then deriving the constraint equation. After obtaining the Jacobian matrix, through the Taylor series expansion, a series of mathematical processing is carried out to clarify the relationship between the deviation of the positioning point and the deviation of the measured part. That is, the positioning deviation of the part caused by the deterministic positioning mode can be calculated. The deviation analysis method based on the deterministic positioning is called the deterministic positioning deviation analysis algorithm. Because of its calculation simplicity, high precision, and that fact that it can be used to analyze the dimensional deviation of the positioning process of complex parts, the deterministic positioning deviation analysis algorithm has become a common algorithm in the field of deviation analysis, which has been widely used in the research of assembly quality control problems in the fields of automobile, aircraft, high-speed train, and other mechanical fields.

For the ring parts, they usually have the characteristics of rotational symmetry. In general, axial and circumferential positioning will be used for positioning, which can match and confirm the parameters given in the design drawings well, and at the same time more conform to the requirements of the structure, so the design of positioning fixture is relatively simple. The positioning method of general annular parts is shown in Fig. 7.22.

In Fig. 7.22, the main positioning surface is perpendicular to the paper surface, controlling the movement of the part in the Z direction and the rotation in X and Y direction. The reference surface controls the movement of the part in X and Y directions, which is actually equivalent to limiting the axial movement of the part. The positioning slot controls the rotation of the part around the Z-axis, which is actually equivalent to limiting the circumferential movement of the part. Due to the special structure of the ring part and the selection of the positioning method, in order

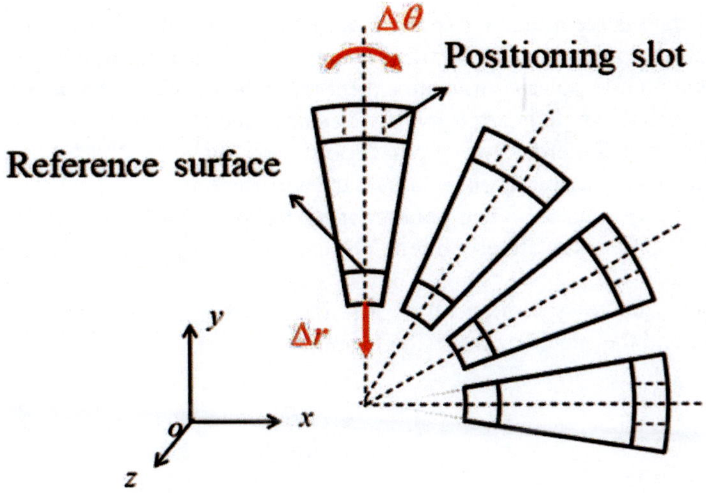

Fig. 7.22 Positioning method of annular parts

to meet the quality accuracy requirements of the part after forming, the positioning process should pay special attention to the axial movement and the circumferential rotation of the part. The axial movement of the part will affect the accuracy of the inner circle diameter and the roundness of each circle after forming, while the angular rotation will affect the coaxiality after forming.

Generally speaking, dimensions can be divided into linear and angular dimensions. For the ring parts, because of the particularity of its rotational symmetry structure, the angular dimension has a great influence on its assembly quality accuracy, while the deterministic positioning deviation analysis algorithm is based on the Cartesian coordinate system, describing the state of linear dimension, which cannot directly and clearly represent the state of the angular dimension of the parts. Even if the linear dimension and the angular dimension can be transformed into each other, the calculation results will deviate due to the coupling of the coordinate system and the calculation process will be relatively complex.

The main content of this section is to solve the problem of how the traditional deterministic positioning deviation analysis algorithm can represent more clearly the angular dimension with better application to the ring parts [11].

1. The establishment of the diagnosis and analysis model of definite positioning deviation in the two-dimensional polar coordinate system

The coordinate system composed of poles, polar axis, and polar diameter in the plane is called the polar coordinate system. A point O in the plane is called the pole. A ray Ox from O is called the polar axis. Then take a length unit, usually, the specified angle is positive in the counterclockwise direction. In this way, the position of any point P in the plane can be determined by the length ρ of the line OP and the angle θ from OX to OP. If there is an ordinal pair (ρ, θ), it is called the polar coordinate of

P point, which is recorded as P (ρ, θ). ρ is called as the polar diameter of P point, θ is called as the polar angle of P point. When $\rho \geq 0, 0 \leq \theta < 2\pi$, every point in the plane has a unique polar coordinate except the pole O, where the pole diameter is zero and the pole angle is arbitrary. If we remove the above limitation, every point on the plane has numerous sets of polar coordinates. In general, if (ρ, θ) is the polar coordinate of a point, then $(\rho, \theta + 2n\pi)$ is the corresponding polar coordinate, where n is an arbitrary integer. When polar coordinates are used for some curves on the plane, the equation is relatively simple. The calculation will also be relatively simple and fast.

Polar coordinate system and the Cartesian coordinate system can be mutually converted, and the specific conversion formula is

$$\begin{cases} x = \rho\cos\theta \\ y = \rho\sin\theta \end{cases} \quad (7.23)$$

In order to establish the model of definite positioning analysis in the polar coordinate system, the vector representation in the polar coordinate system must be definite. As shown in Fig. 7.23, to represent vector AB in the polar coordinate system, we can formulate it according to the basic mathematical relationship:

$$\overrightarrow{AB} = \overrightarrow{AC} + \overrightarrow{CB} \quad (7.24)$$

Fig. 7.23 Vector representation in polar coordinates

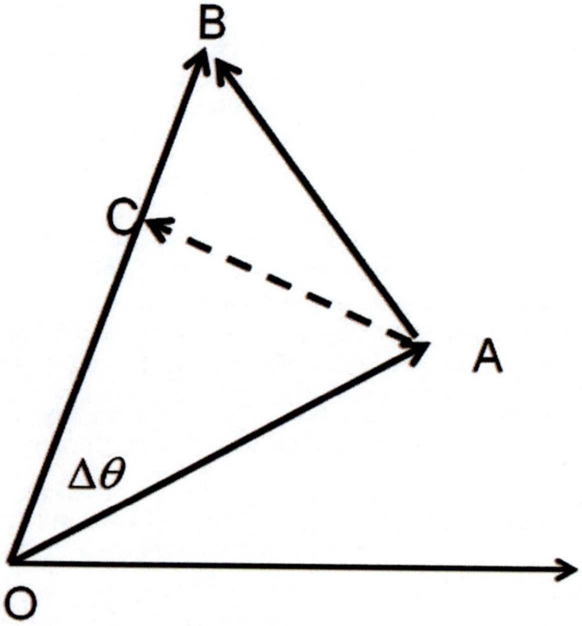

According to the triangle relationship:

$$\overrightarrow{AC} = \overrightarrow{OA} \cdot \Delta\theta \cdot \overrightarrow{\theta} \tag{7.25}$$

$$\overrightarrow{CB} = \overrightarrow{OA} - \overrightarrow{OB} = \Delta\rho \cdot \overrightarrow{\rho} \tag{7.26}$$

Then the vector AB can be expressed as:

$$\overrightarrow{AB} = \overrightarrow{OA} \cdot \Delta\theta \cdot \overrightarrow{\theta} + \Delta\rho \cdot \overrightarrow{\rho} \tag{7.27}$$

According to the above four formulas:

$$\begin{cases} N_{i\rho} = \Delta\rho \cdot \overrightarrow{\rho} \\ N_{i\theta} = \overrightarrow{OA} \cdot \Delta\theta \cdot \overrightarrow{\theta} \end{cases} \tag{7.28}$$

It can be seen from Eq. (7.28) that the representation of vector AB in polar coordinates is related to the coordinates of its endpoints A and B.

Through the transformation of this coordinate form, the expression of the positioning normal vector of the analysis algorithm of definite location deviation in polar coordinates can be obtained. This is also the basis of the analysis model of the definite positioning deviation in the polar coordinate system.

2. The establishment and solution of the algorithm model of definite positioning deviation analysis in the two-dimensional polar coordinate system

(1) Model building

For ring parts, the drawing size is mostly represented by length and angle, so the deviation direction of most key control dimensions is the radius, and the independent variables and control variables of the deterministic positioning analysis model established in a Cartesian coordinate system do not have a one-to-one correspondence. It needs a lot of work on error decomposition and synthesis when solving. This process involves a lot of coordinate transformations. Because of the traditional rounding data processing method, the rounding error [3, 4] will be brought in the calculation process. For this type of problem, the performance of the traditional deterministic algorithm is not good either in terms of calculation accuracy or calculation complexity. In contrast, the use of the polar coordinate system will make deviation calculation more intuitive, which can solve the above problems well.

The specific polar coordinate introduction process is as follows.

As shown in Fig. 7.24, the coordinate system oxy is the global coordinate system and the coordinate system $o'x'y'$ is the local coordinate system.

For location point i: $r_i' = (x_i', y_i')r' = (x', y')$, where (x_i', y_i') is the coordinate of the point i in the coordinate system $o'x'y'$, n_i^T is the normal vector of the curve at the location i, then the tangent equation at the location i is:

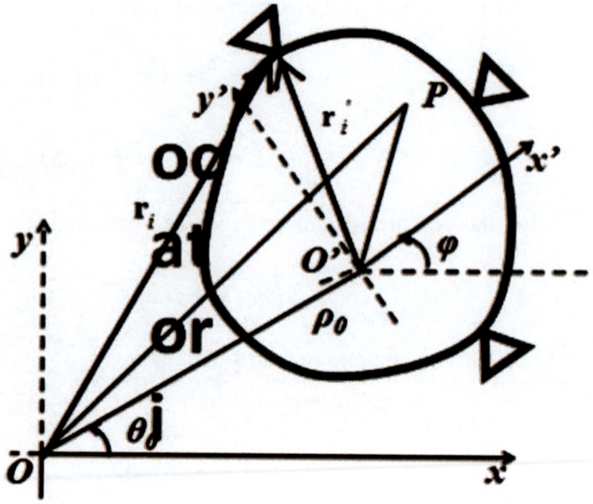

Fig. 7.24 Polar coordinate system transformation and part deterministic positioning model

$$n_i^T(r' - r_i') = 0 \tag{7.29}$$

Let ox denote the original polar coordinate system, $o'x'$ denote the new polar coordinate system. Then the coordinate of pole o' in system ox is (ρ_0, θ_0). The coordinate of any point P in system $o'x'$ is (ρ', θ'). The coordinate in system ox is (ρ, θ). The angle between the two polar axes is φ. The coordinate of point P in system oxy is $(x, y) = (\rho\cos\theta, \rho\sin\theta)$. The coordinate in system $o'x'$ is $(x', y') = (\rho'\cos\theta', \rho'\sin\theta')$. The coordinate of point o' is $(x_0, y_0) = (\rho_0\cos\theta_0, \rho_0\sin\theta_0)$.

It is easy to know from geometric relations that

$$\begin{cases} x = x'\cos\varphi - y'\sin\varphi + x_0 \\ y = x'\sin\varphi - y'\cos\varphi + y_0 \end{cases} \tag{7.30}$$

The transformation formula of polar coordinates is obtained as

$$\begin{cases} \rho'\cos\theta' = \rho\cos(\theta + \varphi) - \rho_0\cos(\theta_0 + \varphi) \\ \rho'\sin\theta' = \rho\sin(\theta - \varphi) - \rho_0\sin(\theta_0 - \varphi) \end{cases} \tag{7.31}$$

The polar coordinates of 6–29 are treated as

$$n_i'^T = \begin{pmatrix} \rho'\cos\theta' - \rho_i'\cos\theta_i' \\ \rho'\sin\theta' - \rho_i'\sin\theta_i' \end{pmatrix} = 0 \tag{7.32}$$

Substituting (7.31) into (7.32) we can obtain the spatial motion equation of workpiece positioning point in the polar coordinate system from moving coordinate system to the fixed coordinate system:

$$\phi(i) = \boldsymbol{n}_i^{\prime T}\begin{pmatrix} \rho\cos(\theta+\varphi) - \rho_0\cos(\theta_0+\varphi) - \rho_i'\cos\theta_i' \\ \rho\sin(\theta-\varphi) - \rho_0\sin(\theta_0-\varphi) - \rho_i'\sin\theta_i' \end{pmatrix} = 0 \tag{7.33}$$

In the case of 3-D, only three points of positioning are needed to meet the positioning requirements. Then the total positioning constraint equation is:

$$\phi(q) = (\phi_1, \phi_2, \phi_3) \tag{7.34}$$

Let $R_i = (\rho_i, \theta_i)$ and $R = (R_1, R_2, R_3)$, then formula (7.10) can be written as

$$\phi(q, R) = 0 \tag{7.35}$$

Equation (7.35) is a deterministic positioning model in a polar coordinate system. It describes the relationship between positioning point deviation and assembly deviation. Given positioning deviation ΔR, assembly deviation Δq can be known. Here $\Delta\rho$ is the deviation of the polar direction, representing the radius deviation.

The total differential form of Eq. (7.35) is:

$$\frac{\partial\phi}{\partial q}\Delta q + \frac{\partial\phi}{\partial R}\Delta R = 0 \tag{7.36}$$

Let $\frac{\partial\phi}{\partial q} = J$

$$\Delta q = -J^{-1}\frac{\partial\phi}{\partial R}\Delta R \tag{7.37}$$

J is Jacobian matrix, $J_i = \left[\frac{\partial\phi_i}{\partial\rho}, \frac{\partial\phi_i}{\partial\theta}, \frac{\partial\phi_i}{\partial\varphi}\right] J = (J_1, J_2, J_3)$, and $\phi_R = (\frac{\partial\phi_i}{\partial\rho_i}, \frac{\partial\phi_i}{\partial\theta_i})$, the derivation of each parameter can be obtained by ϕ_i:

$$\frac{\partial\phi_i}{\partial\rho_0} = -N_{i\rho}\cos(\theta_0-\varphi) - N_{i\theta}\sin(\theta_0-\varphi) \tag{7.38}$$

$$\frac{\partial\phi_i}{\partial\theta_0} = N_{i\rho}\rho_0\sin(\theta_0-\varphi) - N_{i\theta}\rho_0\cos(\theta_0-\varphi) \tag{7.39}$$

$$\begin{aligned}\frac{\partial\phi_i}{\partial\varphi} &= N_{i\rho}[\rho\sin(\theta-\varphi) - \rho_0\sin(\theta_0-\varphi)] \\ &\quad + N_{i\theta}[\rho_0\cos(\theta_0-\varphi) - \rho\cos(\theta-\varphi)]\end{aligned} \tag{7.40}$$

$$\frac{\partial\phi_i}{\partial\rho} = N_{i\rho}\cos(\theta-\varphi) + N_{i\theta}\sin(\theta-\varphi) \tag{7.41}$$

$$\frac{\partial\phi_i}{\partial\theta} = -N_{i\rho}\rho\sin(\theta-\varphi) + N_{i\theta}\rho\cos(\theta-\varphi) \tag{7.42}$$

Substituting the formulas above into Eq. (7.37), we can get the positioning assembly deviation of the parts after the definite positioning. Assuming that the manufacturing deviation of the parts is $\Delta q'$, the assembly deviation of the parts is $S = \Delta q + \Delta q' = -J^{-1}\frac{\partial \phi}{\partial R}\Delta R + \Delta q'$. The polar radius and polar angle are used as parameters to express the model.

(2) Error analysis

When the key control dimension is the radial direction deviation, the flow of the traditional deterministic analysis method and the improved deterministic analysis method in the assembly of the ring parts is shown in Figs. 7.25 and 7.26, respectively.

It can be seen from the figure that the key control dimensions on the design drawing of ring casting are mostly characterized by length and angle. In this section, the polar

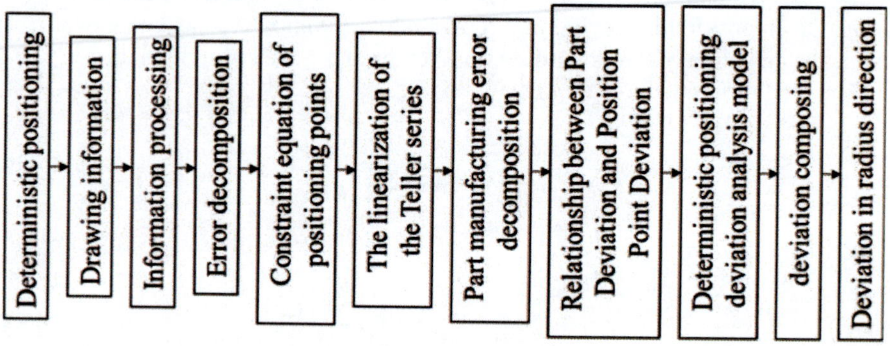

Fig. 7.25 Traditional deterministic positioning algorithm for wax mold assembly process of ring casting

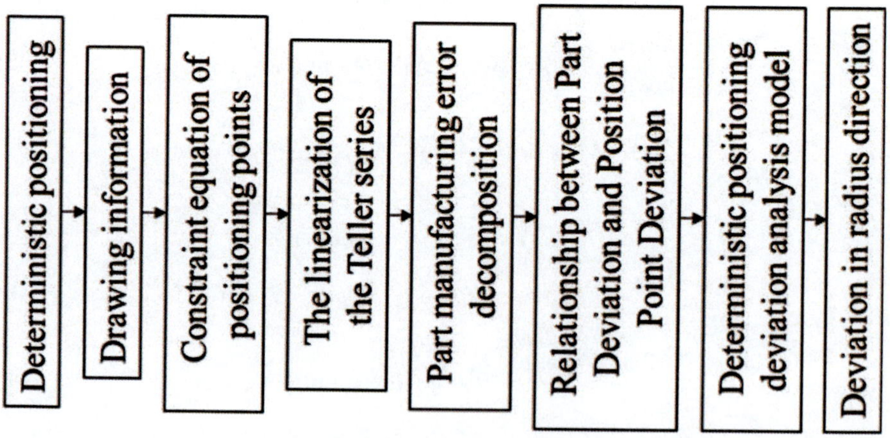

Fig. 7.26 Deterministic positioning algorithm for wax mold assembly process in polar coordinates

coordinates are introduced into the deterministic positioning deviation analysis algorithm to reduce the steps of drawing information processing, error decomposition and synthesis, which makes the calculation process simpler with higher efficiency. It can be seen from the literature that the rounding error generated in the calculation process has a geometric progression growth relationship with the calculation steps. Because of the simplification of the calculation, the improved deterministic positioning deviation analysis algorithm effectively reduces the rounding error caused by rounding in the calculation. In this section, the introduction of polar coordinates in the deterministic positioning analysis model for the wax mold assembly will reduce or even eliminate the influence of rounding errors caused by rounding on the accuracy of calculation results in the calculation process of wax mold assembly deviation analysis.

If the maximum absolute error in the calculation is Δ (the relative error is δ) and the error introduced in the data processing is Δ_k (the relative error is δ_k), then Δ_k (δ_k) must be a micro error of Δ (δ) to ensure that the error caused by data processing does not significantly affect the measurement error.

As mentioned above, the error in data processing is caused by rounding. As a result, its maximum absolute error is half of the last significant digit of the calculation result. If the calculation result A has a total of m significant digits, and the number power of the first digit is 0, then the number power of the last digit is $(-m + 1)$, then the maximum relative error δ_k introduced in data processing is:

$$\delta_k = 0.5 * 10^{(-m+1)}/A \tag{7.43}$$

The first number of A is usually between 1 and 9, so the fluctuation range of δ_k caused by rounding is $\left[0.5 * 10^{-m}, 0.5 * 10^{(-m+1)}\right]$.

It can be seen from the literature that the rounding error generated by rounding in the calculation process has a geometric progression growth relationship with the calculation steps, so the relative error of N calculation steps is:

$$\delta_{kn} = 1 - (1 - \delta_k)^n \tag{7.44}$$

Because the precision of large thin-walled ring casting is not higher than CT6, according to the casting precision class table, we can know that $m = 2$, which suggests that the range of relative error caused by rounding is between 0.5 and 5%. The accuracy of the model in this section can be increased by 5–10% compared with that of the traditional deterministic positioning analysis model, which is very important for precision casting.

3. Case study and validation

As shown in Fig. 7.27, the positioning points P1 and P2 control the radial movement of the parts, and the positioning point P3 controls the rotation of the parts. The positioning deviation of P1 and P2 is 0.2 mm. Along the radius direction, the coordinates of the positioning points are given in the figure. Under different steering errors, the radial deviation and steering deviation of MP is calculated.

Fig. 7.27 Simple part positioning model

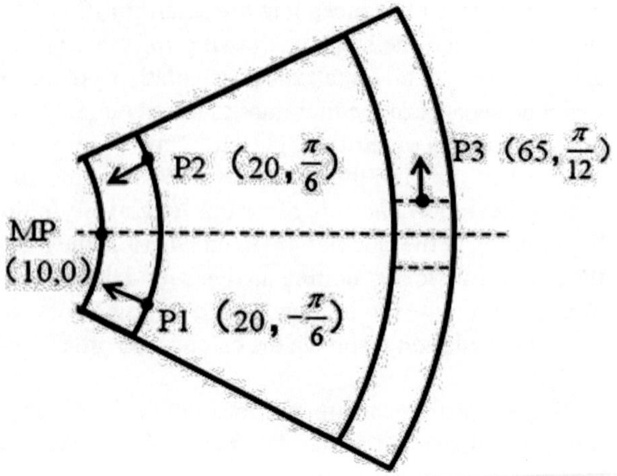

The positioning Jacobian matrix J of parts in Cartesian coordinates and polar coordinates are as follows:

$$J_C = \begin{bmatrix} 0.866 & 0.5 & 0 \\ 0.866 & -0.5 & 17.32 \\ 0 & -1 & 62.785 \end{bmatrix} \tag{7.45}$$

$$J_P = \begin{bmatrix} -1.1547 & 0 & -11.4943 \\ -1.1547 & 0 & 11.4943 \\ 0.0046 & 0.1654 & 0.7929 \end{bmatrix} \tag{7.46}$$

When the angular deviation is 0.2 rad, the deterministic positioning deviation matrix obtained in the polar coordinate system is:

$$\Delta q = \begin{bmatrix} -0.0866 & -0.0866 & 0 \\ 0.0441 & -0.0393 & 0.1163 \\ -0.0087 & 0.0087 & 0 \end{bmatrix} \tag{7.47}$$

The columns of the matrix represent the influence of location points P1, P2, and P3 on the positioning deviation of parts, and the rows of the matrix represent the influence of location points on the parameters of ρ, θ, φ. It can be seen from the definite positioning deviation matrix that the deviation of location point P1 and location point P2 has the same effect on the radial positioning deviation of the part. In addition, the deviation is smaller than that of the location point itself, which proves that the errors offset each other in the process of positioning error transmission. It also proves the correctness of the selection of the main positioning surface. When the radial accuracy of the parts is high, the accuracy requirements of the main positioning point P1 and P2 should be paid more attention.

Table 7.3 Calculation results of the conventional model and the proposed model under different tolerances

	Results\Deviation	0.02	0.04	0.06	0.08	0.1
Conventional model	$\Delta\rho$	−0.173	−0.173	−0.173	−0.173	−0.173
	Δ_θ	0.1155	0.2368	0.3513	0.4687	0.5853
The proposed model	$\Delta\rho$	−0.173	−0.173	−0.173	−0.173	−0.173
	Δ_θ	0.1167	0.2374	0.3536	0.4699	0.5862

The deviation of positioning point P3 has no influence on the radial direction of the part but has a great influence on the circumferential direction of the part, which is related to the control function of the positioning point itself. When the accuracy of the circumferential direction of the part is high, the accuracy requirements of secondary positioning point P3 should be paid more attention.

From Table 7.3, it can be seen that the calculation results of the traditional model and the proposed model in this section are basically the same, which verifies the correctness of the proposed model. Meanwhile, from the error rate, it can be seen that when using the deterministic positioning deviation analysis model under the polar coordinate system to analyze the ring parts, the calculation accuracy has been improved. The efficiency has also been improved correspondingly in the calculation process.

Figure 7.28 is the assembly of a typical flange. Take wax block 1 as the research object. The radius of the inner flange is 1 dm, the radius of the outer flange is 8 dm, the polar axis is *ox* axis, and the whole flange is assembled by four wax blocks. P1, P2, and P3 are positioning points with coordinates of $(2, \pi/6)$, $(2, -\pi/6)$, $(6, \pi/12)$, respectively. P1 and P2 control the centering direction, whereas P3 controls the

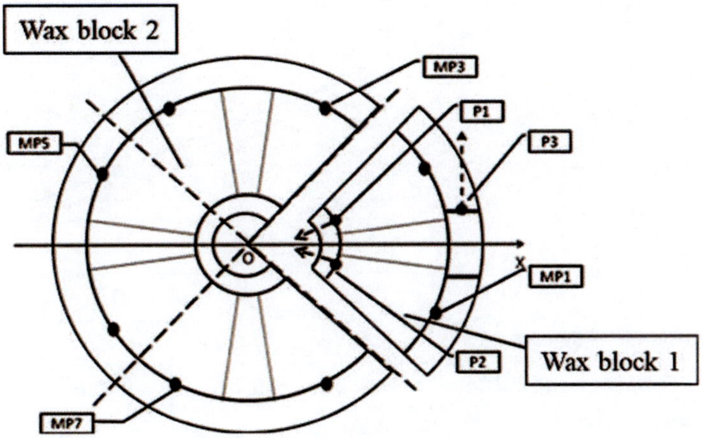

Fig. 7.28 Flange assembly model

direction perpendicular to the polar axis. MP is the measurement point and the corresponding coordinate is $(4, \pi/6)$. The assembly deviation of measurement points in the direction of the flange radius is solved by the method of deterministic positioning deviation analysis. Three coordinate measurement points of wax block 1 under different tolerances are given in Table 7.4.

According to the model in this paper, the corresponding Jacobian matrix $J = \begin{bmatrix} -1.47 & 2.9064 & 1.0936 \\ -1.47 & -2.9069 & -1.0931 \\ -0.5594 & 10.6515 & -7.55 \end{bmatrix}$

Figure 7.29 shows the comparison of the calculation results of the positioning deviation of this model and the traditional deterministic positioning analysis model using different tolerances. The data show that the calculation results of this model are more accurate than those of the traditional deterministic positioning algorithm and are close to the three coordinate measurement values. It can be seen from the calculation results that the larger the positioning error is, the larger the assembly deviation is.

Table 7.4 Three coordinate measurement points of the wax block

	X (Tolerance 0.2)	Y (Tolerance 0.2)	Z (Tolerance 0.2)	X (Tolerance 0.5)	Y (Tolerance 0.5)	Z (Tolerance 0.5)
1	−666.929	392.220	−430.732	−668.098	396.291	−431.062
2	−648.038	260.208	−430.781	−648.067	263.678	−431.109
3	−657.811	159.329	−430.853	−658.141	157.799	−431.185

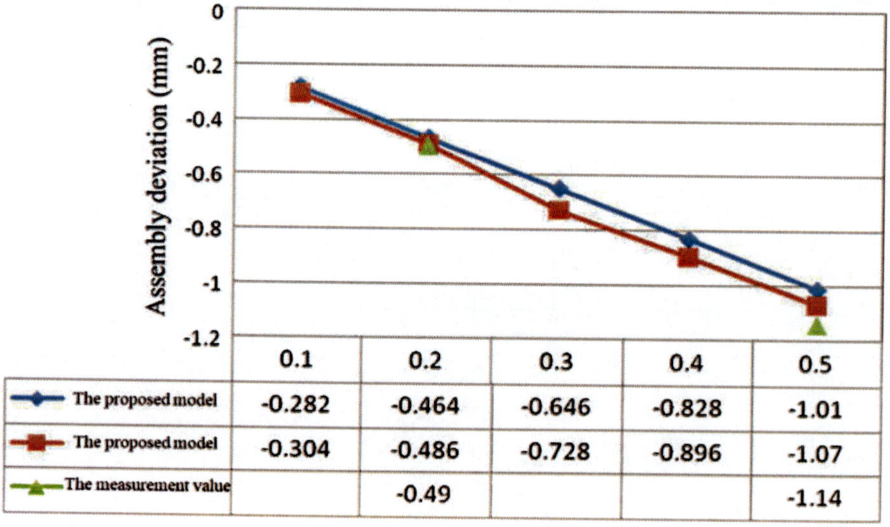

Fig. 7.29 Comparison of assembly deviation results of different models

7.2.3.2 Case Studies for the Rapid Diagnosis of Large Casting Error Source

1. Optimal selection of wax pattern location for large castings

The bottom and top views for a large thin-walled ring casting wax block are shown in Fig. 7.30.

It can be seen that due to the large dimension and thin wall, it is difficult to find suitable positioning features to limit the freedom of wax block. In this case, two schemes can be adopted for positioning. One is that due to the high precision of the outer plane design of the inner flange, according to the "6-6-1" positioning principle, two positioning points are arranged on the outer plane of the inner flange to limit the radial freedom of the wax block. In the aspect of freedom limitation in the circumferential direction, it can be located through the positioning slot. The other scheme is to use the outer flange section as the main locating surface and use the assembly relationship between the parts to limit the circumferential degrees of freedom, and then set a locating point on the outer plane of the inner flange to limit the radial degrees of freedom of the wax block. Both of them are feasible in practice. Next, through the quantitative representation of positioning, the stability and error transferability of the two positioning methods will be compared and analyzed, and the better positioning method will be selected. The two positioning methods are shown in Fig. 7.31. Information of the two positioning methods is listed in Table 7.5.

In order to compare the quality of positioning methods, the stability parameters and pathological index of positioning Jacobian matrix need to be calculated. According to the information on positioning points given in Table 7.4, the results can be calculated by combining Eqs. 7.22 and 7.23. The specific calculation results are listed in Table 7.5.

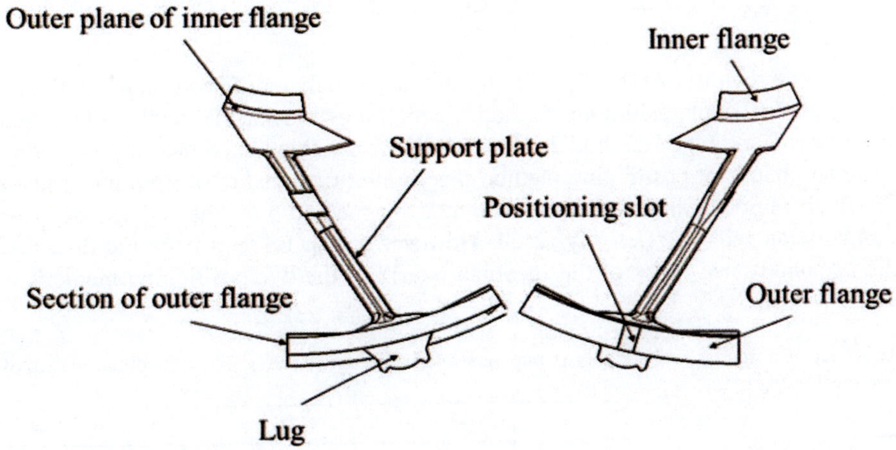

Fig. 7.30 Bottom view and top view of large thin wall ring casting

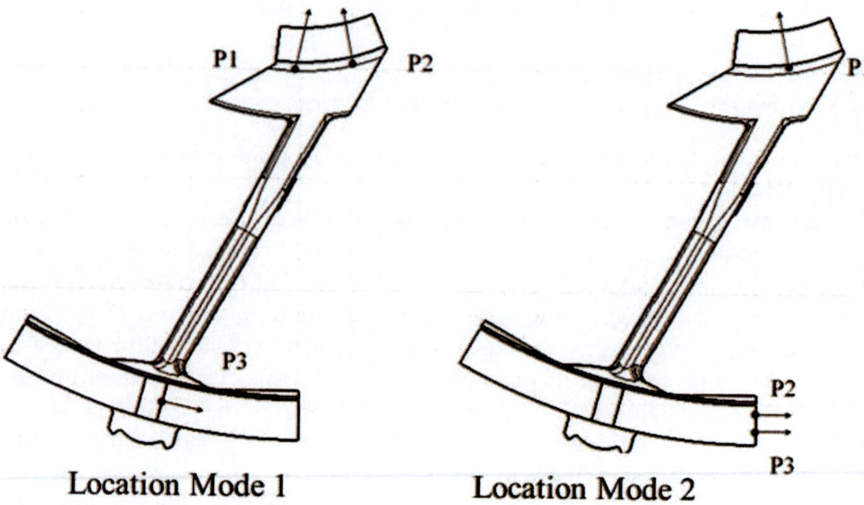

Fig. 7.31 Two positioning methods

Table 7.5 Positioning information of two positioning methods

	Locating points	x	y	n
Positioning method 1	P1	−137.177	137.177	(0.7071, 0.7071)
	P2	137.177	137.177	(−0.7071, 0.7071)
	P3	−120.63	500	(0.9659, −0.2588)
Positioning method 2	P1	0	194	(−0.2588, 0.9659)
	P2	−20	480	(1, 0)
	P3	−20	520	(1, 0)

It can be seen from Table 7.6 that the stability parameter of positioning method 1 is larger than that of positioning method 2, which suggests that the stability of method 1 is better than that of method 2. Meanwhile, the pathological index of positioning Jacobian matrix of positioning method 1 is smaller than that of positioning method 2, which indicates that the error generated by method 1 in the process of error transmission will be relatively small. However, it can be seen from the data that the ill-conditioned index of the Jacobian matrix of the two positioning methods is

Table 7.6 Comparison of stability parameters and pathological index of the Jacobian matrix of two positioning methods

	w	k
Positioning method 1	23.147	50.332
Positioning method 2	14.927	50.515

virtually the same, which also explains why both methods are commonly used in practice. According to the comparison of the two indexes, positioning method 1 is better than positioning method 2.

In order to verify the correctness of the conclusion, this section uses the commercial deviation analysis software VSA to analyze the deviation of the radius on the wax block flange of the two positioning methods. In this case, the other information such as the deviation of the measurement point, the deviation of the fixture and so on are consistent. The specific modeling process and information are shown in Fig. 7.32, and the deviation analysis results of the two cases are shown in Table 7.7.

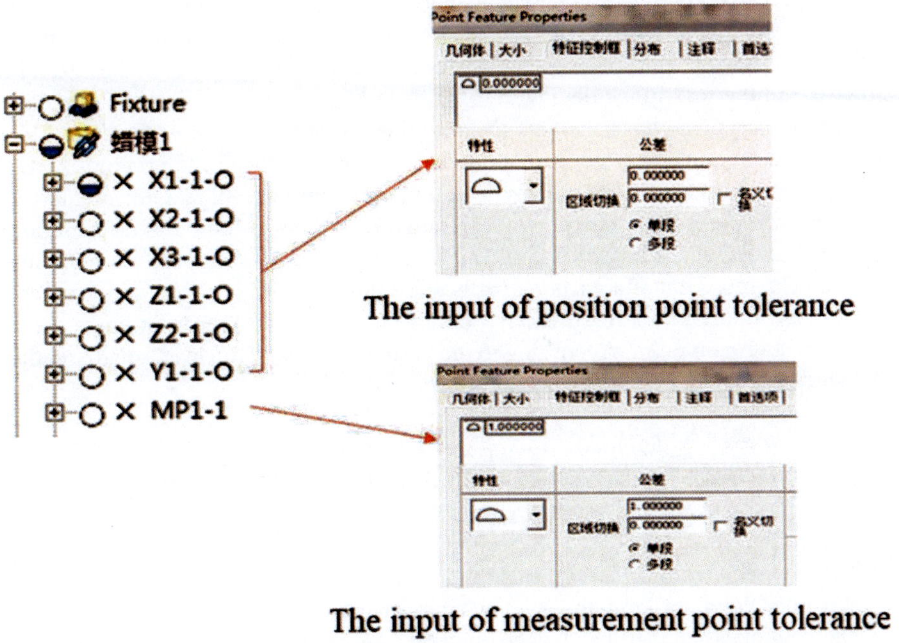

Fig. 7.32 VSA deviation analysis model and related information

Table 7.7 VSA analysis results of two positioning methods

	Object	Mean	±3σ
Positioning method 1	Inner flange radius	194	±0.427
	Outer flange radius	536	±0.702
	Circle radius of the support plate	251	±0.533
Positioning method 2	Inner flange radius	194	±0.828
	Outer flange radius	536	±1.0818
	Circle radius of the support plate	251	±0.9903

It can be seen from the results in Table 7.7 that the positioning deviation caused by positioning method 1 in the assembly process is less than that of positioning method 2, which confirms the correctness of the current conclusions. It can also be seen from this example that in the process of optimal selection of positioning mode, not only the limitation of positioning points on degrees of freedom but also the stability of positioning method is very important, which must be considered.

2. Shape and position tolerance evaluation of the wax block for large castings

(1) Roundness evaluation of the wax mold for large castings

Roundness error is caused by the assembly inconsistency between wax blocks. There are two main reasons for roundness deviation in the process of casting flange assembly. They are the spacing inconsistency between wax blocks in the assembly process and the wax block deviation. Therefore, in order to calculate the roundness deviation of castings, it is necessary to determine the deviation value of the wax block where each measurement point is located. It can be seen from the previous section that in order to obtain the roundness deviation of the research object, the radial deviation of each point must be known. The radial deviation of the measurement point can be obtained according to Eq. (7.37). The fixture positioning error is ±0.5 mm, and the wax mold manufacturing error is ±0.2 mm.

The deviation of the measurement point is considered into the coordinates of the ideal measurement point. According to the basic elements of the least square method of roundness, it can be calculated as follows:

$$\begin{cases} a = \frac{2}{60}\sum(\rho + \Delta\rho_i + \Delta\rho_i^{'})\cos\theta_i = 1.178 \\ b = \frac{2}{60}\sum(\rho + \Delta\rho_i + \Delta\rho_i^{'})\sin\theta_i = 0.581 \\ \rho_0 = \frac{1}{n}\sum(\rho + \Delta\rho_i + \Delta\rho_i^{'}) = 1.533 \end{cases}, (1 \le i \le 60) \tag{7.48}$$

It can be seen from Eq. (7.41) that the roundness error of the least square method for the outer flange of large thin-walled ring casting is 1.643.

The roundness deviation of each method can be obtained by setting different search nodes, and then the roundness of root mean square can be calculated to obtain the roundness value of large-scale thin-walled ring casting. The calculation results of each roundness of different search nodes are shown in Table 7.8. It can be seen from the data in the table that there is no significant difference between the roundness calculation results of the research object and the actual three-coordinate measurement value using the roundness error search optimization algorithm given in this section, which shows that the current model can make an accurate prediction of the roundness error for the large thin-walled ring casting and provide a reliable error source diagnosis.

According to the accuracy control requirements of roundness, the measurement points should be optimized accordingly. When the number of different measurement points is selected, the calculation results of the relative error value of roundness error are shown in Table 7.9.

The relative error value of roundness is defined as:

Table 7.8 Roundness calculation results of the outer flange of large thin-walled ring casting

Search points	Appraising method	Roundness error	Root roundness of mean square	Three coordinate measurement value
$s = 3$	Least square method	1.643	1.6493	1.657
	Minimum area method	1.561		
$t = 100$	Minimum circumscribed circle method	1.724		
	Maximum inscribed circle method	1.665		
$s = 4$	Least square method	1.643	1.6501	
	Minimum area method	1.582		
$t = 200$	Minimum circumscribed circle method	1.702		
	Maximum inscribed circle method	1.653		

$$U = (f_{act} - f_{thero})/f_{thero} \times 100\% \tag{7.49}$$

where U represents the relative error value of roundness, f_{act} represents the actual value of roundness, and f_{thero} represents the theoretical value of casting roundness.

From Table 7.10 and Fig. 7.33, it can be seen that when the number of measurement points is between 20 and 30, the increase of the measurement accuracy is the fastest, with the relative error decreasing to about 10%. Therefore, when measuring the roundness, the number of measurement points can be reduced to between 20 and 30 correspondingly. This provides a basis for the optimization of roundness measurement points for the large ring castings.

(2) Evaluation of angle distribution uniformity of support plate for the wax block

Theoretically, the support plates of the large thin-walled ring casting should be completely and evenly distributed, which is beneficial to the rotational and aerodynamic properties of the casting. However, in the process of the wax mold assembly, due to the errors in the assembly process and due to the influence of human factors, the angular deviation of the support plate is inevitable. This kind of deviation has a great influence on the quality of parts, which needs to be avoided as much as possible. To evaluate the angular deviation of the support plate of the wax mold, we can use the parallelism error evaluation model in this paper. The following is the specific process of model analysis to locate the angular deviation of parts.

Table 7.9 Measurement point information and radial deviation of the outer flange

	X	Y	Z	Radial deviation
1-1	−267.15	462.717	20.917	1.722
1-2	−275.35	476.92	25	1.763
1-3	−283.55	491.123	16.35	1.746
6-1	0	534.3	20.917	1.831
6-2	0	550.7	25	1.822
6-3	0	567.1	16.35	1.859
6-1	267.15	462.717	20.917	1.542
6-2	275.35	476.92	25	1.648
6-3	283.55	491.123	16.35	1.562
4-1	462.717	267.15	20.917	1.73
4-2	476.92	275.35	25	1.77
4-3	491.123	283.55	16.35	1.892
5-1	534.3	0	20.917	1.886
5-2	550.7	0	25	1.873
5-3	567.1	0	16.35	1.692
6-1	462.717	−267.15	20.917	1.707
6-2	476.92	−275.35	25	1.724
6-3	491.123	−283.55	16.35	1.706
7-1	267.15	−462.717	20.917	1.628
7-2	275.35	−476.92	25	1.652
7-3	283.55	−491.123	16.35	1.593
8-1	0	−534.3	20.917	1.556
8-2	0	−550.7	25	1.548
8-3	0	−567.1	16.35	1.637
9-1	−267.15	−462.717	20.917	1.558
9-2	−275.35	−476.92	25	1.637
9-3	−283.55	−491.123	16.35	1.611
10-1	−462.717	−267.15	20.917	1.543
10-2	−476.92	−275.35	25	1.537
10-3	−491.123	−283.55	16.35	1.505
11-1	−534.3	0	20.917	1.661
11-2	−550.7	0	25	1.647
11-3	−567.1	0	16.35	1.766
16-1	−462.717	267.15	20.917	1.598
16-2	−476.92	275.35	25	1.546
16-3	−491.123	283.55	16.35	1.552

Table 7.10 Calculation results of relative roundness errors of castings under different measurement points

The number of measurement points	The outer circle of the inner flange	Connecting circle of the support plate	Inner circle outer flange	The outer circle of the outer flange
5	16.33	18.28	22.43	25.86
10	15.58	16.64	19.23	22.15
15	13.96	15.25	17.94	18.26
20	12.49	13.33	14.67	14.96
25	9.88	10.86	11.2	11.9
30	8.24	8.93	9.8	10.63
35	7.69	8.28	8.67	9.66
40	7.33	7.73	7.92	8.87

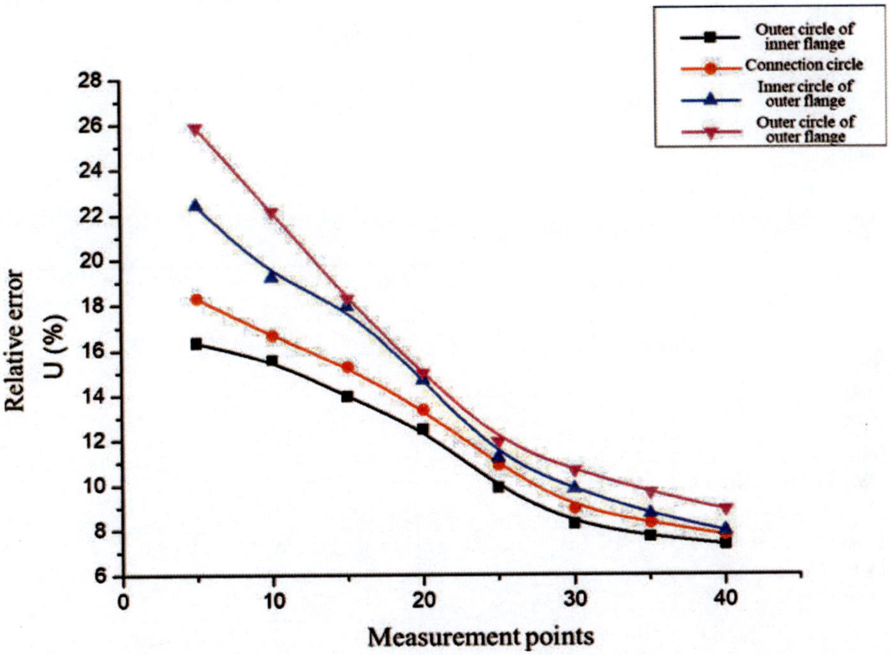

Fig. 7.33 Relationship between different measurement points and roundness of castings

For the ring part, the positioning method 1 is selected for positioning, and the angular deviation generated by the measurement point of the support plate during the positioning process can be calculated according to Eq. (7.37). The specific information and deviation information of measurement points on support plate 1 is shown in Table 7.11.

Table 7.11 Calculated value and the measured value of angle deviation of the large casting support plate

	Calculation deviation (degree)	Measurement deviation (degree)	Deviation percentage (%)
Support plate 6—support plate 3	0.264	0.28	5.7
Support plate 6—support plate 4	0.103	0.11	5.9
Support plate 4—support plate 5	−0.128	−0.12	6.7
Support plate 5—support plate 6	−0.644	−0.66	2.4
Support plate 6—support plate 7	0.046	0.05	8
Support plate 7—support plate 8	−0.158	−0.17	7.05
Support plate 8—support plate 9	0.057	0.06	5
Support plate 9—support plate 10	0.156	0.17	8.2
Support plate 10—support plate 11	0.159	0.17	6.3
Support plate 11—support plate 12	1.216	1.27	4.23
Support plate 16—support plate 1	2.301	2.36	2.5
Support plate 1—support plate 2	0.159	0.17	6.3

From the calculation results in Table 7.12, it can be seen that the deviation calculation of the model in this paper is not much different from the actual measurement value, which shows that the current model can be used to predict and diagnose these errors very well. In addition, it shows that it is effective to use the angular deviation to measure the angle uniformity of the support plate.

Table 7.12 Measurement point information and angular deviation of support plates

	X	Y	Z	Angular deviation
1	−359.767	334.998	−20.1	0.255
2	−360.034	334.843	−20.2	0.243
3	−360.538	334.553	−20.5	0.257
4	−361.038	334.264	−21	0.255
5	−361.357	334.08	−21.5	0.246
6	−361.636	333.919	−22	0.263
7	−361.908	333.762	−22.5	0.264
8	−362.172	333.609	−23	0.262
9	−362.681	333.315	−24	0.26
10	−364.053	332.523	−27	0.255
11	−366.172	331.3	−33	0.257
12	−367.563	330.497	−39	0.245
13	−368.514	329.948	−45	0.262
14	−369.143	329.585	−51	0.238
15	−369.455	329.404	−57	0.267
16	−369.494	329.382	−60	0.265
17	−369.455	329.404	−63	0.262
18	−369.338	329.472	−66	0.265
19	−369.143	329.585	−69	0.267
20	−368.868	329.743	−72	0.253
21	−368.514	329.948	−75	0.249
22	−368.079	330.198	−78	0.258
23	−367.039	330.799	−84	0.259
24	−365.982	331.41	−90	0.273
25	−364.924	332.02	−96	0.269
26	−363.866	332.631	−102	0.273
27	−362.809	333.242	−108	0.268
28	−362.28	333.547	−111	0.262
29	−361.751	333.852	−114	0.255
30	−361.575	333.954	−115	0.259
31	−361.487	334.005	−115.5	0.258
32	−361.399	334.056	−116	0.276
33	−361.311	334.107	−116.5	0.265
34	−361.222	334.157	−117	0.263

3. Study on the transfer law of assembly deviation of large casting wax block

According to the analysis model of definite positioning deviation in a polar coordinate system for the ring parts, it can be seen that the transfer rule of deviation in wax block assembly process of large thin-walled ring castings is affected by many factors such as fixture positioning accuracy, wax block manufacturing accuracy, positioning mode, and assembly mode. The content of this section is to study the influence degree and the law of different influencing factors on the deviation transfer in the assembly process through the deviation analysis model.

(1) Fixture positioning accuracy

Under different fixture positioning accuracy, the assembly deviation of each flange of large thin-walled ring casting is analyzed and solved, and the result is shown in Fig. 7.34.

From Fig. 7.35, it can be seen that the fixture deviation and assembly deviation are nearly linear. The larger the fixture deviation is, the larger the assembly positioning deviation is. At the same time, because the diameter of the outer circle of the outer flange is the largest, when the fixture deviation is the same, the deviation of the outer circle of the outer flange is also the largest among the different flanges. Therefore, in the design process of the fixture, it is necessary to consider the auxiliary positioning,

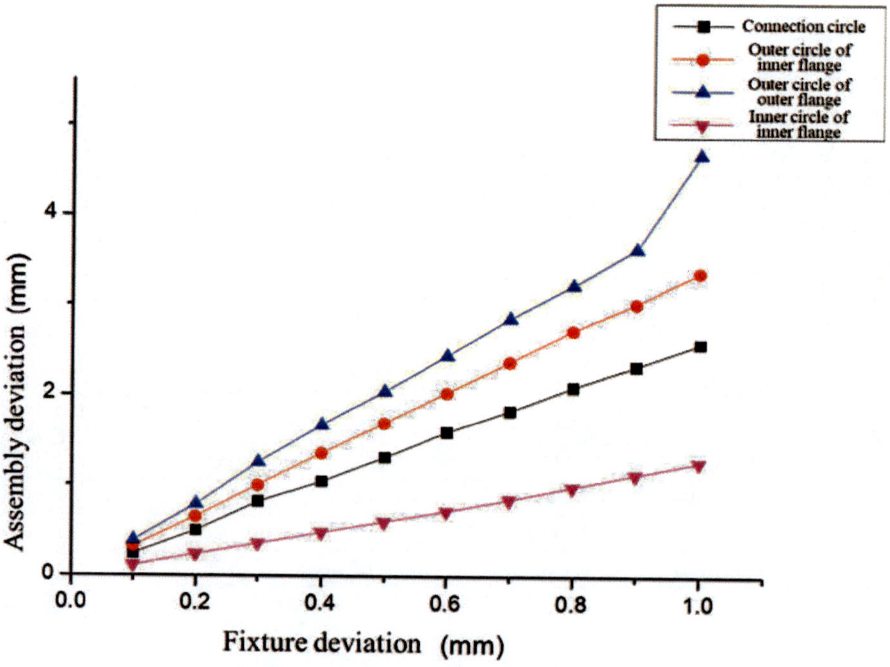

Fig. 7.34 Relationship between the roundness of each flange and assembly accuracy

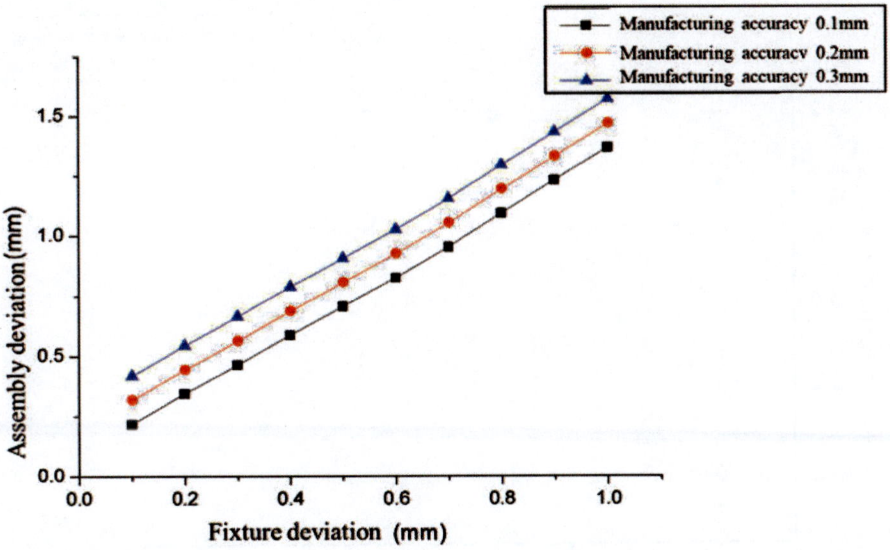

Fig. 7.35 Relationship between diameter deviation and manufacturing accuracy of the inner circle of the inner flange

clamping, and other measures to control the deviation of the flange with a large diameter.

(2) Wax block manufacturing accuracy

Under different wax block manufacturing accuracy, the relationship between the inner diameter deviation of the inner flange and manufacturing accuracy is studied through deviation analysis and the results are shown in Fig. 7.35.

It can be seen from Fig. 7.35 that as the manufacturing accuracy of wax mold decreases, the deviation value also increases. Therefore, controlling the manufacturing accuracy of wax mold is also an effective way to control the assembly accuracy of the wax mold.

(3) Assembly method

In this section, the assembly deviation of each flange diameter under the different assembly modes of a single block and two pairs of wax mold assembly are compared, and the results are shown in Fig. 7.36. From left to right, the midpoint in the figure represents the diameter of inner circle of upper inner flange, outer circle of upper inner flange, inner circle of upper outer flange, outer circle of upper outer flange, inner circle of lower inner flange, outer circle of lower inner flange, inner circle of lower outer flange, and straight diameter of outer circle of lower outer flange, respectively.

From Fig. 7.36, we know that the deviation of the assembly method after two bonding is significantly smaller than that of the single block assembly. This is due to

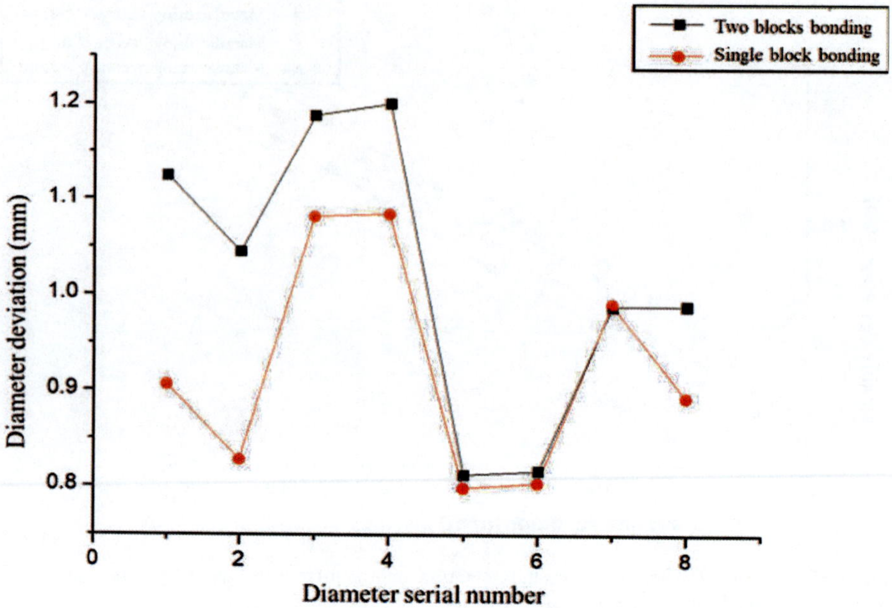

Fig. 7.36 Relationship between flange diameter deviation and assembly method

the mutual coupling of the positioning modes during the assembly of the two pairs, which leads to a decrease of the deviation.

7.2.4 Variation Stream Modeling and Process Optimization

7.2.4.1 Dimensional Variation Stream Modeling of Investment Casting Process Based on State Space Method

In most cases, the final dimensions of the wax mold and the casting, which are constrained by the core and the mold, are determined by the viscoelasticity of the wax mold, the viscoplastic creep of the shell, and the plastic deformation of the alloy. The mold design becomes more difficult because of the complex shape of the casting and the precise dimensional accuracy requirements. The requirement of the casting to the dimensional accuracy cannot be met when the deformation lows for the wax mold, the shell, the metal material, and the casting are not mastered. In order to improve the precision of the final casting, the variation transfer and accumulation in the manufacturing process must be taken into account, and the whole process variation transfer model of the complex casting system must be established, which can then enable us to realize the description and control of the dimensional variation in the whole process of the complex investment casting [12].

1. **Determination of State vector**

The investment casting process (ICP) is a process involving multiple stages to produce the final casting product, that is, it is a complex multistage manufacturing process (MMPs). In the ICP, the linear dimensions, such as length, width, radius and so on, change at each stage After wax injection, the wax will shrink and thus the dimension of a wax part is smaller than that of the part die. When the shell is fired before casting, the inner dimension of the shell will expand. At the casting stage, the solidification of the metal leads to the contraction of the dimension directly. If the dimension of a casting part is rejected, then the heat treatment for the dimension is needed. Therefore, the final linear dimension and dimensional variation of final parts depend on the cumulative and transitive effects of the linear contractions or expansions in every individual stage.

In order to study the linear dimensional change in the whole ICP and develop a mathematical model to describe it, the dimensions of the part at different stages can be written by:

$$X(k) = (x_1(k), x_2(k), \cdots, x_m(k))^T, k = 1, 2, 3 \cdots n \tag{7.50}$$

where $x_j(k)$ is the j-th key control dimension of the part at state k, which is the k-th stage in the ICP, and m is the total of the key control dimensions.

Definition 1 State vector: A state vector is defined to represent the dimensional variations of the part to the theoretical value or target value in different stages. According to Eq. (7.1), it is described as:

$$\Delta X(k) = (\Delta x_1(k), \Delta x_2(k), \cdots, \Delta x_m(k))^T \tag{7.51}$$

where $\Delta x_j(k)$ describes the j-th dimensional variation at state k.

2. **Determination of Process Parameter Vector**

(1) Analysis of factors affecting dimensions

The dimensional changes in ICP result from the multi-field coupling, in particular, the thermo-mechanical coupling in each stage. It is also affected by time-dependent processes such as viscoelastic deformation of the wax, viscoplastic creep, and plastic yield of the shell and alloy. The material properties and process parameters determine the dimension in each stage directly and their variations lead to the dimensional variations in the individual stage. At each stage, the variation of parameters generates a dimensional variation, which is transferred to the next stage. Hence, the dimensional variations at a current stage are not only controlled by the variations of the parameters in this stage but also affected by the accumulated variations transmitted from previous stages. As the geometrical dimension of the part going through all the ICP, a stream of geometrical dimension flow, as well as an associated flow of variation, will be generated. Here, it is named as the Dimensional Variation Stream (DVS) for the ICP.

In order to identify the process parameters that affect the dimension of the part processed by ICP, an Ishikawa cause-effect diagram is constructed, as shown in Fig. 7.4. The Ishikawa cause-effect diagram depicts that the following process parameters may affect the dimensional stability of the part in different stages.

i. People-based variables: laziness, fatigue, and absent-mindedness, all of which may occur in the whole ICP.
ii. Wax characteristics-based variables: thermal conductivity, the composition of the wax, rheology, and viscosity.
iii. Die based variables: venting, material, and durability.
iv. Wax injection-based variables: packing pressure, holding time, injection velocity, and temperature.
v. Wax cooling-based variables: cooling time, the thermal conductivity of the die, and the ambient temperature of the die.
vi. Wax storage-based variables: ambient temperature, time, and humidity.
vii. Coating- and stuccoing-based variables: the sum of the layers, uniformity of the thickness, and slurry viscosity.
viii. Shell drying based variables: ambient temperature, time, and humidity.
ix. Material characteristics of the shell-based variables: deformability, permeability, and composition.
x. Dewaxing-based variables: temperature, external pressure, heat input direction, and heating speed.
xi. Roasting-based variables: temperature, heating speed, and soaking time.
xii. Material characteristics of the metal-based variables: thermal conductivity, melting point, rheology, and composition.
xiii. Geometric characteristics-based variables: uniformity of the dimension and Gating and Risering system.
xiv. Casting-based variables: temperature, casting velocity, and retention time.

It is worth mentioning that there are many variables that affect the final dimensions of the casting, and this Ishikawa cause-effect diagram just depicts main factors which affect the final dimensions of the casting (Fig. 7.37).

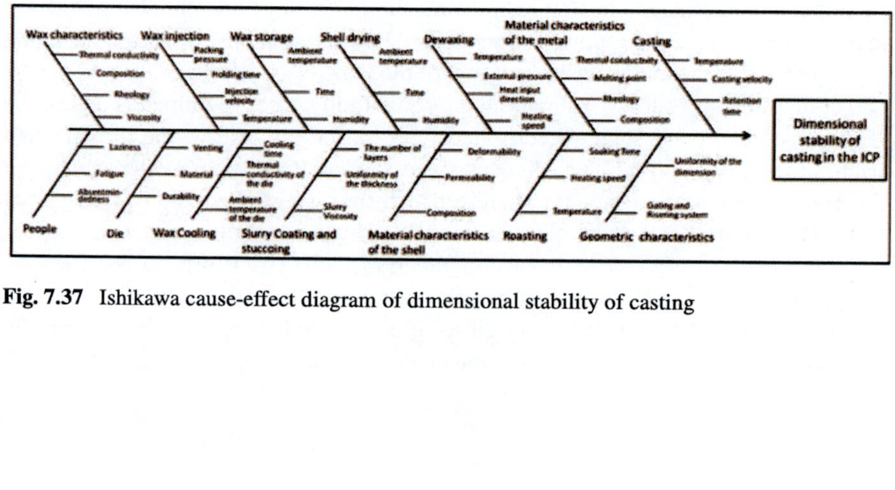

Fig. 7.37 Ishikawa cause-effect diagram of dimensional stability of casting

(2) Process Parameter Vector

In this section, a State Space Model will be developed to describe the DVS for the ICP. It is a mathematical model that links the key control dimensions with key control variables (process parameters) and expresses the complex production stream and data stream involved in modeling and analysis of variation together with its propagation in the whole ICP. Hence, the process parameter vector and process parameter variation vector should be defined rigorously.

Definition 2 Process parameter vector: In the ICP, when the material is selected, the dimensions of the part in each stage mainly depend on different process parameters on the corresponding stage. Accordingly, the process parameter vector is defined as

$$\boldsymbol{u}_j(k) = (\alpha_{j.1}(k), \alpha_{j.2}(k), \cdots, \alpha_{j.l}(k))^T \tag{7.52}$$

where j is the order number of the dimension and $\alpha_{j.i}(k)$ represents the i-th process parameter which affects $\Delta x_j(k)$ at stage k (such as packing pressure, temperature, and holding time at wax injection stage, as Fig. 7.4 illustrated), and l is the total number of process parameters at stage k.

With the definition of the process parameter vector, the process parameter variation vector at the current stage k can be given as

$$\Delta\boldsymbol{u}_j(k) = (\Delta\alpha_{j.1}(k), \Delta\alpha_{j.2}(k), \cdots, \Delta\alpha_{j.l}(k))^T \tag{7.53}$$

where $\Delta\alpha_{j.i}(k)$ is the variation of the process parameter to the target value.

3. **State Space Equation**

The linear dimension of the ICP changes as a result of the linear contractions or expansions that occur in every individual stage. In order to describe the relative dimensional difference between the current and the previous stages, the following definition is adopted.

Definition 3 Dimensional change rate (DCR) is defined as the following relationship.

$$s_j(k) = \frac{x_j(k)}{x_j(k-1)} \times 100\%,\ k = 1, 2, 3 \cdots n \tag{7.54}$$

where k is the order number of process stage in the ICP, $s_j(k)$ is the DCR of the k-th stage for $x_j(k)$, and $x_j(k)$ represents the j-th actual linear dimension in the k-th stage, such as width, length, and radius and so on.

According to Eq. (7.54), $x_j(k)$ can be represented by $x_j(k-1)$ as

$$x_j(k) = s_j(k).x_j(k-1) \tag{7.55}$$

As for $x_j(k)$, a functional relationship between DCR and process parameters always exists. This is because process parameters determine the dimension in each stage and their variations lead to the dimensional variations directly in each individual stage. The implicit function between DCR and process parameters is represented as Eq. (7.55) so that the model of the ICP for dimension variation can be conducted.

$$s_j(k) = f_{k.j}(\boldsymbol{u}_j(k)) \tag{7.56}$$

Here, $f_{k.j}(\boldsymbol{u}_j(k))$ is a vector function related to the process parameter vector $\boldsymbol{u}_j(k)$. According to the variation definition, the variation of $x_j(k)$ can be expressed by:

$$\Delta x_j(k) = x_j(k) - x_j^t(k) \tag{7.57}$$

where $x_j^t(k)$ is the target value of the j-th dimension at stage k.

Based on the Eqs. (7.55) and (7.57), one of the dimensional variations of the part at stage k can be described as

$$\Delta x_j(k) = s_j(k).x_j(k-1) - s_j^t(k).x_j^t(k-1) \tag{7.58}$$

where $s_j^t(k)$ is the target value of DCR to the j-th dimension when the process parameter vector gets the target value.

According to Eq. (7.56), if $f_{k.j}(\boldsymbol{u}_j(k))$ is derivable at the point $\boldsymbol{u}_j^t(k)$, the Taylor series expansion of Eq. (7.9) can be given as

$$\Delta x_j(k) = f_{k.j}(\boldsymbol{u}_j^t(k)).\Delta x_j(k-1) + x_j^t(k-1).\boldsymbol{f}'_{k,j}(\boldsymbol{u}_j^t(k)).\Delta \boldsymbol{u}_j(k) + h(\Delta \boldsymbol{u}_j(k)) \tag{7.59}$$

In this equation,

$$\boldsymbol{f}'_{k.j}(\boldsymbol{u}_j^t(k)) = \left(\frac{\partial \boldsymbol{f}_{k.j}(\boldsymbol{u}_j^t(k))}{\partial \alpha_{j.1}(k)}, \frac{\partial \boldsymbol{f}_{k.j}(\boldsymbol{u}_j^t(k))}{\partial \alpha_{j.2}(k)}, \cdots, \frac{\partial \boldsymbol{f}_{k.j}(\boldsymbol{u}_j^t(k))}{\partial \alpha_{j.l}(k)} \right) \tag{7.60}$$

$$\begin{aligned} h(\Delta \boldsymbol{u}_j(k)) &= r(\Delta \boldsymbol{u}_j(k)).(x_j^t(k-1) + \Delta x_j(k-1)) \\ &\quad + \boldsymbol{f}'_{k.j}(\boldsymbol{u}_j^t(k)).\Delta \boldsymbol{u}_j(k).x_j(k-1) \end{aligned} \tag{7.61}$$

where $\boldsymbol{u}_j^t(k)$ is the target value of the process parameter vector and $r(\Delta \boldsymbol{u}_j(k))$ is the higher-order remainder term of $\Delta \boldsymbol{u}_j(k)$.

Considering the fact that there are some random factors affecting the whole procedure, such as people-related factors (laziness, fatigue, and absent-mindedness), Eq. (7.59) can be written as:

$$\begin{aligned} \Delta x_j(k) = {} & f_{k.j}(\boldsymbol{u}_j^t(k)).\Delta x_j(k-1) + x_j^t(k-1).\boldsymbol{f}'_{k,j}(\boldsymbol{u}_j^t(k)).\Delta \boldsymbol{u}_j(k) \\ & + h(\Delta \boldsymbol{u}_j(k)) + \varphi_j(k) \end{aligned} \tag{7.62}$$

In order to simplify the form of Eq. (7.62), it can be rewritten as

$$\Delta x_j(k) = a_j(k).\Delta x_j(k-1) + \boldsymbol{b}_j(k).\Delta \boldsymbol{u}_j(k) + w_j(k) \tag{7.63}$$

where

$$a_j(k) = f_{k,j}(\boldsymbol{u}_j^t(k)) \tag{7.64}$$

$$\boldsymbol{b}_j(k) = x_j^t(k-1).\boldsymbol{f}'_{k,j}(\boldsymbol{u}_j^t(k)) \tag{7.65}$$

$$w_j(k) = h(\Delta \boldsymbol{u}_j(k)) + \varphi_j(k) \tag{7.66}$$

Due to the fact that there are different key dimensions in each stage, Eq. (7.63) can be rewritten in a matrix form:

$$\Delta X(k) = A(k).\Delta X(k-1) + B(k).\Delta U(k) + W(k) \tag{7.67}$$

In Eq. (7.67), $A(k)$ and $B(k)$ are diagonal matrices, $\Delta U(k)$ is the control vector, and $W(k)$ is a noise vector, which are given by:

$$A(k) = \begin{bmatrix} a_1(k) & & & \\ & a_2(k)) & & \\ & & \ddots & \\ & & & a_m(k) \end{bmatrix}_{m \times m} \tag{7.68}$$

$$B(k) = \begin{bmatrix} \boldsymbol{b}_1(k) & & & \\ & \boldsymbol{b}_2(k) & & \\ & & \ddots & \\ & & & \boldsymbol{b}_m(k) \end{bmatrix}_{m \times ml} \tag{7.69}$$

$$\Delta U(k) = \begin{bmatrix} \Delta \boldsymbol{u}_1(k) \\ \Delta \boldsymbol{u}_2(k) \\ \vdots \\ \Delta \boldsymbol{u}_m(k) \end{bmatrix}_{ml \times 1} \tag{7.70}$$

$$W(k) = \begin{bmatrix} w_1(k) \\ w_2(k) \\ \vdots \\ w_m(k) \end{bmatrix}_{m \times 1} \tag{7.71}$$

Here, the coefficient matrices $A(k)$ and $B(k)$ are determined by the process parameter vector. The former describes the impact of the variation transition from stage k-1 to stage k, called system matrix, and the latter represents the impact of the process parameter variation on the dimensional variations. On the other hand, $\Delta U(k)$ is the control vector, which includes the process parameter variations. $W(k)$ is the noise term, which represents the imperfections in the modeling, such as the lack of representation of the random factors at stage k, the disregard of the higher-order remainder term of $\Delta \boldsymbol{u}_j(k)$.

According to the linear control theory, the relationship between the k-th state vector and control vectors at different stages can be expressed as:

$$\Delta X(k) = \boldsymbol{\phi}(n, 0).\Delta X(0) + \sum_{k=1}^{n} \boldsymbol{\phi}(n, k).B(k).\Delta U(k) + W(k) \tag{7.72}$$

where the state transition matrix $\boldsymbol{\phi}(\cdot, \cdot)$ is adopted from linear control theory and defined as:

$$\boldsymbol{\phi}(n, i) = \mathbf{A}(n-1)\mathbf{A}(n-2)\ldots\mathbf{A}(i) \text{ for } n > i \text{ and } \boldsymbol{\phi}(i, i) = \mathbf{I} \tag{7.73}$$

4. **Observation Equations**

During the measurement, the measurement noise is inevitable. If all the key control dimensions are directly measurable, the j-th measured value of the variation can be written as

$$\Delta y_j(k) = \Delta x_j(k) + v_j(k) \tag{7.74}$$

Therefore, the observation equation can be expressed by:

$$\Delta Y(k) = \Delta X(k) + V(k) \tag{7.75}$$

where $\Delta Y(k)$ is an observed vector related to all the measured value at stage k and represents the information of $\Delta X(k)$, which dimension is $m \times 1$. $V(k)$ represents the measurement noise, they are given by:

$$\Delta Y(k) = \begin{bmatrix} \Delta y_1(k) \\ \Delta y_2(k) \\ \vdots \\ \Delta y_m(k) \end{bmatrix}_{m\times 1} \tag{7.76}$$

$$V(k) = \begin{bmatrix} v_1(k) \\ v_2(k) \\ \vdots \\ v_m(k) \end{bmatrix}_{m\times 1} \tag{7.77}$$

According to Eq. (7.72), the observation equation can be rewritten as:

$$\Delta Y(k) = \phi(n,0)\Delta X(0) + \sum_{k=1}^{n} \phi(n,k).B(k).\Delta U(k) + W(k) + V(k) \tag{7.78}$$

5. **Quantitative calculation of the Stream of Variation for the whole process**

(1) Design of feature parts

In this section, for the purpose of making a good interpretation of the state space modeling process, a simple case is adopted, which is the ICP of a rectangular-shaped specimen. The dimensions of the die are 400 ± 0.01 mm, 100 ± 0.01 mm, and 2.5 ± 0.01 mm (as illustrated in Fig. 7.38).

Figure 7.1 shows that the whole ICP consists of so many stages. In this case, main stages such as wax injection, wax cooling, and storage, the roasting of the

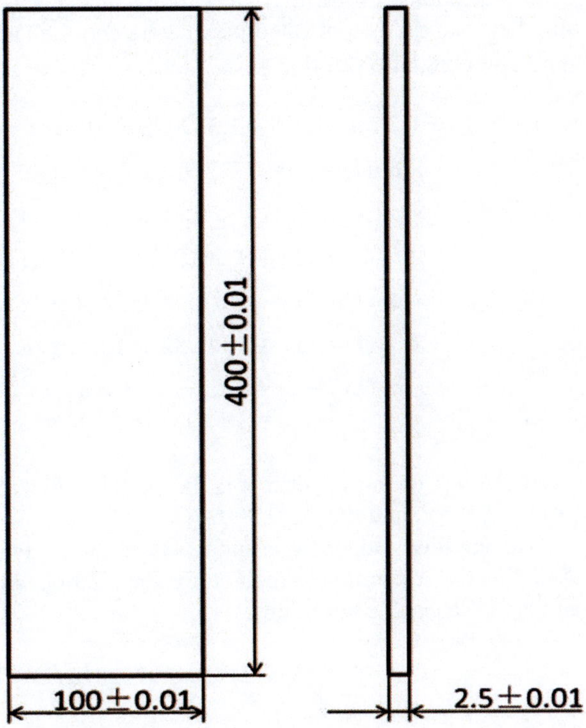

Fig. 7.38 The dimension of the specimen

shell and casting are abstracted, and other stages such as coating and stuccoing, shell drying, and dewaxing are neglected owing to the reason that the dimensions of the part experiences little changes in these stages. When setting up the state space model of these abstracted stages, some empirical formulas or fitting formulas in others' previous studies are applied.

(2) Variation Modeling

Die: The dimensions of the die are the initial dimensions of the part. The thickness of the specimen is selected and according to the machining accuracy, the tolerance of the die is ± 0.01 mm that is looked at the variation range of stage 1. Since only one dimension is selected, the state vector degenerated into one dimension and the order number j can be deleted in all the stages. The dimension $x(0)$ is 2.5 and the state vector degenerate as $\Delta x(0) = \pm 0.01$.

Wax injection: In this stage, a series of experiments are conducted using KCRO17B material. The details are described in Appendix. The process parameters include mold temperature ($\alpha_1(1)$), melt temperature ($\alpha_2(1)$), packing pressure ($\alpha_3(1)$), and holding time ($\alpha_4(1)$), while other parameters maintain constant. As a result, the process parameter vector can be written as

$$\boldsymbol{u}(1) = (\alpha_1(1), \alpha_2(1), \alpha_2(1), \alpha_4(1))^T \tag{7.79}$$

The Response Surface Methodology analysis is used to model the relationship between these controlled parameters and the DCR via regression models. The regression model is developed as

$$\begin{aligned} f_1(\boldsymbol{u}(1)) = {} & 0.776684357 - 1.352023 \times 10^{-3} \times \alpha_1(1) + 6.336718 \times 10^{-3} \\ & \times \alpha_2(1) - 1.897728 \times 10^{-3} \times \alpha_3(1) - 6.65351 \times 10^{-5} \\ & \times \alpha_4(1) + 1.375 \times 10^{-5} \times \alpha_1(1)\alpha_2(1)_m + 2 \times 10^{-6} \\ & \times \alpha_2(1)\alpha_3(1) + 1 \times 10^{-6} \times \alpha_1(1)\alpha_4(1) + 2.1375 \times 10^{-5} \\ & \times \alpha_2(1)\alpha_3(1) - 2.875 \times 10^{-6} \times \alpha_2(1)\alpha_4(1) - 3.5 \times 10^{-7} \\ & \times \alpha_3(1)\alpha_4(1) + 1.00982 \times 10^{-5} \times \alpha_1(1)^2 - 4.5636 \times 10^{-5} \\ & \times \alpha_2(1)^2 + 2.69825 \times 10^{-6} \times \alpha_3(1)^2 - 2.70175 \times 10^{-6} \times \alpha_4(1) \end{aligned} \tag{7.80}$$

According to the experiment, the target value of the process vector $\boldsymbol{u}^t(1)$ is $(30, 74, 25, 5)^T$ and $a(1)$ is 0.989346.

Wax cooling and storage: In this stage, the study results of Bonilla et al. [13] is used directly. According to their study, the relationship between storage time ($\alpha(2)$) and the DCR can be rewritten as:

$$f_2(\boldsymbol{u}(2)) = 0.996281\alpha(2)^{-0.0025} \tag{7.81}$$

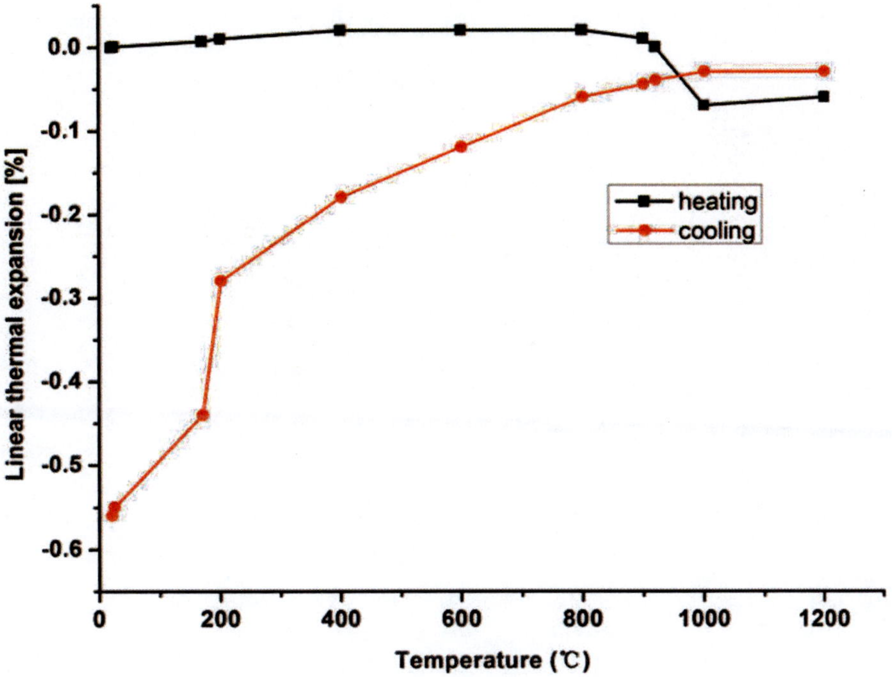

Fig. 7.39 Measured linear thermal expansion for the shell mold (Snow 1995)

In this case, only storage time is considered, and which unite is minute. Suppose the wax part will be used to make a shell as a core 180 min late after injection. Then $a(2)$ is 0.98343.

Shell roasting: The main determinant of shell expansion is roasting temperature, which can be seen from the study of Sabau and Porter [14]. From Snow's study (as illustrated in Fig. 7.39), roasting temperature ($\alpha(3)$) is the main variable affecting the dimensional variation of the shell. In addition, the relationship between the roasting temperature and the DCR at the heating process of the shell is given by

$$f_3(\boldsymbol{u}(3)) = 0.99995 + 1.42454 \times 10^{-6}\alpha(3) - 1.70489 \times 10^{-9}\alpha(3)^2 \quad (7.82)$$

By setting the target temperature to be 1200 °C, $a(3)$ is then determined to be 0.997783.

Coasting: Since the shell is much stiffer than the melting steel, the effects of the contraction of the shell to the dimension of the casting are insignificant and can be neglected at casting contraction stage. Based on the study of Sabau and Porter [14], the thermal expansion of 17.4 PH alloy in the 20–1360 °C temperature range are obtained through experiments (as illustrated in Fig. 7.40).

From these results of this previous work, the final cooling temperature ($\alpha(4)$) is the main variable and an abrupt expansion at approximately 265 °C during cooling is

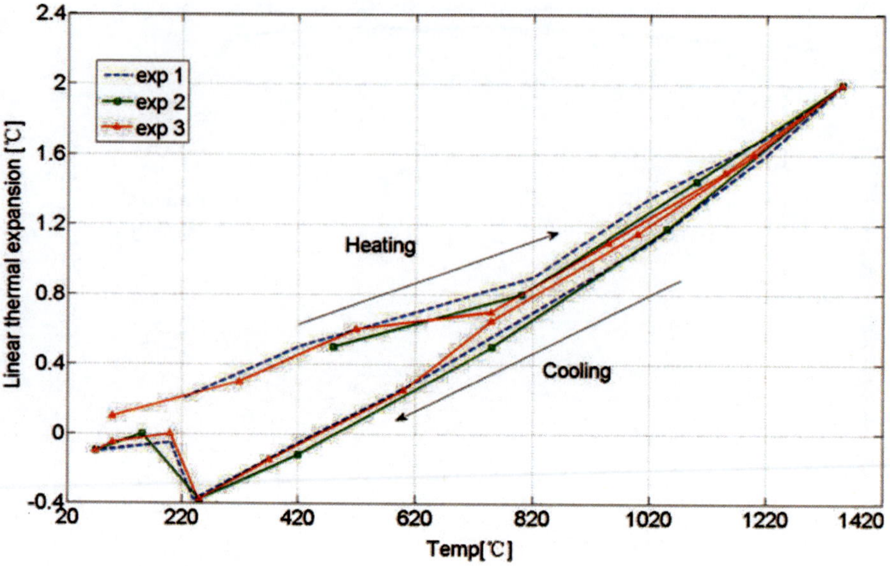

Fig. 7.40 Measured linear thermal expansion for 17-4PH alloy on heating and cooling (Adrian S. Sabau and Wallace D. Porter, 2008)

shown. Using the result of the linear thermal expansion in the 20–265 °C temperature range, a regression model is obtained:

$$f_4(\boldsymbol{u}(4)) = 0.99818 + 4.432027 \times 10^{-5}\alpha(4) - 7.33681 \times 10^{-7}\alpha(4)^2 + 6.88197 \times 10^{-9}\alpha(4)^3 - 2.62351 \times 10^{-11}\alpha(4)^4 + 3.0152 \times 10^{-14}\alpha(4)^5,\ (20 \le \alpha(4) \le 265) \quad (7.83)$$

If the final cooling temperature is set to 25 °C, then the target value of $a(4)$ would be 0.998972.

(3) Simulation

Since the process parameter vectors ($\boldsymbol{u}^t(k)$) at every stage are set up, and $f_k(\boldsymbol{u}(k))$ at different stages are obtained via experiments or some previous works, if the process parameter vectors at each stage (the inputs) are set up, the variation simulation can be carried out.

A group of process parameter variation ranges are shown in Table 7.13, based on which the target thickness ($x^t(k)$) and variation range ($\Delta x(k)$) can be calculated via Eqs. (7.55) and (7.63), respectively. These results are also shown in Table 7.13.

In this simulation, the worst case is considered, that is, the limiting values of process parameter variations at every stage are taken into account to maximize the part variation range such that all the probable values are covered. From this simulation, the changes of the nominal dimension of the thickness and the variation range of

Table 7.13 The simulation of variation range

Stage NO./k	Stage	Process parameter vector/$u^t(k)$				Process parameter variation vector $\Delta u(k)$				$a(k)$	$x^t(k)$	$b(k)$				$\Delta x(k)$
		$\alpha_1(k)$	$\alpha_2(k)$	$\alpha_3(k)$	$\alpha_4(k)$	$\Delta\alpha_1(k)$	$\Delta\alpha_2(k)$	$\Delta\alpha_3(k)$	$\Delta\alpha_4(k)$							
0	Die										2.5					±0.01
1	Wax injection	30	74	25	5	±0.3	±0.74	±0.25	±0.05	0.989346	2.473365	8.16E−004	0.0013	−3.07e−004	−7.13e−004	±0.011213
2	Wax storage	180				±20				0.98343	2.432381	−3.38E−05				±0.011703
3	Shell roasting	1200				±20				0.999204	2.430445	−6.49E−06				±0.011823
4	Casting	25				±2.5				0.998972	2.427947	4.6E−05				±0.011926

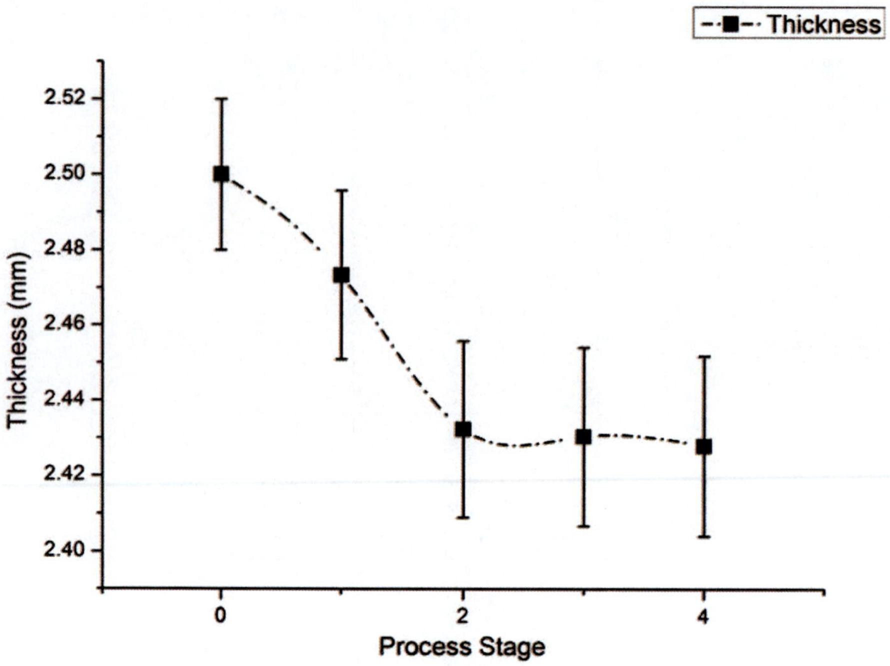

Fig. 7.41 The dimensional changes and the variation range

each stage can be seen in Fig. 7.41. Figure 7.42 shows us the accumulation and transmission of linear dimension variation of the thickness.

7.2.4.2 Sensitivity Analysis and Optimization of Process Parameters

The response surface methodology (RSM), introduced by Box and Wilson [15], is very effective to quantify the relationship between multiple inputs and obtained responses surfaces [16]. Consequently, it has been widely used in different fields to evaluate and improve the quality of products or process parameters [17–19], especially injection molding process. Chuang et al. discussed the variation of warpage and the tensile stress properties depended on process parameters via the Taguchi approach and RSM [20, 21].

A genetic algorithm, as a random search algorithm without limitation in the search space to optimize the real-world problem, is developed by Holland. Largely because of Goldberg's [22] studies, it has been widely applied to solving complex optimization problems. Changyu et al. [23] proposed an integrated artificial neural network and the GA method to optimize the injection modeling process. However, these studies just focus on the plastic injection molding process. With the thickness of the final casting wall required thinner and the accuracy requirements for higher in the ICP,

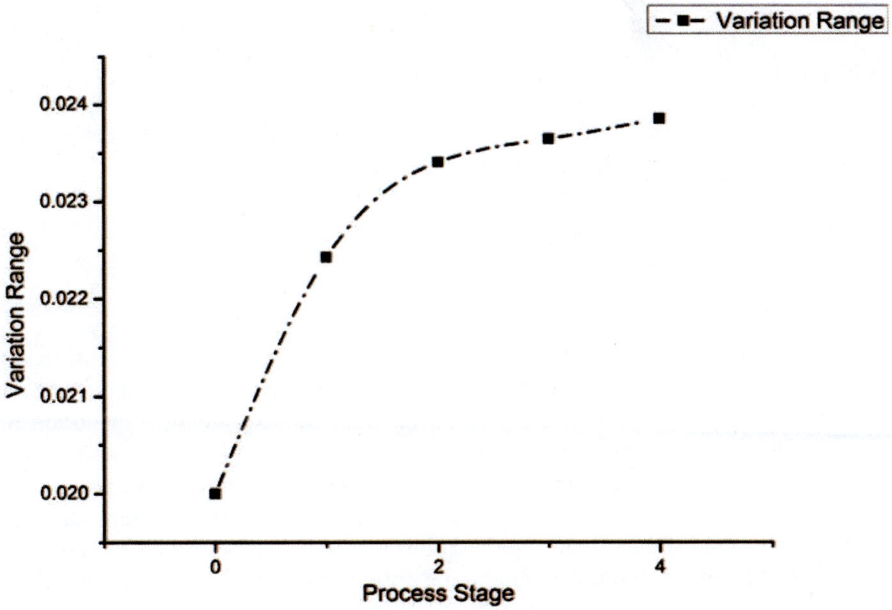

Fig. 7.42 The accumulation and transmission of the variation range

the quantitative investigation about the interaction of the process parameters on the shrinkage of the thin-walled wax part is increasingly important.

In this study, a Box-Behnken design based on RSM was applied to designing the experiments [24]. Four process parameters of mold temperature, melt temperature, packing pressure, and holding time with three levels for each factor were selected to investigate their effects on the dimensional shrinkage variation of a thin-walled wax part in the injection process. Then, the regression model was obtained by implementing the RSM analysis and the validity and accuracy of the model were confirmed through the analysis of variance (ANOVA). Furthermore, the significant factors to the dimension shrinkage were found out and their contribution was calculated. Finally, the optimization of process parameters was carried out by GA and the desirability function respectively. These two predicted optimum injection conditions were compared and validated with experimental results.

1. **RSM-GA method**

(1) Response surface methodology

As a statistical approach, RSM integrates the design of experiment (DOE) technique and regression analysis. With its advantage to describe the multi-input problem, it is very effective to find the best range of design space for performance. The quadratic response surface is always described as

$$y = b_0 + \sum_{i=1}^{n} b_i x_i + \sum_{i=1}^{n} b_{ii} x_i^2 + \sum_{i<j}^{n} b_{ij} x_i x_j \quad (7.84)$$

where b_0, b_i, and b_{ij} are coefficients of the linear term, quadratic term, and interactive term, respectively, and n is the number of design variables.

The basic steps of RSM consist of the determination of the controlled parameters and response, DOE, regressive model construction, analyses of variance (ANOVA), and confirmation tests.

(2) The GA Approach

In order to find the global optimization or nearly optimal solution, GA is commonly adopted to solve complex non-linear problems on the basis of a randomized search and Darwin's theory of survival of the fittest. It simulates biological evolution by natural selection. In general, a set of potential solutions (i.e., chromosomes) is randomly generated and comprises a population. The GA operators such as selection/reproduction, crossover, and mutation are performed at each step based on different criteria [25] that the next generations (i.e., offsprings) can be created from the parent population. Moreover, due to the elitism, the genetic drift is reduced and the number of individuals that can survive to the next generation is guaranteed.

In order to make up the next generation, the selection operator selects the solutions to build up a breeding pool. In this work, the roulette wheel method was used such that weaker genes stand a chance to be selected, as they might contain some good features [25]. The main purpose of the selection is to choose better parents for the next generation.

Subsequently, the parts of two-parent solutions are mixed by the crossover operator so that two different offspring solutions are created. Crossover fraction specifies the fraction of the next generation that is produced by crossover. The fraction value ranges between 0 and 1. Generally, a crossover operator is applied on the chromosome with single-point, two-point, or scattered crossover approaches, and the crossover fraction is larger than 0.5. In this study, the scattered crossover approaches were used.

The mutation operation makes some small random changes in the solutions, maintaining the diversity of the population to prevent premature convergence to a local optimum solution. In this study, a uniform mutation is used.

The outline of the above optimization algorithm is given in Fig. 7.43. The optimization scheme of RSM-GA model is summarized as follows:

i. Based on RSM design and experiment, a correct and reasonable RSM model is obtained.
ii. Generating an initial population of chromosomes (candidate solutions to the problem) from the first generation via the combined RSM-GA model.
iii. Calculating the fitness $F(x)$ of each of the chromosomes in the population
iv. Creating a population for the next generation by genetic operations (crossover and mutation). These operations are applied to the chromosomes in a formal generation with the probabilities based on their fitness.

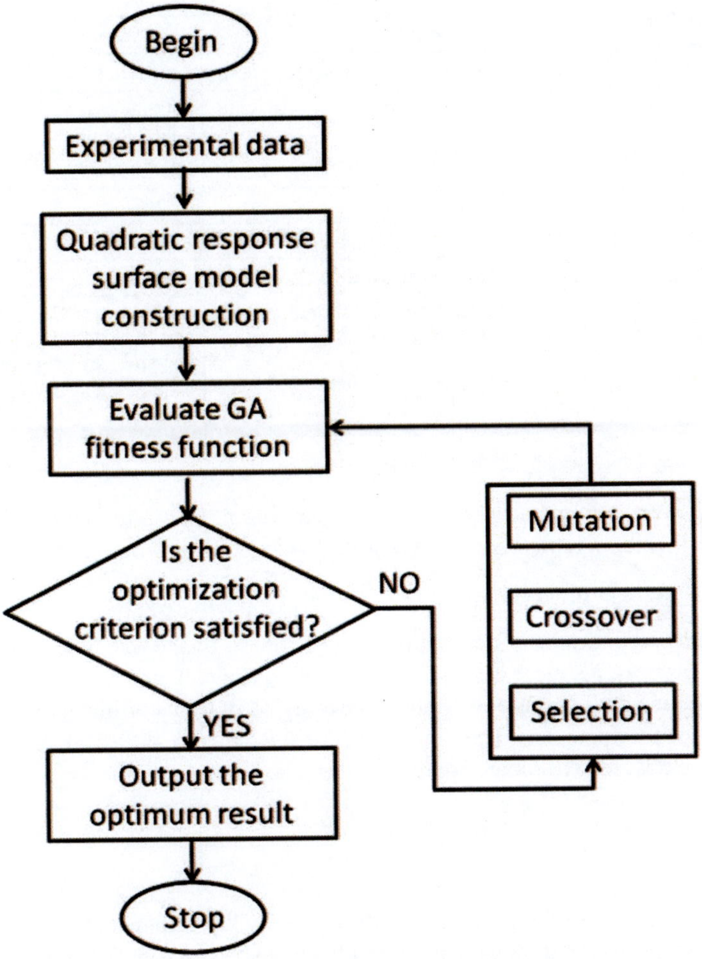

Fig. 7.43 The basic procedure of the optimization scheme based on the combined RSM-GA

v. Generating a new generation of injection conditions. This algorithm is repeated to breed the next generation.
vi. Repeating the above steps until the termination criterion has been satisfied.

2. **Design of Experiments**

(1) Materials and equipments

In this study, a crystalline wax named KC4017B that was manufactured by Kindt-Collins was used. Table 7.14 lists its fundamental physical properties. The experiments were carried out on the MPI55 injection molding machine, with the technical capability of a maximum injection rate of 350 cm^3/s, a maximum injection pressure

Table 7.14 The basic physical property of KC4017B

Properties	Specification
Ring and Ball softening point	148 °F ≈ 64.4 °C
Penetration	450 gm load 17.2 dmm; 100 mg load 3.8 dmm
Ash content (ICI1.0)	<0.015%
Filler content	≈32%
Specific gravity	1.002
Flow test (ICI2.0)	>50%@44 °C
Viscosity (Brookfield thermosel Spindle)	160 °F ≈ 71 °C 680 100; 170 °F ≈ 76 °C 480 100; 180 °F ≈ 82 °C 350 100;
Expansion	48 °C 2.7%; 60 °C 5.3%; 72 °C 6.7%;

of 70 bar, nozzle diameter of 10 mm and a maximum clamp force of 100 tons, and the experimental parameters can be adjusted via a control program.

(2) Measurement of shrinkage

In order to investigate the effects of process parameters on the wax products in the wax injecting process, a simple case is adopted, which is a rectangular-shaped specimen. All the dimensional variations have been evaluated in terms of linear changes in the thickness of the specimen. The dimensions of the die were 400, 100, and 2.5 mm.

In this study, the shrinkage is defined as:

$$S = \frac{L}{L_0} \tag{7.85}$$

where L_0 is the dimension of the cavity and L is the dimension of the wax part after injection. All the dimensional variations have been evaluated in terms of linear change of the thickness in the A, B, C, D, E and F locations, which were, respectively, 10, 95, 200, 295, and 390 mm away from the gate (as illustrated in Fig. 7.44). In this experiment, a CMM with 0.1 um resolution was used, and the mean of the shrinkage of the five different locations was taken as the shrinkage of the thickness of the specimen.

(3) Experimental scheme

The series of experiments were carried out to figure out the relative influence of the injection molding conditions on the dimensional stability of the wax pattern. To this end, process parameters, which includes mold temperature (T_d), melt temperature (T_m), packing pressure (P) and holding time (t), were investigated and selected as inputs and other parameters were maintained constant. Then the shrinkage of the thickness of the specimen was considered as responses. The process factors and factor levels of the injection molding conditions are listed in Table 7.15. The coded

Fig. 7.44 The geometry of the die and wax pattern

Table 7.15 Experimental factors and factor levels

Levels of experiments	Mold temperature/(°C)	Melt temperature /(°C)	Packing pressure /(bar)	Holding time /(s)
1	30	74	35	15
0	25	72	30	10
-1	20	70	25	5

value +1, 0, −1 represented a high level, middle level, low level of the factors, respectively. The coded variables are calculated as

$$A = \frac{T_d - T_{d0}}{\Delta T_d}, B = \frac{T_m - T_{m0}}{\Delta T_m}, C = \frac{P - P_0}{\Delta P}, D = \frac{t - t_0}{\Delta t} \tag{7.86}$$

where A, B, C, and D are the coded values of the factors T_d, T_m, P, and t, respectively. T_{d0}, T_{m0}, P_0, and t_0 are the values of mold temperature, melt temperature, packing pressure, and holding time at zero level, respectively.\DeltaT$_d$,\DeltaT$_m$, ΔP, and Δt are the intervals of T_d, T_m, P, and t, respectively.

The experiment plan, generated in accordance with faced Box-Behnken design, consists of 30 runs. The factorial portion is a full factorial design with all combinations of the parameters at three levels. All the corresponding results of the shrinkage are given in Table 7.16.

Table 7.16 Results of experimental runs

Runs.	Factors				Response (Shrinkage)
	A	B	C	D	
1	0	0	0	0	0.98503
2	1	-1	1	1	0.9832
3	0	0	0	0	0.98503
4	1	-1	1	−1	0.98423
5	0	0	0	−1	0.9862
6	0	0	1	0	0.98487
7	−1	−1	−1	−1	0.9846
8	0	0	0	0	0.98503
9	1	−1	−1	1	0.98463
10	1	0	0	0	0.98707
11	1	1	1	1	0.98643
12	0	−1	0	0	0.9837
13	0	0	0	0	0.98503
14	−1	1	−1	1	0.98533
15	1	1	1	−1	0.98857
16	0	0	0	1	0.9846
17	0	1	0	0	0.98687
18	−1	0	0	0	0.98437
19	−1	1	1	1	0.98387
20	1	1	−1	−1	0.98906
21	−1	−1	1	1	0.98093
22	−1	−1	−1	1	0.98317
23	1	−1	−1	−1	0.98697
24	−1	1	−1	−1	0.9872
25	0	0	−1	0	0.9862
26	0	0	0	0	0.98503
27	−1	1	1	−1	0.9858
28	1	1	−1	1	0.98747
29	−1	−1	1	−1	0.9832
30	0	0	0	0	0.98503

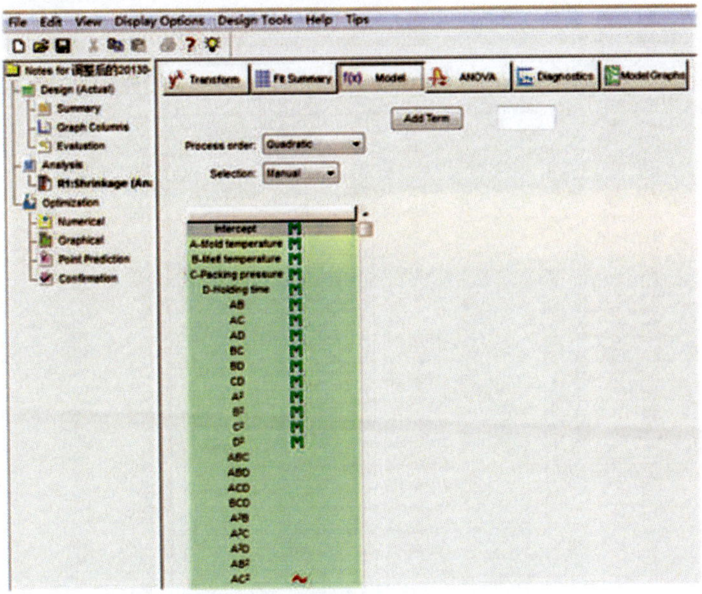

Fig. 7.45 Multinomial fitting of Experimental data

3. **Result discussion and optimization**

(1) RSM model

In this section, Design Expert 8.0.6 is used to fit the mathematical model. Design Expert is one of the most commonly used experimental design software, which integrates many functions, such as experimental design, data fitting, variance analysis, model suitability detection and so on. After importing the experimental data into Design Expert, the polynomial fitting is selected in the model fitting term, as shown in Fig. 7.45. Finally, the fitted mathematical model can be viewed in the model fitting results, as shown in Fig. 7.46.

The response equations in terms of coded facts and actual facts are depicted as follows:

$$
\begin{aligned}
S = {} & 0.776684357 - 1.352023 \times 10^{-3} \times T_d \\
& + 6.336718 \times 10^{-3} \times T_m - 1.897728 \times 10^{-3} \times P \\
& - 6.65351 \times 10^{-5} \times t + 1.375 \times 10^{-5} \times T_d T_m + 2 \times 10^{-6} \\
& \times T_d P + 1 \times 10^{-6} \times T_d t + 2.1375 \times 10^{-5} \times T_m P \\
& - 2.875 \times 10^{-6} \times T_m t - 3.5 \times 10^{-7} \times Pt + 1.00982 \\
& \times 10^{-5} \times T_d^2 - 4.5636 \times 10^{-5} \times T_m^2 + 2.69825 \times 10^{-6} \\
& \times P^2 - 2.70175 \times 10^{-6} \times t^2
\end{aligned} \tag{7.87}
$$

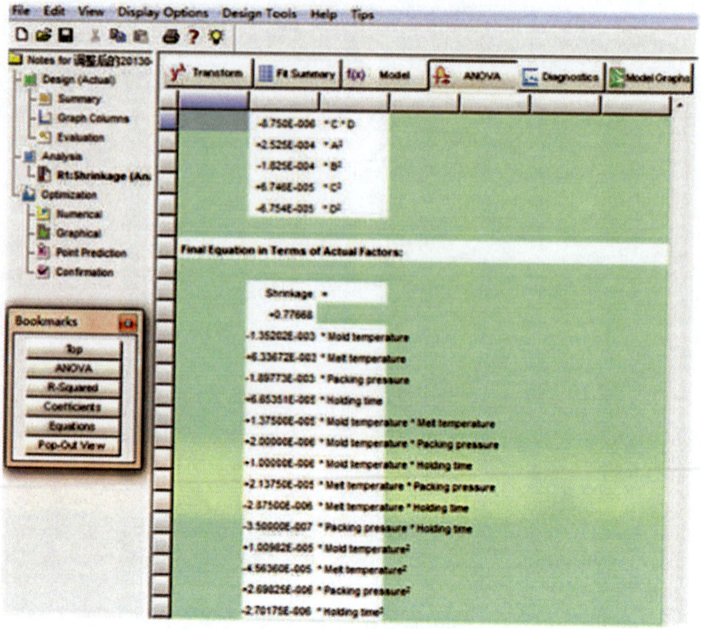

Fig. 7.46 RSM model

(2) ANOVA Results

In order to compare the residual variance with the term variance, the "Model F value" is calculated from a model mean square which is divided by a residual mean square. If the factors do not have any significant effects on the response, the ration will be closed to 1. On the contrary, the "Model P Value" smaller (less than 0.05) the effect of the terms in the model is more significant on the response. On the other hand, as a test to compare the model variance with the residual variance, the "F Value" on any individual factor terms should be not close to 1, which means the variances are close to the same and the effects of the term on the response are insignificant. Furthermore, if a "P Value" of any model terms is less than 0.05, the effect on the response of the individual terms is a significant effect on the response.

The ANOVA results of the shrinkage are shown in Table 7.17. The "Model F value" of 48.58 with a "Model P value" of less than 0.0001, implies the model is significant. The "Model P Value" of A, B, C, D are all less than 0.0001, which means that they are significant for the response. The model term BC is also significant for its "Model P Value" 0.0279 is less than 0.05. In contrast, other terms are not significant, since all of their "Model P Values" are more than 0.05.

Furthermore, in order to test the rationality and acceptability of the model, an investigation about the correlation coefficients (R^2), the adjusted R-squared ($Adj.R^2$), the predicted R-squared ($\Pr ed.R^2$), and the Adequate Precision is indispensable. They are explained by Eq. (7.88)

Table 7.17 Analysis of variance table

Source	Sum of Squares	df	Mean Square	F- value	P-value Prob > F	
Model	8.39E−05	14	6E−06	48.57794	<0.0001	Significant
A	2.04E−05	1	2.04E−05	165.222	<0.0001	
B	3.75E−05	1	3.75E−05	303.5434	<0.0001	
C	1.02E−05	1	1.02E−05	82.38963	<0.0001	
D	1.46E−05	1	1.46E−05	118.1155	<0.0001	
AB	3.02E−07	1	3.02E−07	2.450614	0.1383	
AC	4E−08	1	4E−08	0.324048	0.5776	
AD	1E−08	1	1E−08	0.081012	0.7798	
BC	7.31E−07	1	7.31E−07	5.922181	0.0279	
BD	1.32E−08	1	1.32E−08	0.107138	0.7479	
CD	1.23E−09	1	1.23E−09	0.009924	0.9220	
AA	1.65E−07	1	1.65E−07	1.337746	0.2655	
BB	8.63E−08	1	8.63E−08	0.699417	0.4161	
CC	1.18E−08	1	1.18E−08	0.095509	0.7615	
DD	1.18E−08	1	1.18E−08	0.095758	0.7612	
Residual	1.85E−06	15	1.23E−07			
Lack of Fit	1.85E−06	10	1.85E−07			
Pure Error	0	5	0			
Cor total	8.58E−05	29				

$$R^2 = 1 - \left[\frac{SS_{residual}}{SS_{residual} + SS_{model}}\right] \tag{7.88}$$

where *SS* is the abbreviation of "Sum of Squares".

$$Adj.R^2 = 1 - \left[\left(\frac{SS_{residual}}{df_{residual}}\right) \Big/ \left(\frac{SS_{residual} + SS_{model}}{df_{residual} + df_{residual}}\right)\right] \tag{7.89}$$

In the above equation, *df* is the abbreviation of the degree of freedom.

$$\Pr ed.R^2 = 1 - \left[\frac{PRESS}{SS_{residual} + SS_{model}}\right] \tag{7.90}$$

where *PRESS* is the sum of squares of the predicted data.

$$\begin{cases} Adeq.\Pr ecision = \frac{\max(\hat{Y}) - \min(\hat{Y})}{\sqrt{\bar{\nabla}(\hat{Y})}} \\ \bar{\nabla}\left(\hat{Y}\right) = \frac{p\sigma^2}{n} \end{cases} \tag{7.91}$$

where p is the number of model parameters (including intercept and any block coefficients), δ^2 is the residual mean square from ANOVA table, and n is the number of experiments

The ability of the correlation coefficients (R^2) is to measure the amount of variation of the model to the experimental data. It should be as close to 1 as possible. The difference between the adjusted R-squared measurement the amount of variation around the mean and the predicted R-squared measurement the amount of variation in new data should be within 0.20. The Adequate Precision, a signal-to-noise ratio, comparing the range of the predicted values at the design points to the average prediction error, should be greater than 4 if the model discrimination is adequate.

In this study, the correlation coefficient $R^2 = 0.97842$ is close to 1, so the model is reasonable and acceptable. The "Pred R-Squared" of 0.958279 is in good agreement with the "Adj R-Squared" of 0.874245. "Adeq Precision" with a ratio of 33.48089 indicates an adequate signal. In conclusion, this model can be used to explore the design space.

(3) Sensitivity Analysis of process parameters

According to the results of ANOVA, a sensitivity analysis for process parameters on the dimensional shrinkage variation can be conducted. The results are shown in Fig. 7.47. It clearly shows that the most significant factor is melt temperature (B), followed by mold temperature, holding time, and packing pressure. Their contribution percentages are 43.67%, 23.77%, 16.99% and 11.85%, respectively.

The effects of mold temperature and melt temperature on the dimension shrinkage variation while keeping the packing pressure at 25 bar and holding time at 5 s are shown in Fig. 7.48a. When the mold temperature increases from 20 to 30 °C and the melt temperature increases from 70 to 74 °C, the dimension shrinkage variation increases. Figure 7.48b shows the effects of packing pressure and holding time on

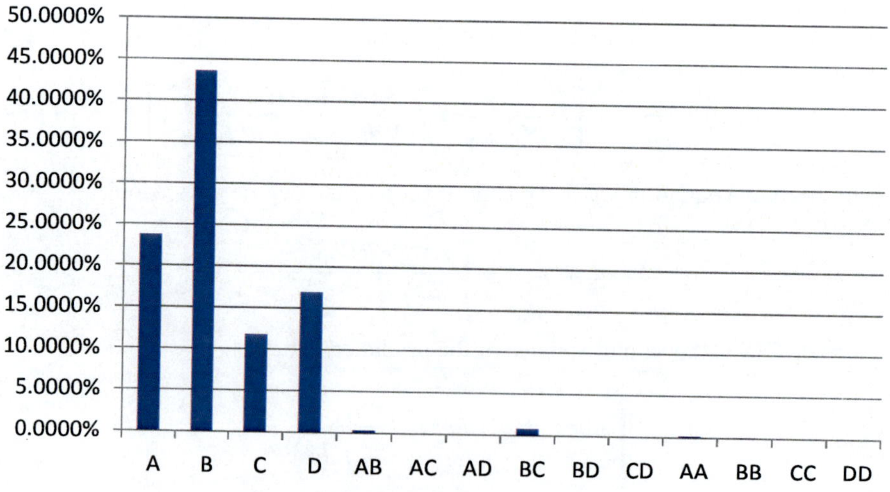

Fig. 7.47 The sensitivity of factors (coded factors) on the shrinkage

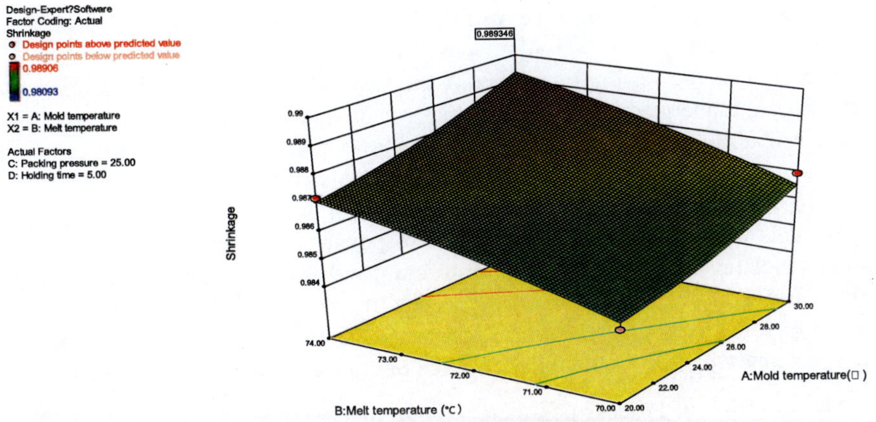

(a)

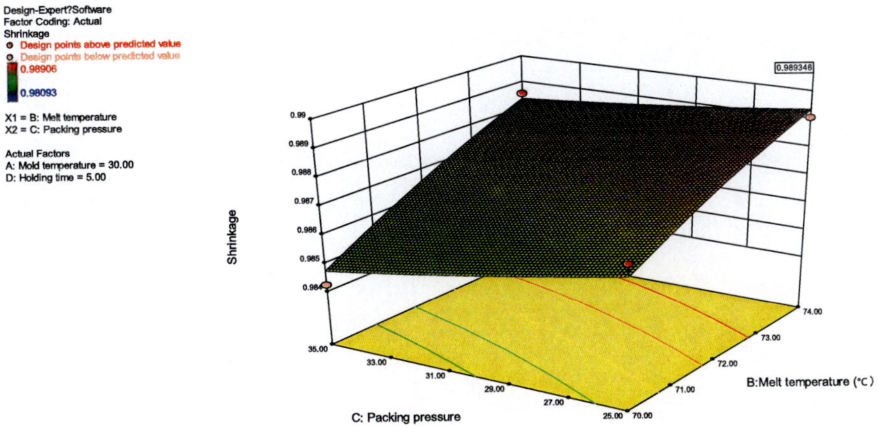

(b)

Fig. 7.48 **a** Response surface and contour plot of the shrinkage between the effect of packing pressure and holding time at 25 bar and 5 s, respectively. **b** Response surface and contour plot of the shrinkage between the effect of mold temperature and melt temperature at 30 °C and 74 °C, respectively

the responses while keeping the mold temperature at 30 °C and melt temperature at 74 °C. It clearly indicates that by decreasing the packing pressure and holding time, the shrinkage is maximized.

(4) Regression Models Adequacy Check

In order to confirm that all relevant information from the experimental data has been used by the mathematical predictive models, the regression model adequacy should be checked. Figure 7.49 shows the results of the actual value and predicted value of the experiment run.

According to the residual analysis, if the distribution of residuals is normally distributed, the adequacy of the mathematical predictive model is sufficient. The normal probability plot (Fig. 7.50) clearly shows that the points in experiments follow a straight line, which indicates that residuals follow a normal distribution. Furthermore, as a testing tool of the assumption of constant variance, the plot of the residuals versus the ascending predicted response values (Fig. 7.51) is a random scatter, which revealed that any unusual structure and obvious pattern do not exist for the reason that the plot is a random scatter. It tests the assumption of constant variance. These imply that the regression models proposed are adequate and there is no reason to suspect any violation of the independence or constant variance assumptions.

(5) Optimization of Process parameters based on GA.

In this study, GA is adopted to find the optimal process parameters for the minimal shrink of the thin-wall wax part, namely, the shrinkage should be as close to 1 as possible. Therefore, the GA optimization task is to minimize the objective function:

$$f(z) = -S \tag{7.92}$$

where

$$z = (T_d, T_m, P, t) \tag{7.93}$$

The varied range of the process parameters is:

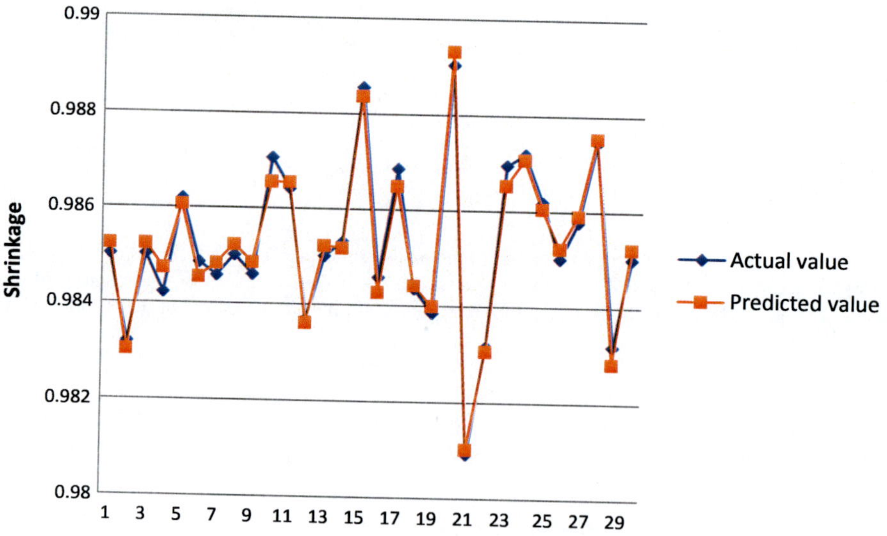

Fig. 7.49 The actual value and the predicted value of the experiment runs

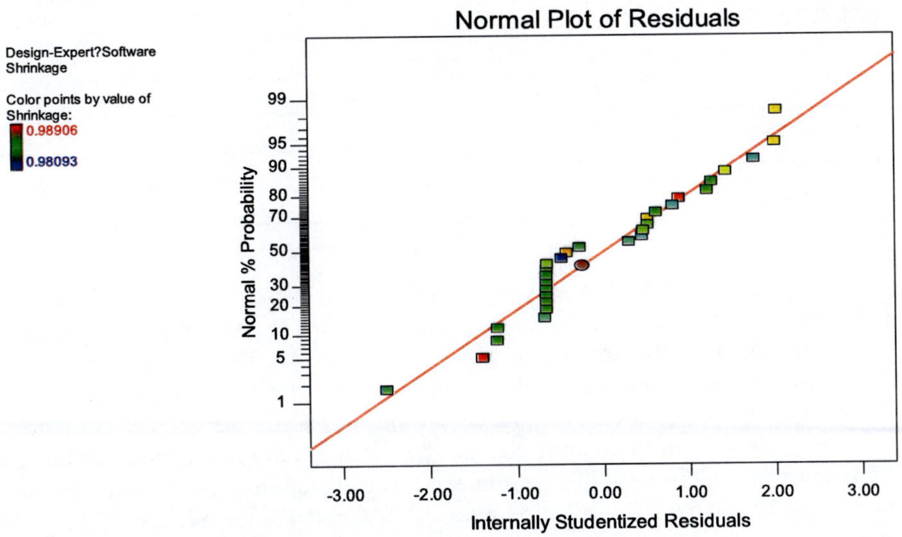

Fig. 7.50 Normal probability plot for the residual of shrinkage

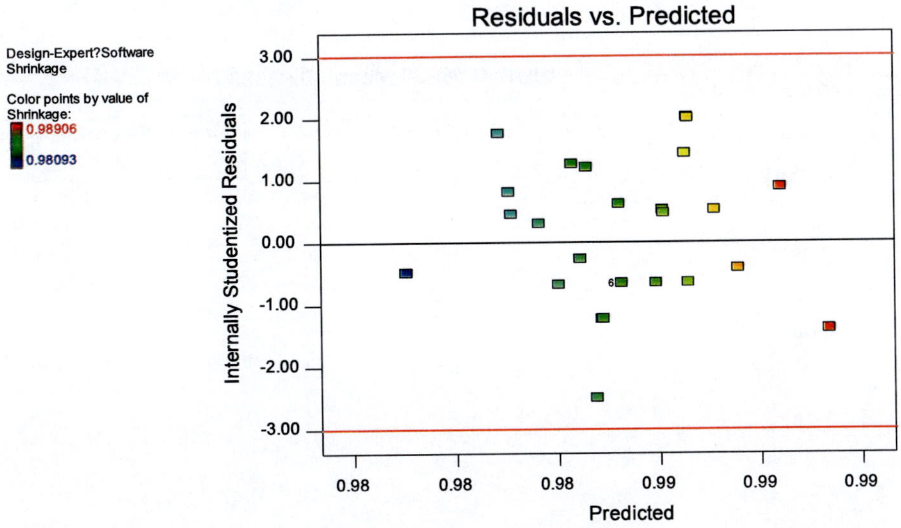

Fig. 7.51 Studentized residual plot for the predicted of shrinkage

$$\begin{cases} 20\,^{\circ}\mathrm{C} \le T_d \le 30\,^{\circ}\mathrm{C} \\ 70\,^{\circ}\mathrm{C} \le T_m \le 74\,^{\circ}\mathrm{C} \\ 25bar \le P \le 35bar \\ 5s \le t \le 15s \end{cases} \tag{7.94}$$

Table 7.18 GA parameters set-up

Parameter	Value at the following level		
	Level 1	Level 2	Level 3
Population size	30	60	90
Crossover rate	0.7	0.8	0.9
Mutation rate	0.01	0.1	0.2

Because the GA solution may be sensitive to its set-up, 27 experiments were performed to represent the combination of different levels for population size, crossover rate, and mutation rate, as shown in Table 7.18. The values in Table 7.18 are based on computational experiences and the corresponding literature review.

There were no significant differences in these 27 GA experiments. The effects of population size and crossover rate on the shrinkage are not significant, but the mutation rate exhibits a significant influence. The evolution of generations is shown in Figs. 7.52 and 7.53 for mutation rates of 20% and 1% respectively. Table 7.19 shows the corresponding values for the GA and injection parameters. It is verified that high mutation rates create offspring that push parameter values to the bounds.

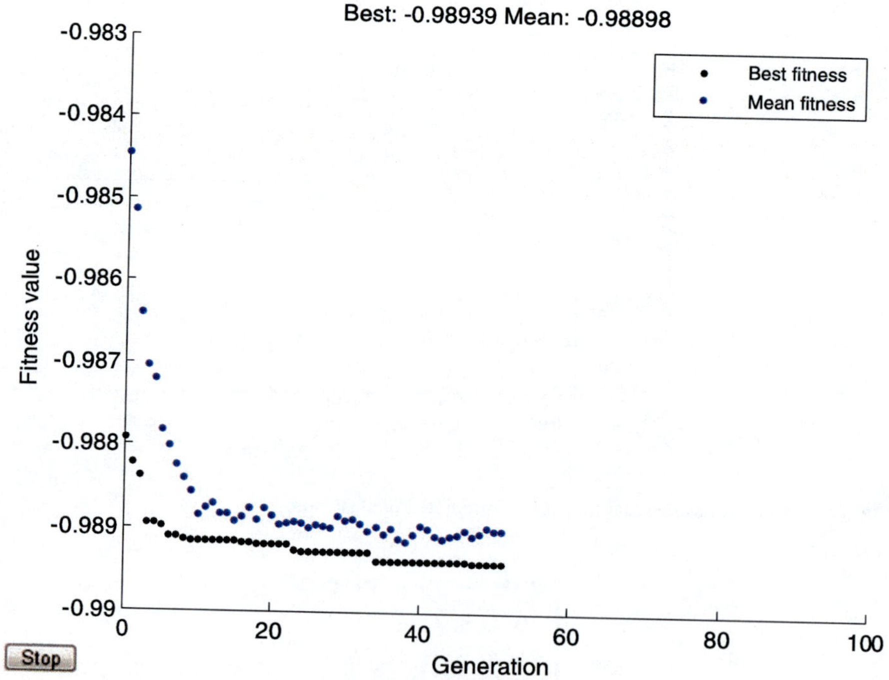

Fig. 7.52 Variation of shrinkage with number of generations (mutation rate, 0.2)

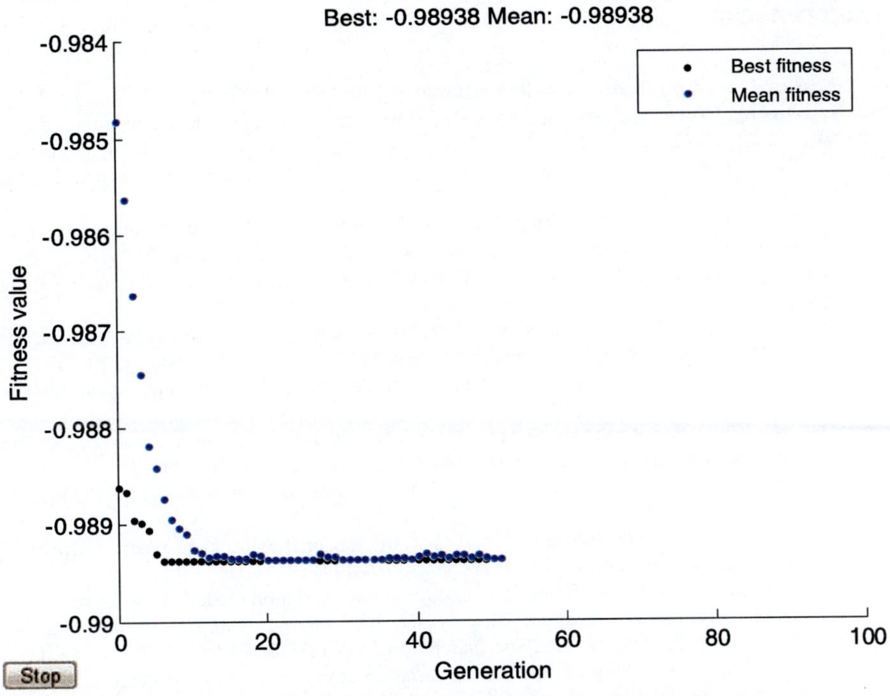

Fig. 7.53 Variation of shrinkage with number of generations (mutation rate, 0.01)

Table 7.19 Effect of mutation rate on the values of the injection parameters

	Value at the following levels	
	Level 1 (Fig. 7.9)	Level 2 (Fig. 7.10)
GA parameters		
Population size	60	60
Crossover rate	0.8	0.8
Mutation rate	0.2	0.01
Mean shrinkage	0.98898	0.98938
Injection parameters		
Mold temperature (°C)	29.998	29.868
Melt temperature (°C)	73.958	73.903
Packing pressure (Pa)	25.156	25
Holding time (s)	5.227	5.09

References

1. M. Sun, *Investment Casting*. (Beijing Institute of Technology Press, 2009)
2. L. Zhigang, History and development trend of investment casting industry in China. Casting **61**(4), 347–356 (2012)
3. L. Zhang, *Theory and Practice of Near Net-shape Investment Casting* (National Defense Industry Press, 2007)
4. Lamei Cao, Xing Tang, Yong Zhang et al., Progress of advanced near net-shape investment casting technology of superalloys. J Aeronaut Mater **26**(3), 238–243 (2006)
5. B. Jiang, *Practical Casting Technology* (Liaoning Science and Technology Publishing House, 2008)
6. M. Guo, *Moldflow Analysis on Big-sized Injection Molding Parts of Composite Materials and Their Optimization of Process Parameters* (Zhejiang University of Technology, 2012)
7. Chao Luo, Kan Long, Qiang Wang, Application of CAE technology into injection molds. Coal Min. Mach. **32**(5), 195–197 (2011)
8. Xiaoping Chen, Yun Wang, Hu Shugen et al., Optimization of injection molding processing based on the flow simulation technology. Mach. Des. Manuf. **2**, 68–69 (2005)
9. Q. Gao, Y. Chang, Q. Liang, Development of industrial computed tomography. Mod. Sci. Instrum. (4):5–8 (1993)
10. C. Liu, S. Jin, X. Lai, et al., Influence of complex structure on the shrinkage of part in investment casting process[J]. Int. J. Adv. Manuf. Technol. **77**(5–8), 1191–1203 (2015)
11. C. Liu, S. Jin, X. Lai, J. Luo, B. He, F. Li, The assembly variation modeling for the rear casing wax part based on polar coordinate[C], in *ASME 2015 International Manufacturing Science and Engineering Conference*, Am. Soc. Mech. Eng. V001T02A083-V001T02A083 (2015)
12. C. Liu, S. Jin, X. Lai, et al., Dimensional variation stream modeling of investment casting process based on state space method[J], in *Proceedings of the Institution of Mechanical Engineers*, Part B: J. Eng. Manuf. **229**(3), 463–474 (2015)
13. W. Bonilla, S.H. Masood, P. Iovenitti, An Investigation of Wax Patterns for Accuracy Improvement in Investment Cast Parts[J]. Int. J. Adv. Manuf. Technol. **18**(5), 348–356 (2001)
14. A.S. Sabau, W.D. Porter, Alloy shrinkage factors for the investment casting of 17-4PH stainless steel parts. Metall. Mater. Trans. B **39**, 317–330 (2008)
15. G.E.P. Box, K.B. Wilson, On the experimental attainment of optimum conditions. J. Roy. Stat. Soc.: Ser. B (Methodol.) **13**(1), 1–38 (1951)
16. R.H. Myers, D.C. Montgomery, C.M. Anderson-Cook, *Response Surface Methodology: Process and Product Optimization using Designed Experiments* (Wiley, 2016)
17. V. Beal, P. Erasenthiran, N. Hopkinson, P. Dickens, C. Ahrens, Optimisation of processing parameters in laser fused H13/Cu materials using response surface method (RSM). J. Mater. Process. Technol. **174**(1), 145–154 (2006)
18. L. Wu, K.-L. Yick, S.-P. Ng, J. Yip, K.-H. Kong, Parametric design and process parameter optimization for bra cup molding via response surface methodology. Expert Syst. Appl. **39**(1), 162–171 (2012)
19. S. Rajakumar, V. Balasubramanian, Establishing relationships between mechanical properties of aluminium alloys and optimized friction stir welding process parameters. Mater. Des. **40**, 17–35 (2012)
20. M.T. Chuang, Y.K. Yang, Y.H. Hsiao, Modeling and optimization of injection molding process parameters for thin-shell plastic parts. Polym.-Plast. Technol. Eng. **48**(7), 745–753 (2009)
21. C.P. Chen, M.T. Chuang, Y.H. Hsiao et al., Simulation and experimental study in determining injection molding process parameters for thin-shell plastic parts via design of experiments analysis. Expert Syst. Appl. **36**(7), 10752–10759 (2009)
22. D.E. Goldberg, J.H. Holland, Genetic algorithms and machine learning (1988)
23. C. Shen, L. Wang, Q. Li, Optimization of injection molding process parameters using combination of artificial neural network and genetic algorithm method. J. Mater. Process. Technol. **183**(2–3), 412–418 (2007)

24. C. Liu, S. Jin, X. Lai, D. Wang, Y. Wang, Experimental study of injection conditions for a thin-walled wax pattern using response surface methodology[C], in *Applied Mechanics and Materials*, Vol. 607, pp. 185–192, (Trans Tech Publications, 2014)
25. C. Reeves, J.E. Rowe, Genetic Algorithms: Principles And Perspectives: A Guide to GA Theory (Springer Science & Business Media, 2002)

Chapter 8
Advanced Adjusted Pressure Casting Process

Anping Dong, Dafan Du, Hui Xing, and Guoliang Zhu

With the continuous advancement of the aviation and aerospace industry, higher requirements are desired for the structure and size of casting parts, e.g., lightweight, holistic, and precision. In line with this development trend, the demand for precision forming of complex thin-walled casting parts is getting stronger and stronger. Defects, such as porosity, cold shut, and misrun for thin-walled casting parts, are easy to form during the filling process with a fast cooling rate that cannot guarantee the fluidity and feeding ability of the liquid metal. Besides, the larger Laplace force and viscous force caused by the thinner thickness of the mold cavity wall may influence the flow state of the front edge of the liquid metal during the filling process of the casting parts, which has a great impact on the precision forming of these thin-walled casting parts. Therefore, it becomes more and more important to explore new technologies of precision forming for complex thin-walled casting parts. Especially, the precision investment casting under anti-gravity condition has gradually become a research hotspot. The anti-gravity precision casting mainly includes low-pressure casting, counter pressure casting, adjusted pressure casting, etc. The study of low-pressure casting and counter-pressure casting has been relatively mature, whereas the adjusted pressure casting is mainly limited to the casting of light alloys such as aluminum and magnesium, which has not been applied in superalloy casting.

B. Sun et al., *Precision Forming Technology of Large Superalloy Castings for Aircraft Engines*, https://doi.org/10.1007/978-981-33-6220-8_8

8.1 Overview of Adjusted Pressure Precision Casting Technology for Complex Thin-Walled Casting Parts

8.1.1 Casting Method of Complex Thin-Walled Casting Parts

8.1.1.1 Complex Thin-Walled Casting Parts of Aluminum Alloys

In the early stage, complex thin-walled casting parts of aluminum alloys were mostly produced using sand mold gravity casting. The filling of liquid metal completely depends on gravity, and the filling speed is relatively slow and difficult to control. There are too many uncontrollable factors in the casting process. The dimensional precision of sand mold and sand core, resulting in a low surface finish. As a result, the quality of the casting parts is poor. There are many defects such as slag, porosity, shrinkage cavity, surface cracks, and shrinkage porosity, which can severely limit the production. After multiple studies and research, it was decided to solve the quality problems of complex thin-walled casting parts of aluminum alloy by adopting more advanced casting method [1].

During the World War II, the Britain Birmingham Cast Aluminum Co., Ltd. successfully developed the aero-engine cylinder block casting parts using the low pressure casting technology, which attracted a great attention of the casting industry in the world and motivated many countries to begin the study of anti-gravity casting method of aluminum alloys [2].

Recently, developed countries/regions such as Europe and the United States have been widely using anti-gravity casting technology to produce complex thin-walled casting parts of aluminum alloys with complex structure and excellent performance, which can be used in key parts such as aircraft and space shuttles, engine and cabin of cruise missiles, the engine of high-performance armored vehicles, and high-end electronic equipment. The resulting performance and dimensional precision exceed that of alloy forging parts and sheet metal parts of the same type [3].

8.1.1.2 Complex Thin-Walled Casting Parts of Magnesium Alloys

Compared with aluminum alloys, magnesium alloys are chemically very active. The liquid metal is easy to be oxidized in the filling process, and the generation of shrinkage porosity and shrinkage cavity will increase the contact area between magnesium alloy and the environment, making it vulnerable to electrochemical corrosion. Besides, the density of magnesium alloy is low, the crystallization temperature interval is large, and the feeding channel is easy to be blocked, which leads to the generation of defects such as shrinkage porosity and shrinkage cavity in the casting parts. The casting of magnesium alloys is much more difficult than that of aluminum alloys. The traditional gravity casting method is difficult to satisfy the growing demand for complex thin-walled casting parts of magnesium alloys [4–6].

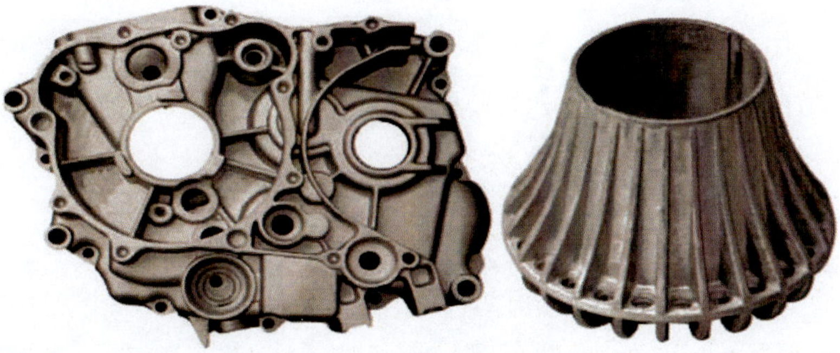

Fig. 8.1 Typical precision casting parts of magnesium alloy used in aero-engine [4]

At present, there are two main forming methods of complex thin-walled casting parts of magnesium alloy: die casting and anti-gravity casting.

The anti-gravity casting technology of magnesium alloy has emerged recently. Compared with die-casting technology, the anti-gravity casting method becomes more and more popular in the field of complex thin-walled casting parts of magnesium alloy because its characteristics of filling, solidification, and forming process under pressure are controllable. Developed countries in Europe and the United States have developed the integrated gas protection technology of magnesium alloys melting, transportation, and anti-gravity casting through in-depth research on the casting technology of complex thin-wall casting parts of magnesium alloys. The mass production of complex thin-wall anti-gravity casting parts of magnesium alloy is achieved by using the advanced process equipment integrated with computer-aided process design methods. Figure 8.1 shows typical precision casting parts of magnesium alloy used in aero-engine [4].

8.1.1.3 Complex Thin-Walled Casting Parts of Titanium Alloys

Complex thin-walled casting parts of titanium alloy exhibit typically features such as large overall dimensions, thin wall, complex structure with high precision requirements, together with strong chemical activity at high temperature and a wide crystallization temperature range. As a result, they are difficult to produce using the conventional casting method. At present, the centrifugal casting process is often used to produce the casting parts which have the characteristics of high metal utilization and fine structure, with a relatively low cost [7–10]. The centrifugal casting method is relatively mature in China. Chen Yuyong, et al. of Harbin Institute of Technology has successfully prepared Ti-Al-Zr alloy parts by centrifugal casting method (as shown in the Fig. 8.2) [11].

Although the forming problems during the complex thin-walled casting of titanium alloys can be solved using the centrifugal casting method [10, 11], the existence of centrifugal force leads to the uneven distribution of chemical composition of the

Fig. 8.2 Complex thin-walled casting parts of titanium alloy used in the engine [11]

casting parts, which makes the metallurgical quality, microstructure, and mechanical properties of the casting parts location-dependent. To get better quality casting parts, researchers began to explore to produce thin-walled casting parts of titanium alloy by using anti-gravity casting.

The anti-gravity casting technology was first applied successfully to the precision casting of TiAl alloy by Noda [12], who obtained the turbine with excellent performance. The maximum rotation speed of the turbine can break through the limit of the traditional one. Researchers from Harbin Institute of Technology of China [13, 14] investigated the influence of various process parameters and process factors on the casting quality of complex thin-walled casting parts of titanium alloy, put forward the corresponding casting plan, and completed the preparation of small complex thin-walled casting parts. Compared to the previous method, the casting parts perform some advantages: higher density, lower composition segregation, finer structure, fewer shrinkage cavity, no interface reacts, and lower cost. This technique reflecting the advantages of the anti-gravity casting method in titanium alloy casting.

8.1.1.4 Complex Thin-Walled Casting Parts of Superalloys

Complex thin-walled casting parts of superalloys are widely used in the aerospace industry. According to the requirements of aero-engine that is the core equipment and the heart of aircraft. The combat capability and transportation capacity of military aircraft depend on the design and manufacturing level of the aero-engine. To improve the performance of the aircraft, it is necessary to reduce the weight as much as possible and to improve the thrust specific and structural efficiency. 70–80% of the realization of this high-performance engine technology depends on the material and structure that are manufactured using the thin-walled casting parts, such as the turbine rear frame (TRF), the pre-swirl nozzle, the floating wall tile of the engine combustion chamber, the pre-diffuser case, the turbine interstage case and so on. The integral precision casting technology has become the mainstream route of the manufacturing technology of aero-engine structural parts.

At present, large complex thin-walled precision casting parts of superalloy has been successfully applied to many kinds of military and civil engines such as CFM56 series, JT90D, PW4000, RB211, EJ200, etc. [15]. In 1984, Howmet Company applied the integral casting technology of K4169 superalloy to the production of the exhaust guide ring of a high-pressure compressor of the CFM56 engine, and successfully manufactured the ring complex casting parts with diameter more than 500 mm. The integral casting of the compressor case and combustion chamber shell of the engine was completed on this basis [16]. After that, many large-scale complex thin-walled fine-grained integral casting parts were completed using MX fine-grained casting process and the complex thin-walled casting parts of superalloy were successfully applied to the diffusion cases of a rocket engine, such as the CF680 engine turbine rear frame of superalloy with casting parts size more than 1000 mm [17]. GE Company and PCC Company also have a very high level in the field of large-scale complex thin-walled casting parts superalloy. They have been able to produce complex engine cases with a diameter of more than 1100 mm, a minimum wall thickness of less than 2 mm, and a large number of hollow support plates with different structures.

Currently, the superalloy precision casting technology of China is relatively backward, and the early structural parts of the superalloy case have adopted the technology of "Plate Bending Forming-Forging Rough Machining-Welding Machining Forming". This process involves many processing procedures and a long manufacturing cycle, with a low material utilization rate and high manufacturing cost. Besides, it is easy to cause serious deformation of components due to many and dense welds and it is difficult to control the deformation and defects in the heat-affected zone of the welding site, which is easy to cause poor rigidity of the overall structure. In recent years, some research institutes, such as Shanghai Jiao Tong University, have carried out a lot of studies on the integral precision casting parts of large-scale complex thin-walled components of superalloy and have made certain progress. In the past, gravity casting was used to develop such casting parts. Because the thin-walled area of casting parts is getting larger, the structure is becoming more and more complex, and the surface quality requirements are getting better. The design of the casting system under the gravity casting process has certain limitations, the yield is only 20%, and some complex structures of casting parts are prone to quality defects such as shrinkage porosity, shrinkage cavity, deformation, and size out of tolerance. The weight of the alloy used in the casting system is about 4 times that of the casting itself, which greatly increases the manufacturing cost of the casting. It is thus urgent to develop new technologies and equipment to solve the casting and forming problems of this kind of superalloy components.

8.1.2 *Precision Forming Technology of Adjusted Pressure Casting*

The adjusted pressure casting is a complex thin-walled casting forming technology based on the traditional anti-gravity casting technology. In 1987, Prof. Yaohe Zhou and Jianmin Zeng proposed a method of adjusted pressure casting based on anti-gravity differential pressure casting to solve the problem of complex thin-walled casting forming. In addition to the advantages of traditional anti-gravity casting technology, it also has its unique characteristics. First of all, vacuum degassing shall be carried out before mold filling and the negative pressure shall be maintained for a period, which is conducive to the gas precipitation in the liquid metal, making the liquid metal purer and removing the gas and water adsorbed on the mold surface to prevent defects. Secondly, when filling, negative pressure filling can avoid the phenomenon of aspiration and air entrainment of liquid metal, and there is no blocking effect of air back pressure in the cavity on filling, which not only enhances the filling ability but also maintains the stability of filling. Consequently, this stable filling process provides favorable conditions for the sequential solidification of casting parts. Finally, after filling, the liquid metal solidifies under positive pressure, which prevents the liquid metal from flowing back and is conducive to the feeding of the liquid metal. Meanwhile, this positive pressure solidification environment reduces the occurrence of shrinkage porosity and shrinkage cavity, restrains the precipitation of residual gas to reduce the formation of vacuum defects, improves the density of the casting, and results in a good performance of the casting parts. The advantages of adjusted pressure casting improve the strength and plasticity of the casting parts and change the microstructure and properties of the casting parts, which is suitable for the forming of complex thin-walled casting parts with a thickness of 1–5 mm.

Additionally, due to the effect of vacuum degassing and negative pressure filling, the exhaust volume of the mold is very low during the filling process, which reduces the requirements for the permeability of the mold. Moreover, the pressure difference between the upper and lower chambers follows the principle of the minimum pressure difference and the effective pressure difference on the mold is not very large. As a result, the strength of the mold is not required to be high. In this way, the adjusted pressure casting has strong adaptability to the mold, which can be applied to various casting methods such as metal mold casting, sand mold casting, and investment casting to produce complex, thin-walled and integral metal casting parts that are difficult to be poured by other methods, to solve the filling problem in the pouring of complex thin-walled casting parts.

The main part of the adjusted pressure casting device is shown in Fig. 8.3, which mainly includes two isolated pressure chambers with controllable gas pressure and the control equipment for regulating the pressure. The upper and lower pressure chambers are connected with the positive pressure and negative pressure control system, and the positive pressure and negative pressure are controlled at the same time. Figure 8.4 shows the pressure variation curve with time in the whole process of adjusted pressure casting.

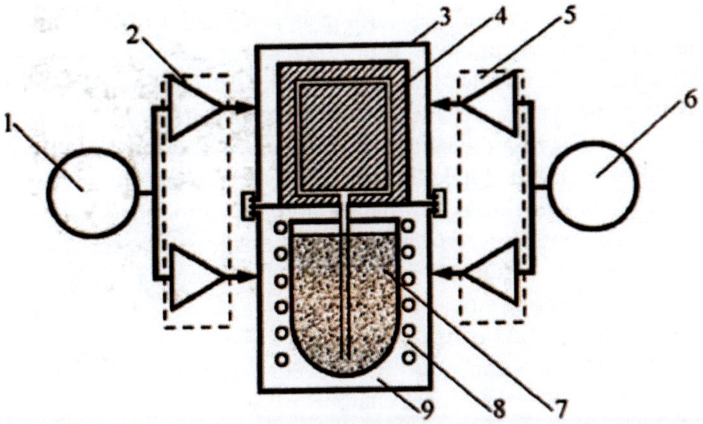

Fig. 8.3 Schematic diagram of adjusted pressure casting equipment for aluminum alloys: *1* Pressure tank *2* Positive pressure control system *3* Upper pressure chamber *4* Casting mold *5* Negative pressure control system *6* Vacuum tank *7* Molten metal *8* Holding furnace *9* Lower pressure chamber

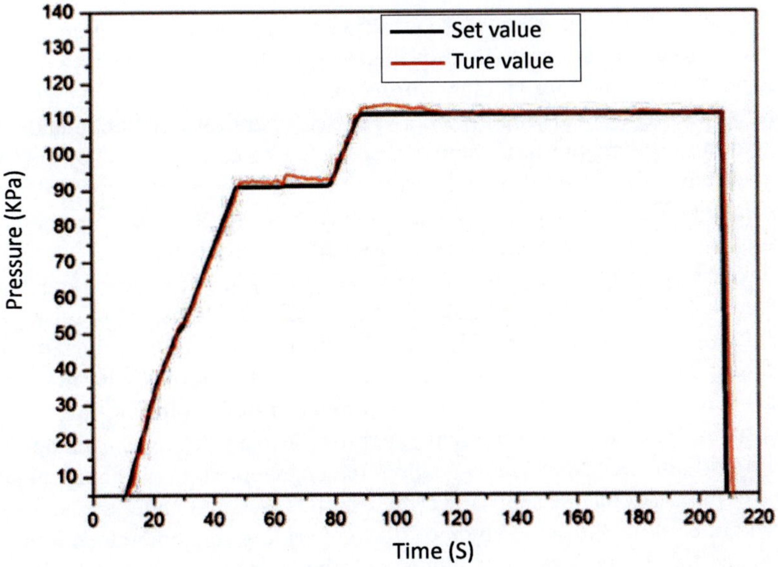

Fig. 8.4 Curve of pressure versus time during adjusted-pressure cast

As a kind of anti-gravity casting technology used in the precision casting of complex thin-walled casting parts of aluminum and magnesium alloys, the adjusted pressure casting has a great advantage in solving the problem of thin-walled casting. At present, with the further development of major equipment such as aero-engine, aerospace vehicle, and ground gas turbine in China, the demand for large-scale complex ultrathin-walled components are increasing. Adopting the gravity casting method for casting parts with wall thickness greater than 1.5 mm, through the special design of the casting system, the area of the thin-walled area less than 100 × 100 mm can be achieved. If the area of the thin wall is further enlarged, there will be looseness, incomplete filling, cold shut, and other casting defects. However, for the thin-walled casting parts with wall thickness less than 1.5 mm, it is difficult to complete the entire filling of the melt in the thin-walled area by adopting the gravity casting process, even if the area is slightly less than 100 mm × 100 mm. Therefore, it is very necessary to explore and establish the adjusted pressure casting device and technology for the complex thin-walled casting parts of the superalloy.

Pressure change in the upper and lower chambers during the adjusted pressure casting process where P1 is the upper chamber pressure and P2 is the lower chamber pressure.

Based on the processing characteristics of adjusted pressure casting, the physical properties, fire resistance, pressure regulating, and casting system of superalloy must be considered comprehensively if the process is used for the precision casting of superalloys.

(1) The density of superalloy is about three times that of aluminum alloy. It is necessary to solve the problem of metal liquid filling under external pressure. Superalloys need more pressure during the filling process. However, due to the difference in physical properties between the superalloy and light metals such as aluminum and magnesium, there will be the backpressure caused by the Laplace effect when the superalloy melt rises along the riser, together with the impact effect of liquid alloy on the inside of the shell. As a result, we should seek solutions from two aspects of dynamics and hydrodynamics.
(2) The casting temperature of superalloy is about 1500 °C, which is much higher than that of common aluminum and magnesium alloys. This raises higher requirements for the riser and other refractories used in the adjusted pressure casting. For example, the riser should use special materials with high thermal shock resistance to replacing the traditional graphite pipe or cast iron pipe, to avoid the erosion and thermal shock of the riser during the rising process of liquid alloy. For superalloy sealing, the engagement between the upper and lower tanks adopts the snap ring type, which is fastened under the action of the hydraulic pump and sealed by the silica gel sealing ring embedded in it.
(3) The melting process of superalloy should be carried out under a high vacuum condition, which is another important feature different from the adjusted pressure casting of aluminum and magnesium alloys. At present, the aluminum alloy adjusted pressure casting device adopts the structure of upper and lower tanks. There is a partition in the middle to separate the upper and lower pressure tanks.

The shell and riser are baked and preheated in an external oven and then put into the upper-pressure tank. If the superalloy also adopts this design, the vacuum degree will be greatly affected during the installation of the riser and the shell. Inert gas protection should be considered in the design to avoid more casting defects because of strong oxidation.

(4) Requirements for shell: in common precision casting, the choice of shell material and dip-coating process plays a decisive role in the surface quality and internal quality of casting parts. In particular, the surface layer and back layer slurry and the composition and physical properties of the molding sand have a crucial impact on the quality of casting parts. The application of these materials is required to be higher by the adjusted pressure precision casting, and the shell is required to bear higher pressure. Generally, the packing pressure of the adjusted pressure casting is about 0.6–1.0 Mpa.

(5) Special design of casting system: the key of adjusted pressure casting is to obtain the differential pressure so that the liquid metal can smoothly reach the complex ultrathin wall area and complete the filling procedure. It is undoubtedly a significantly important topic on how to design a reasonable casting system and to choose the appropriate casting parameters to realize the smooth filling of metal and the specific solidification sequence.

8.2 Adjusted Pressure Casting Equipment of Superalloys

In the 1980s, Prof. Yaohe Zhou and Jianmin Zeng established the principle prototype of the adjusted pressure casting in the School of Materials Science and Engineering of Northwestern Polytechnic University, and operated it and produced on a small scale under laboratory conditions for a long time. As early as around 2002, the automation controlled adjusted pressure casting engineering application equipment was developed by Weidong Huang, Wanqi Jie, Qitang Hao, etc., and the mass production was carried out in AVIC Jiangxi Hongdu Aviation Industry Group Company Ltd. Jianfei Sun of Harbin Institute of Technology had also carried out the development and technology research of the integrated equipment of adjusted pressure/differential pressure/low pressure and developed various types of integrated casting equipment of anti-gravity adjusted pressure/differential pressure for many research institutions and production enterprises. At present, the adjusted pressure casting technology has been successfully applied to the engineering production of aluminum alloy and magnesium alloy casting parts. Based on the principle of generality and technological advantages of the adjusted pressure casting technology, and the research and development requirements for large complex thin-walled casting parts in key industries such as aerospace and ground gas turbine in recent years, Shanghai Jiao Tong University with the assistance of Shanghai Huizhong Electric Furnace Co., Ltd. and Prof. Jianfei Sun's research group of Harbin Institute of Technology, has designed and manufactured the superalloy precision casting voltage regulating device.

Figure 8.5 is the schematic view of a device of superalloy adjusted pressure casting. The adjusted pressure casting device is composed of the upper and lower chamber vacuum pressure tanks, induction heating device, vacuum system, riser, pressure control device, and other main components. The device can control not only the positive pressure but also the negative pressure. The device includes two isolated chambers with independent and controllable internal air pressure and control equipment for gas pressure regulation. The upper and lower cavities are separated by a partition, while the isolation between the two cavities and the external environment is ensured at the same time. An induction heating coil and a crucible are installed in the lower pressure chamber to melt and hold the superalloy liquid. The power is regulated by the medium frequency power supply and thermocouple to control the temperature of the alloy liquid. The shell is placed in the upper tank, one end of the shell opening is facing down and connected with the riser, and the bottom end of the riser is inserted into the molten metal liquid level. The upper and lower pressure chambers are respectively connected with the pressure control system through pipes, and the precise control of the air pressure in the pressure chamber from negative pressure to positive pressure is realized by leading the gas into or out of each pressure chamber. Figure 8.6 is the picture of the device.

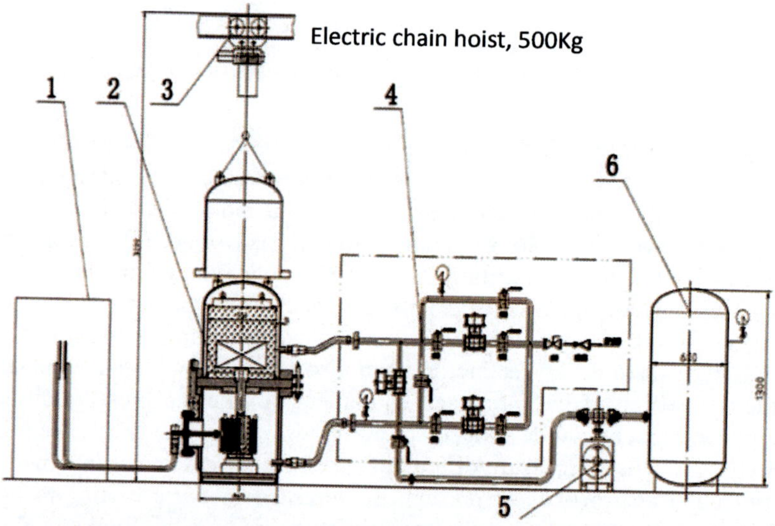

Fig. 8.5 Schematic diagram of the superalloy adjusted pressure casting equipment: *1* Medium frequency power supply *2* The main body of adjusted pressure casting furnace *3* Lifting vehicle *4* Pneumatic control system *5* Vacuum line *6* Vacuum tank

Fig. 8.6 Adjusted pressure precision casting equipment of complex thin-walled superalloy parts

8.2.1 Structural Design

The lower pressure tank is of flat sandwich structure, and the material is 16 MnR special steel for pressure vessel. The structural pressure is designed and manufactured according to 1.0 MPa pressure. The lower tank is equipped with an electromagnetic induction heating coil and crucible. The upper-pressure tank is of dome-roof structure, and the material is 16 MnR special steel for pressure vessel. The structural pressure is the same as the lower pressure tank, and the shell is placed in the upper tank. The intermediate partition is mainly used to isolate the upper and lower pressure tanks, place the riser and the shell. The material of the intermediate partition is 16 MnR special steel for pressure vessels. The riser hole in the center of the intermediate partition is provided with the connecting thread for installing the sealing plate flange, which is used for the sealing during the equipment commissioning. The upper and lower pressure tanks and intermediate partitions are clamped by a movable joint screw rod. When installing the mold and riser, we use the chain hoist to lift the upper tank.

8.2.2 Pneumatic Control and Seal Design

In the process of adjusted pressure casting, the vacuum degree and sealing of the upper and lower tanks are the key control parameters of the pressure regulating device. Consequently, it is difficult to accurately control the pressure difference between the upper and lower pressure tanks. To solve this problem, the upper and lower seal chambers are connected through pipelines with the vacuum pressure control device and the protective gas pressure control device, and the gas storage tank of the protective gas pressure control device is connected with sensors, all of which are connected with the control system.

The liquid metal is pressed smoothly into the mold cavity by the inert gas or compressed air that has been fully dried and filtered. Through the air distribution mechanism, the gas enters the upper and lower pressure chambers, respectively, and the pressure of the lower chamber is fed to the casting through the infusion pipe. The pressure of the upper chamber acts on the mold to ensure that at the same time of solidification under high pressure, the force on the mold shell itself is very small.

The initial pressure in the cavity before filling should be considered comprehensively according to the mold material, the complexity of the casting structure, and the physical parameters of the superalloy. If the air permeability of the mold is poor and if the structure of the casting is complex, the initial pressure should be taken as a relatively low value; otherwise, it should be taken as a relatively high value. o meet this requirement, the pressure distribution in the system shall meet:

$$\begin{cases} dP_1/dt = dP_2/dt \\ P_1 = P_2 + \Delta P \end{cases} \tag{8.1}$$

where P_1 is the lower chamber pressure value, P_2 is the upper chamber pressure value, ΔP is filling pressure, and t is filling time.

The initial pressure in the cavity before filling should be considered comprehensively according to the mold material, the complexity of the casting structure, and the physical parameters of the superalloy. If the air permeability of the mold is poor and the structure of the casting is complex, the initial pressure should be taken as a relatively low value; otherwise, it should be taken as a relatively high value.

The seal between the upper and lower tanks is O-ring. The seal between the riser and the middle partition as well as between the mold and the riser depends on the seal of graphite packing, fire-resistant cotton, and the mold itself (pressing iron). The density of superalloy is large, the pressure required by the adjusted pressure casting is large, and the equipment manufacturing is difficult. To solve this problem, the way of increasing large capacity gas storage tank and short-time three-dimensional pressurization is adopted. According to the principles of fluid mechanics and thermodynamics, combined with the coupling phenomenon of filling process parameters, the filling equation under the condition of adjusted pressure casting is given as

$$\dot{P}(t) = \Delta P \cdot \alpha / \left[R \cdot \rho \cdot C \cdot \ln\left(T_\rho - T_0 / T_l - T_0 \right) \right] \tag{8.2}$$

where α is pressure speed, P(t) is pressure function, ΔP is pressure difference, $T\rho$ is casting temperature, T_0 is mold temperature, T_1 is liquidus temperature of the alloy, C is the specific heat of alloy, and R is 0.5 times the wall thickness of the casting parts.

8.2.3 Design of Riser

In the structure of the riser, the first thing is to ensure that the liquid metal flow in the riser is laminar, so the filling speed and time of anti-gravity casting should be reasonably selected. The filling time and speed can be determined according to the size and shape parameters or experience of the casting parts. When the filling time and speed are determined, the radius of the riser can be determined using

$$\mathrm{R} = \sqrt{\frac{V}{tvt}} \tag{8.3}$$

where R is the radius of the riser, V is the volume of mold, v is filling speed, and t is filing time.

In most anti-gravity castings, the outlet area of the riser is larger than the calculated value to ensure smooth filing and the outlet of the riser matches the cast gate. At present, the inner diameter of the riser tube commonly used in the pressure regulating casting of Al-Mg alloy is generally (60–80) mm, and the diameter at the outlet is generally greater than 30 mm. If the alloy has a high melting point, the riser should be thicker. For superalloys with a high melting point, the diameter of the riser should be larger than the calculated value to avoid the blockage caused by the rapid solidification of the melt.

The top of the riser is tapered to facilitate the backflow of the molten metal. The area of the riser outlet is the area of the straight runner. It should be larger than the maximum thermal joint area of the casting to ensure that the casting is fully compensated. In order to ensure the smooth flow of the molten metal, the ratio of the pressure between the molten metal surface in the crucible to the riser tube is preferably controlled at 20 –45.

As for the material selection of the riser tube, the pouring temperature of the superalloy pressure regulating casting is up to about 1500 °C, which requires that the riser tube has good thermal shock resistance and high-temperature corrosion resistance. Ordinary alumina corundum tubes cannot meet the technical requirements. Therefore, a special refractory material of Si_3N_4 combined with SiC is planned to be used in the superalloy pressure regulating casting device. This material has good thermal shock resistance, high-temperature flexural strength, wears resistance, together with a series of advantages such as low linear expansion rate and high thermal conductivity. The short-term use temperature can reach 1600 °C. Table 8.1 list the physical properties of the material.

Table 8.1 The physical properties of the special refractory material of Si_3N_4 combined with SiC

Name of product		Si_3N_4 combined with SiC
Composition (%)	Si	≥72.0
	Si_3N_4	≥22.0
	Fe_2O_3	≤0.3
Apparent porosity (%)		≤16.0
Bulk Density (g/cm3)		2.68
Compressive strength at room temperature (Mpa)		≥180.0
Flexural strength (Mpa)	Room temperature	≥45.0
	1400 °C	≥55.0
Thermal conductivity (W/mK)	650 °C	19.5
	1000 °C	17.2
Temperature coefficient $\alpha/(cm^2/s)$	650 °C	0.064
	1000 °C	0.053
Linear expansion coefficientα/K^{-1}	30–1000	4.02×10^{-6} (K^{-1})
	30–1100	4.02×10^{-6} (K^{-1})

8.2.4 *Main Parameters of Superalloy Pressure Regulating Casting Device*

(a) Maximum temperature: 1650 °C; intermediate frequency power: 60 KW; intermediate frequency voltage: 375 V; heating frequency: ~4000 Hz; crucible capacity: 20 kg.
(b) Multi-line air inlet, the minimum pressure required for the lower tank of the filling molten metal: 0.05 MPa, the pressure of the lower tank at the end of filling is 1.0 MPa, and the pressure of the upper tank during solidification and filling is 0.8 MPa.

8.3 Process Design of Pressure Regulating Casting

8.3.1 *Selection of Casting Solidification Mode*

The choice of the solidification mode of the casting is the guide for the determination of casting parameters. Only after the solidification mode of casting is determined can the casting system be determined. The choice of casting solidification method relies on many factors, such as the casting structure (shape, wall thickness, and distribution complexity), casting properties (thermal conductivity, thermal storage coefficient, wall thickness), casting alloy properties (crystallization temperature interval, thermal conductivity), Shrinkage rate), the external environment during casting (casting method, pressure, gravity, and whether there is forced cooling), etc.

For superalloy castings under the condition of anti-gravity pressure regulated casting, the molten metal is poured from the bottom to top during casting, so the solidification mode of the casting should generally be top-down, and the gate finally solidifies, which is the sequential solidification method.

8.3.2 Design of Superalloy Pressure Regulating Casting Pouring System

During the anti-gravity pressure regulating the casting process, in addition to the requirements of sequential solidification of the casting system, it should also ensure that the molten metal flows smoothly and removes the slag. Meanwhile, the casting system should improve production efficiency, save molten metal, and be easy to remove the riser after pouring. There are four main types of anti-gravity casting gating systems:

(1) Thick wall and box castings: For thick wall or box castings, several internal runners are often used to ensure the short filling process and can also play a role of shrinkage.
(2) Cylinder castings: The diameter of cylinder castings can vary significantly. For castings with a diameter of less than 400 mm, one riser is enough. For cylinders with a diameter of more than 400 mm, however, two risers are often used. The molten alloy liquid enters the gate sleeve through the riser tube, fills the confluence ring, and then enters the cavity from the gap-type inner runner.
(3) Casting system with a thick underside and a larger casting area: When it is difficult to place multiple casting runners in this type of casting, a riser tube can be used to branch into the casting.
(4) Large and complex thin-walled ring and collar structure: In recent years, due to the rapid growth in the application of this structure in key industries such as aerospace and gas turbines, more and more attention has been paid to the research of such large and complex thin-walled ring and ring structure castings.

8.3.3 Design of Molten Metal Surface Pressure Process Parameters for Pressure Regulating Casting

According to the technical principle of pressure regulating casting and the control requirements of the equipment, the parameters such as the ascending speed, ascending pressure, filling speed, filling pressure, solidification pressure, solidification pressure time, and the pouring temperature must be determined.

8.3.3.1 Lifting Speed and Lifting Pressure

Ascending speed refers to the flow speed of the molten alloy in the riser tube. One of the important characteristics of anti-gravity casting is that the molten alloy flows smoothly during the ascent and filling process, which is not likely to cause turbulence.

Lifting pressure refers to the pressure when the molten alloy reaches the outlet of the riser tube or the bottom of the cavity. At the beginning of lifting, the molten metal in the riser tube and the crucible are on the same level, and the molten metal in the riser tube can be raised only by a small increment of pressure on the liquid surface of the crucible. As shown in Fig. 8.7, during the lifting process, when the liquid level in the crucible drops by ΔH, the height of the liquid level in the riser tube is,

$$\Delta \mathrm{h} = \left(\frac{d1}{d2}\right)^2 \Delta H \tag{8.4}$$

$$\text{If } \mathrm{H}' = \Delta \mathrm{H} + \Delta \mathrm{h} = \left[1 + \left(\frac{d1}{d2}\right)^2\right] \Delta \mathrm{H} \tag{8.5}$$

According to $P = \rho gh$, it can be known that the liquid lift pressure required at this time is,

$$P_{lift} = K\rho g H' = K\frac{\rho H'}{1033.6} \times 0.098 \tag{8.6}$$

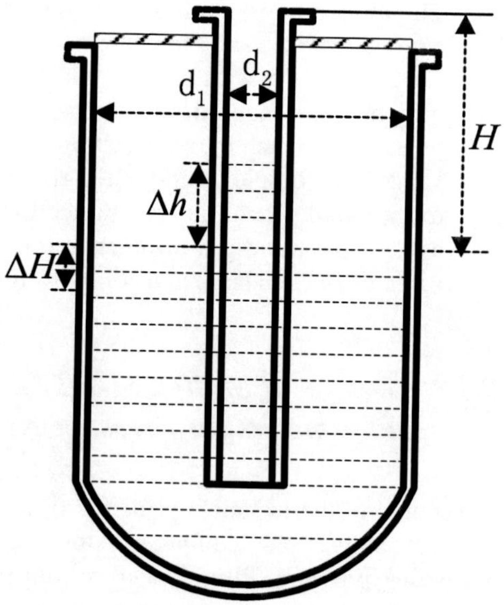

Fig. 8.7 Schematic diagram of the calculation of the pressure to elevate the liquid level during the anti-gravity casting

In the above equation, P_{lift} is the gas pressure acting on the molten metal surface (MPa); K is the resistance coefficient, generally 1.0–1.5, lower limit for thick wall castings, upper limit for thin wall castings; ρ is the density of molten metal (g/cm^3); H′ is the distance (cm) between the molten metal level in the riser tube and the molten metal level in the crucible; 0.098 is conversion coefficient between kg/cm^2 and MPa; 1033.6 is pressure conversion unit.

The resistance coefficient K includes the backpressure of the gas in the cavity, the friction between the alloy and the inner wall surface of the mold, the friction in the molten alloy, the resistance of the surface oxide film, the resistance of the alloy to decrease in temperature and the viscosity to increase, and so on. Therefore, the resistance coefficient is related to the height of the casting, wall thickness, mold material, pouring temperature, exhaust conditions of the mold, and alloy composition.

It can be seen from Eq. (8.6) that with the progress of the liquid lifting process, the required liquid lifting pressure increases continuously. When the liquid lifting ends, the liquid level in the crucible drops, and the liquid lifting pressure required reaches the maximum, i.e.,

$$P_{lift} = \mathrm{K}\rho g\left[1 + \left(\frac{d1}{d2}\right)^2\right] H \times \frac{0.098}{1033.6} \tag{8.7}$$

where d_1 is the diameter of the crucible (m), d_2 is the diameter of the riser tube (m), and H is the distance (m) between the nozzle of the riser tube and the unpressured free liquid level in the crucible.

8.3.3.2 Filling Speed

Filling velocity refers to the rising velocity of molten metal in the mold cavity, which is one of the important technical parameters of counter casting. When the section area of the riser outlet is fixed, the filling speed depends on the increasing speed of the gas pressure acting on the surface of the molten alloy. The faster the pressure speed is, the faster the mold filling speed is. On the contrary, the slower is the pressure speed, the slower the filling speed of the molten alloy is. It is very important to choose the filling speed reasonably in the pressure regulating casting. If the mold filling speed is too fast, the molten alloy will be splashed and billow when it enters the mold cavity, resulting in oxidized slag inclusion and porosity defects. On the other hand, if the filling speed is too slow, the casting may have cold insulation and under-casting defects.

8.3.3.3 Pressure Rate

The pressure rate is determined by the thickness of the casting wall, the complexity of the casting, and the cooling rate of the mold. For thick-walled simple parts, filling a bit slower is also not likely to produce cold insulation and under-casting defects. However, for complex thin-wall castings, if the mold filling speed is too slow and the cooling rate of the mold shell is relatively high, defects such as cold insulation and under-casting are likely to form.

8.3.3.4 Filling Pressure

Filling pressure refers to the gas pressure required when the metal liquid rises to the top of the mold cavity during the filling stage. The filling pressure can be calculated according to Eq. (8.7), except that "*H*" in the equation should now correspond to the distance between the top of the casting and the free liquid level without pressure in the crucible.

8.3.3.5 Solidification Holding Pressure

The solidification holding pressure refers to the boosting value provided to crystallize the casting under a larger pressure. The purpose of crystallization pressurization is to make the casting solidify at a higher pressure, improve the shrinkage ability, increase the microstructure density of the casting, and reduce casting defects. On the premise that the shell strength is high enough, the higher the solidification holding pressure is, the more favorable the filling is, and the denser the casting is.

However, the method of increasing the holding pressure is not suitable in all cases. The crystallization pressure is related to casting type, casting structure, and other factors. For example, the crystallization pressure of low-pressure casting in wet sand mold should not be too high. Excessive pressure will not only affect the surface finish and geometric dimensions of the casting but also seriously lead to sand adhesion and expansion defects. Another example is the casting of thin wall metal castings. Since the casting heat dissipates quickly in thin wall metal castings, the molten alloy will soon solidify after filling the mold cavity. In this case, even if the solidification pressure is increased, the shrinkage effect is small.

8.3.3.6 The Holding Time

The holding time refers to the time required from the maximum holding pressure of the molten metal to the complete solidification of the casting under constant pressurization. During this period, the molten alloy in the cavity will complete the transformation from liquid to solid. If the holding time is not long enough, the solidification of the casting in the mold cavity will not be filled and porosity and shrinkage defects will be formed. Besides, when the molten alloy in the casting mold and spout is not fully solidified, the pressure on the liquid surface of the crucible is removed prematurely and the unsolidified alloy liquid will fall back to the crucible, causing the casting to be "hollow" and discarded. If the holding time is too long, the solidification length of the casting sprue will increase and extend into the riser tube, resulting in the reduction of the metal liquid flowing back into the crucible. This will further affect the separation between the riser tube and the shell and will even damage the seal ring at the middle separator in the worst cases. The selection of casting holding time should consider many factors, such as casting wall thickness, structure, alloy composition, casting temperature, and shell cooling conditions, etc. In the actual casting production, the holding time is often determined empirically.

The holding time can be calculated by the following equation:

$$\sqrt{t} = K\frac{(t_{melt} - t_{mold})bV}{L + \left(t_{filling} - t_{melt}\right)g\gamma S} \tag{8.8}$$

where K is the specific heat coefficient, t_{melt} is the melting point of alloy (°C), t_{mold} is the average temperature of mold during pouring (°C), $t_{filling}$ is filling temperature (°C), b is heat storage coefficient of mold, V is volume of casting, L is the latent heat of crystallization, C is the heat capacity of the molten metal (kJ/(kg • K)), γ is the specific gravity of the metal after solidification, S is the surface area of the casting (m^2), and t is the holding time (h).

When the casting process parameters such as alloy material, mold material, and pouring temperature are determined, Eq. (8.8) can be converted to:

$$\mathrm{t} = \mathrm{const} \times \left(\frac{V}{t}\right)^2 \tag{8.9}$$

It can be seen from the above equation that under the conditions determined by the casting process, the holding time of the casting is only related to the structure of the casting itself.

8.3.4 Pouring Temperature and Mold Temperature

The pouring temperature and mold temperature can directly influence the forming and crystallization structure of the casting, which should be as low as possible on the premise of ensuring the casting forming. If the pouring temperature and mold temperature are too high, the molten alloy will heavily aspirate in a non-vacuum environment, which will tend to the reduction of mechanical properties due to the shrinkage and coarse crystallization structure. On the other hand, if the pouring temperature and mold temperature are too low, the shrinkage tendency will be reduced, but the molten alloy flow performance will be reduced, and the filling capacity will be reduced as well. Therefore, reasonable pouring temperature and mold temperature should be selected according to the characteristics of the casting, the types of alloy materials and molds, and the cooling and heat dissipation conditions. The pouring temperature and shell temperature of thin-walled parts should be higher to facilitate the filling and forming of metal melt; the pouring temperature and mold temperature of thick wall parts can be slightly lower to reduce shrinkage.

The pouring temperature of anti-gravity casting is generally 10–20 °C lower than that of gravity casting. Under the condition of ensuring complete filling, the lower the pouring temperature is, the better the casting properties are.

8.4 Melt Filling and Solidification Under Pressure

8.4.1 *Effect of Pressure Regulating Casting Method on Mold Filling Capacity of the Alloy*

Mold filling capacity refers to the fluidity of molten metal under the influence of casting mold, casting temperature, casting speed, and other factors, namely, the ability of the liquid metal to fill the casting mold and obtain the complete and accurate casting shape. The main factors influencing the filling capacity of liquid metal include: (1) the fluidity of liquid metal; (2) casting conditions; (3) filling conditions of cast mold; (4) casting structure, etc. The improvement of liquid metal fluidity is of great significance to reduce the inner defects of castings (such as underwatering, cold insulation, porosity, etc.).

The filling capacity of molten metal is also affected by the filling speed (pouring speed) [18–20]. The faster is the filling speed, the less heat is dissipated by the molten metal in the mold cavity. The higher is the temperature, the better the flow performance can achieve. However, a filling speed too high may cause molten metal entrapment, spatter, etc., affecting the casting quality. In the filling process of pressure regulating casting method, the flow of liquid metal is controlled by adjustable pressure, and the filling speed of liquid metal can be changed in real-time by selecting different boost speeds at different stages, thus effectively controlling the filling cavity of liquid metal.

Meanwhile, during the filling process of the pressure regulating casting method, the backpressure in the shell is low, the resistance during the filling process of the liquid metal is small, and the possibility of gas entrainment in the casting is reduced. Consequently, a higher boosting speed can be selected to filling the mold. Also, the flow and heat preservation ability of molten metal can be further improved and the filling ability of the molten metal improves at the same time, leading to a more complete and clearer casting contour shape [21–24].

Jiang et al. [25] concluded that there was a linear relationship the filling capacity of the molten metal with boost velocity, pouring temperature, gas pressure, and cavity vacuum. For a 3 mm-thick thin wall plate, the relationship among its filling capacity, boost speed, pouring temperature, gas pressure, and cavity vacuum is

$$L = -192.405 + 0.406T + 3.381Q + 733.333P_V + 450P_g \tag{8.10}$$

where L represents the filling length, T, Q, P_V, and P_g represent, respectively, pouring temperature, boost speed, cavity vacuum, and gas pressure.

However, it should be noted that the thickness of the casting wall will also have a certain impact on the filling process [10]. Under the action of pressure difference Δp, the molten metal fills the cavity layer by layer along the anti-gravity direction. The filling speed of pressure regulating casting method is generally fast, with the increase of wall thickness, the filling process appears obvious reverse filling characteristics, which affects the filling process. While the superalloy density is relatively high, the casting quality is more easily affected by the wall thickness.

8.4.2 Effect of Pressure Regulating Casting on the Solidification Process of Casting

Solidification is one of the most important processes in casting. The change of the solidification process will have different effects on casting structure and properties. As for the pressure regulating casting, the adjustment of pressure in different technological stages will certainly affect the solidification process, thus changing the structure and performance of the casting.

Conventional pressure has almost no effect on the surface energy of the crystal. According to the classical nucleation theory, the effect of pressure on the nucleation rate and growth rate of the alloy crystal can be deduced as [26]:

$$\frac{I(P)}{I(P_0)} = exp\frac{\Delta G'(P_0) - \Delta G'(P)}{RT} \tag{8.11}$$

$$\frac{U(P)}{U(P_0)} = exp\frac{\Delta G''(P_0) - \Delta G''(P)}{RT} \tag{8.12}$$

where $I(P)/I(P_0)$ represents the influence of pressure on the nucleation rate and $U(P)/U(P_0)$ represents the influence of pressure on the crystal growth rate. $\Delta G'(P)$ and $\Delta G''(P)$ are the nucleation activation energy and growth activation energy of the alloy, respectively, under standard conditions; $\Delta G'(P_0)$ and $\Delta G''(P_0)$ are the nucleation activation energy and crystal growth activation energy of the alloy under pressure, respectively.

When the pressure is large enough, the presence of pressure will lower the nucleation activation energy, reduce the difficulty of crystal nucleation, and thus increase the crystal nucleation rate. At the same time, the presence of pressure will increase the density of molten metal, inhibit the diffusion of the heteroatom in the alloy, and increases the activation energy of atomic diffusion. The increase of the diffusion activation energy will reduce the diffusion ability of the atoms and then reduce the crystal growth rate, leading to the refinement of the crystal grains. At the same time, the decrease of the atomic diffusion ability of the atoms will affect the solute redistribution during the solidification process of the alloy and reduce the segregation of the elements. Under the effect of pressure, the grains may undergo elastic or plastic deformation, which can promote the melt to flow between the solidified dendrites, improve the ability of the liquid metal to shrink, and change the shape and distribution of the metal structure [27].

A large number of experimental results [28–32] show that the structure and performance of pressure casting castings are significantly better than gravity casting. For Al alloy [30], the tensile strength was increased by 8–10%, and the elongation was increased by more than 50%.

To improve the microstructure and nucleation rate of the casting, it is necessary to study how to maximize the pressure value during solidification.

8.4.3 *Effect of Pressure Regulating Casting Method on Alloy Casting Defects*

For alloys with a large solidification interval, the casting can easily form microscopic porosity, shrinkage porosity, and shrinkage cavity, which can seriously affect the performance of the casting.

For the improvement of the filling ability of castings, people usually consider the driving force and resistance to filling. The main driving force of shrinkage includes the surface tension of molten metal, static pressure generated by its gravity, and additional pressure on the liquid surface. The resistance to the shrinkage mainly includes the viscous friction force encountered when the molten metal flows, the local resistance generated by the intergranular channel and the gravity of the liquid metal itself, etc.

Based on the analysis of the force involved in the flow of molten metal, the shrinkage and relaxation criterion can be obtained [33]:

$$\frac{G_{SC}\sqrt{P_{SC}}}{\sqrt{P_{SC}}} < K_C \tag{8.13}$$

In this equation, Gsc, Rsc, and Psc correspond to the temperature gradient, cooling rate, and pressure of the critical solid rate, respectively. Kc is the critical value of the shrinkage criterion. When the calculated value is greater than Kc, shrinkage is not easy to occur; when the calculated value is less than Kc, shrinkage is easy to occur.

With the same process condition, the temperature gradient and cooling rate at each position of the casting do not fluctuate significantly. As a result, the tendency of shrinkage of the casting is mainly affected by the pressure. Increasing the pressure can improve the filling capacity of the molten metal, which theoretically reflects the advantages of the pressure regulating casting method in solidification and filling. Increasing the pressure holding pressure will improve the filling speed and filling ability of the casting, resulting in denser and more uniform microstructure, effectively reducing the content of micro-porosity in the casting.

Also, negative pressure is maintained in the pressure chamber during the initial stage of pressure regulating casting to reduce the possibility of gas entrapment, porosity, oxidation, and other defects in the casting, and to reduce the tendency of casting cracks, achieving the purpose of improving the reliability of parts.

While the superalloy pressure regulating casting technology has not been proposed for a long time in China, the design and understanding of the casting process need more exploration. But we believe that with the development of theory and equipment, superalloy pressure regulating casting technology will form a complete technical system and achieve low cost, high reliability, and high qualified rate production.

8.5 Examples of Pressure Regulating Casting for Complex Thin-Walled Castings

Nickel-base superalloy K4169 was selected as the raw material for precision casting. At present, the wall thickness of large complex thin-wall castings of superalloy is mainly concentrated in 1–2 mm. In order to improve the applicability of large complex thin-wall superalloy castings fabricated by the pressure regulating casting, 1 mm or 2 mm thin plates were used as the main design features of the casting.

For thin plate castings without a riser, inclined pouring is beneficial to reducing the impact of alloy liquid level fluctuations on the quality of the casting. But in the investment casting method, the strength of the 1 mm-thick wax mold is not sufficient, so the reinforcement plate is needed. It is easy to cause the wax molded formation or break during the dipping process if the wax mold is set obliquely. As a result, it is placed vertically. The feature structure is shown in Fig. 8.8.

One piece of $100 \times 100 \times 1$ mm thin plate is placed on the right side of the feature piece and 5 mm-thick wax plates were used around the thin plate as reinforcement plate. Two pieces of thin plate with a tilt of $100 \times 100 \times 2$ mm on the right side, with a dip of 15°, connected with the runner through a 10 mm lug, and define the farthest part of the thin plate from the runner as the top, and the bottom near the runner.

The size of the runner is $150 \times 50 \times 30$ mm. The larger size and thickness ensure that the molten metal in the runner can flow back to the crucible after the solidification of the sheet casting, which can lower the production cost. A dark riser is provided at the top of the wax mold to preserve the contaminated molten metal and to improve the metallurgical quality of the sheet castings. A tapered boss is set at the top of the wax mold, which is convenient to cooperate with the riser.

The casting mold is made up of a fused quartz shell and the casting alloy is K4169. The thermal conductivity between the casting and the mold is selected empirically. The mold heat dissipation coefficient is 1000 W/m^{2}K. The casting is strictly insulated with thermal insulation cotton except for the riser part and the heat conduction of the riser tube is air cooling.

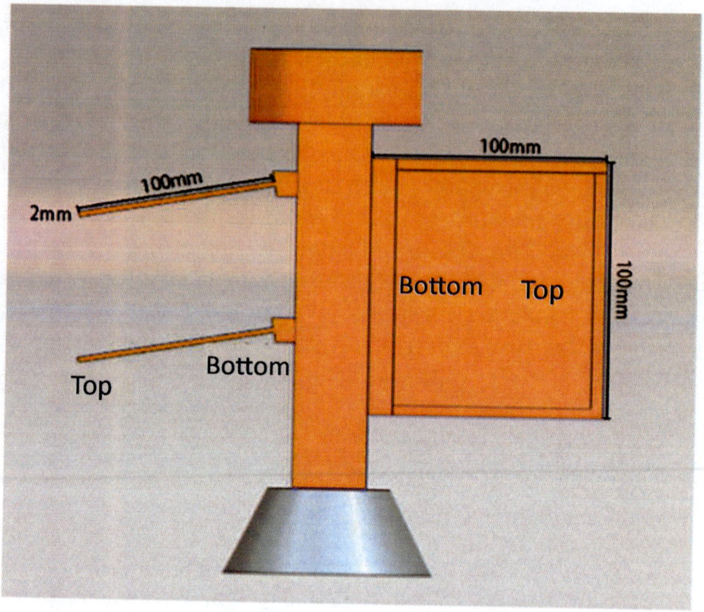

Fig. 8.8 3D model of feature parts

8.5.1 Pressure-Regulating Precision Casting of Superalloy Characteristic Parts

8.5.1.1 Experiment Scheme

According to the actual capacity of the pressure adjusting casting equipment, the process parameters for the experiment are chosen as follows. The casting temperature was 1560 °C, the shell preheating temperature was 955 °C, the boost velocity was set as 3, 4, and 5 kPa/s, and the crystallization holding pressure was 300 KPa to investigate the influence of the boost velocity on the casting structure and performance. Based on the boost speed of 5 KPa/s, the pressure of 100 and 600 KPa were selected for the experiment to study the effect of crystallization pressure on casting structure and performance.

To compare the mold filling capacity of the molten metal under different casting methods, a gravity comparison experiment was prepared. Considering the cooling condition of the mold shell, the casting temperature of gravity casting was 1560 °C, the preheating temperature of the mold shell was 850 °C, and the filling time of the mold cavity was 5 s to observe the filling conditions of different thickness plates.

The selection of process parameters of pressure regulating casting experiment is listed in Table 8.2.

Table 8.2 The process parameters of pressure regulating casting experiment

Number	Boost velocity/(kPa/s)	Differential pressure/kPa	Crystallization holding pressure/kPa	Holding time/s
1	3	90	300	300
2	4	90	300	300
3	5	90	300	30
4	5	90	100	300
5	5	90	600	300

8.5.1.2 Effects of Process Parameters on Mold Filling of Castings

Figures 8.9, 8.10, and 8.11 show the filling conditions of thin plate castings under different process parameters. It can be seen that the 1 mm-thick thin plate cast the gravity casting method reveals the worst filling ability and only a small amount of molten metal flows into the cavity, whereas the 2 mm-thick thin plate is filled. For pressure regulating casting, when the boost velocity is low, there is a large amount of cold insulation and insufficient pouring defects in the 1 mm-thick thin plate. However, the defects can be effectively controlled by increasing the boost velocity. Similarly, the 2 mm-thick thin plate is filled.

For the filling process of thin plate castings, the Laplace force produced by the surface tension hinders the flow of the molten metal and affects the filling quality of the casting. At the same time, Molten metal dissipates heat quickly and the flow capacity decreases rapidly, which leads to the formation of defects such as cold insulation and insufficient pouring.

Under gravity condition, the driving force of filling which is determined by the pouring position of molten metal is gravity; the filling force of adjusted pressure casting is gas pressure, which can be adjusted as required. For a 1 mm sheet, it is

(a) (b) (c)

Fig. 8.9 The pressure curve of the adjusted-pressure cast. **a** 3 kPa/s, **b** 4 kPa/s, **c** 5 kPa/s

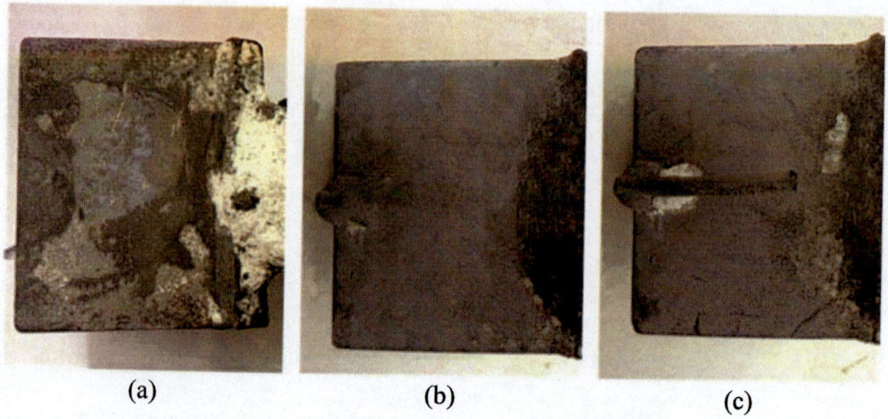

Fig. 8.10 Filling condition of 2 mm thin plate samples under different gas flowrate: **a** 3 kPa/s, **b** 4 kPa/s, **c** 5 kPa/s

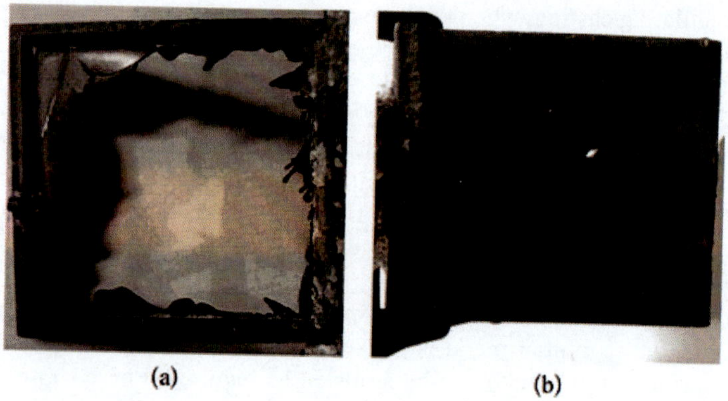

Fig. 8.11 Filling condition of gravity casting samples **a** 1 mm, **b** 2 mm

difficult for gravity casting to provide sufficient driving force, resulting in the poor filling condition. However, adjusted pressure casting can effectively control the filling process and guarantee the filling quality of castings by changing the gas flow rate.

8.5.1.3 Metallographic Analysis of Castings

In the adjusted pressure casting process of superalloys, the selection of different technologies affects the filling and solidification process of castings and has a certain impact on the casting structure. At the same time, the cooling rate and solidification time are different in different parts, which will lead to different structures at different locations.

Figure 8.12 is metallographic of casting microstructure. It can be seen that the alloy matrix is white γ' phases and a large number of black "island" segregation zones are distributed in the dendrite matrix. A certain amount of fine network structure exists inside the "island" structure, which can be judged as the precipitated Laves phase. Besides, there are particle MC carbides in the segregation zone.

Alloying elements continuously segregate after the formation of the casting matrix. If the content of Ti, Mo, Nb, and other elements are greater than their saturated solubility in Laves phases, the supersaturated Laves phases will precipitate between dendrites or grain boundaries as the casting temperature decreases, forming the brittle Laves phases as shown in the figure.

Figure 8.13 shows the distribution of microstructures at different positions of a 2 mm sheet sample. It can be seen that fine equiaxed grains form at the top. Equiaxed grains with increased grain size are in the center part and uniformly oriented columnar grains are in the bottom.

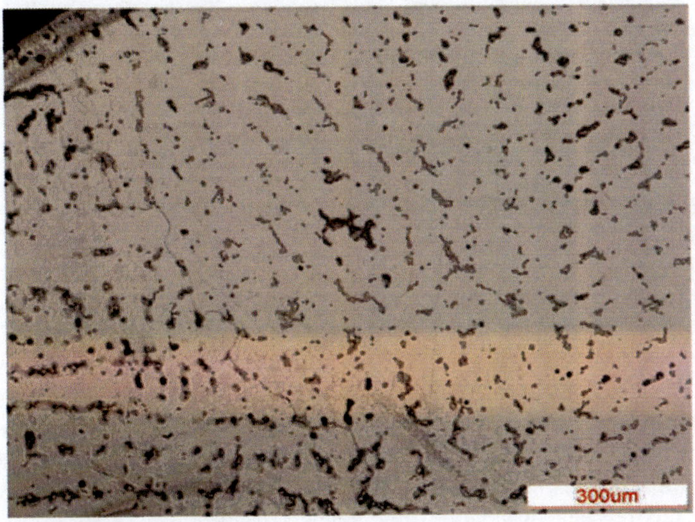

Fig. 8.12 Metallograph of the casting

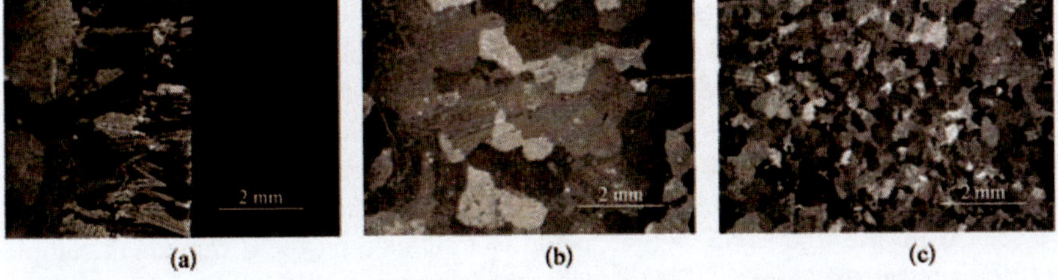

Fig. 8.13 Distribution of microstructures of 2 mm sheet sample at different positions **a** bottom, **b** middle, **c** top

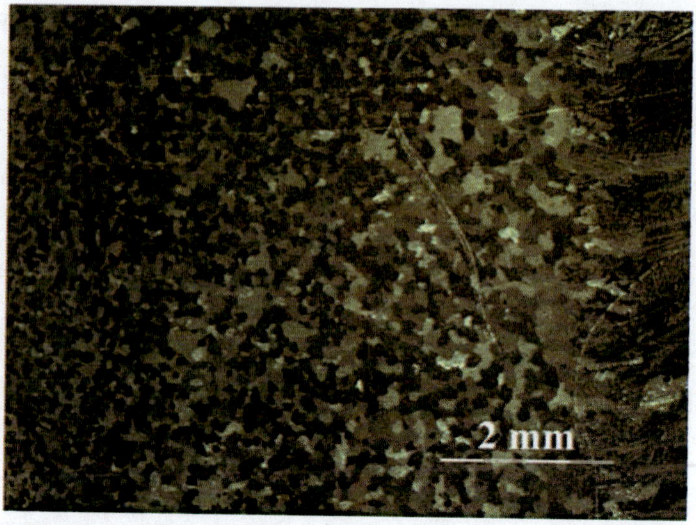

Fig. 8.14 Grain structure of 1 mm sheet sample

Figure 8.14 shows the microstructure distribution of a 1 mm sheet sample along the metal flow direction. Similar to the microstructure of the case of 2 mm, the microstructure transits from fine equiaxed grains to uniformly oriented columnar grains, but grains are finer.

During the filling process, as the contact time between the molten metal flow front and the mold is longer, the temperature drops obviously and the solidification time is shorter. The top position of the sheet mold experiences faster heat dissipation and possesses lower temperatures. The melt nucleates a lot under the strong cooling effect of the mold, forming a non-directional fine and dense equiaxed grain structure. As the solidification layer moves towards the sprue, the heat dissipation capacity of molten metal decreases and the temperature gradient tends to be smooth. As the temperature decreases continuously, the melt begins to nucleate and grow. Since there is no obvious heat dissipation direction during solidification, the microstructure is composed of non-directional equiaxed grains. However, due to a small degree of supercooling, there are fewer nucleation cores in the melt and the resulting grain structure is relatively coarse. The bottom of the sheet is connected to thick runners and the cooling rate is slow and the solidification time is long. Grains in the central equiaxial region extend and grow in one direction, which is parallel to the direction of heat flow and towards the casting system.

Figure 8.15 shows the relationship between the average grain size of the 2 mm sheet sample and its horizontal height under the same process parameters. For convenient comparison, only the representative central part is selected for analysis. As can be seen from the figure, the height has little influence on the 2 mm sheet sample and the central structure is all equiaxed coarse grains at different heights, with no significant changes in grain size.

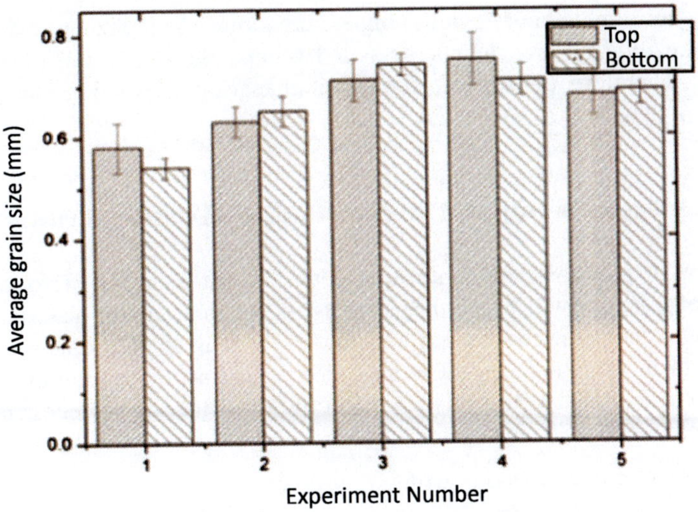

Fig. 8.15 Average grain size versus the height of 2 mm sheet sample

Figure 8.16 shows the trend of the average grain size with a height of 1 mm sheet sample. Similar to the microstructure of 2 mm one, the microstructure which is composed of large equiaxed grains is relatively stable without an obvious change in the grain size.

During the adjusted pressure casting process, the metal flow is controlled by the pressure difference, and the filling process is stable without obvious spatter or turbulence. The temperature distribution of molten metal in the casting system is

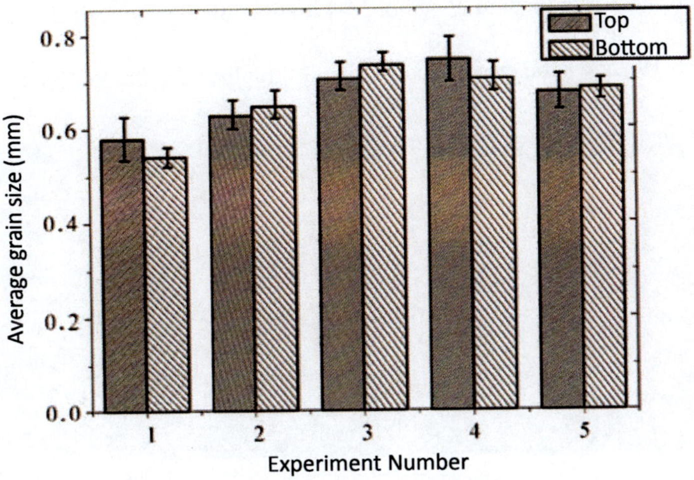

Fig. 8.16 Average grain size versus the height of 1 mm sheet sample

relatively uniform without obvious temperature jump. The temperature difference is small while filling different height sheets and its impacts on the casting microstructure are small. Therefore, there are no significant changes in the casting microstructures at different heights.

8.5.1.4 Influence of Adjusted Pressure Casting on Casting Structure

The grain size plays an important role in determining the casting performance. The increase of grain size will reduce the plasticity and toughness of the material, lowering the medium temperature strength and low-cycle fatigue performance. However, grains with too small sizes will reduce creep and fracture properties. Therefore, it is of great significance to study the effect of different casting process parameters on the grain size of castings.

(1) Influence of thickness on castings

Figure 8.17 shows the metallographic structure of 1 and 2 mm sheet castings. The sampling site is the center of the sheets. The adjusted pressure casting process parameters are a pressuring rate of 3 kPa/s and a crystallization holding pressure of 300 kPa. It can be seen that the solidification structures of different thickness sheets are similar and the grains are composed of a large number of cross dendrite structures. The dendrites have different directions and there is no uniform distribution, which is consistent with the heat dissipation condition. The grain size in the 2 mm sheet sample is larger than the 1 mm one.

For all experimental groups, the grains at the center of different thickness sheets are statistically calculated: the average grain diameter in 2 mm sheet sample is 1.22 mm, with a grain size of about M-9.5, while the average grain diameter in the 1 mm one is 0.64 mm, with a grain size of about M-12, showing a grain size change by more than 40%.

Figure 8.18 shows the cross-section metallographic of different thickness sheets.

(a)

(b)

Fig. 8.17 Microstructure of 1 mm and 2 mm sheet samples: **a** 1 mm, **b** 2 mm

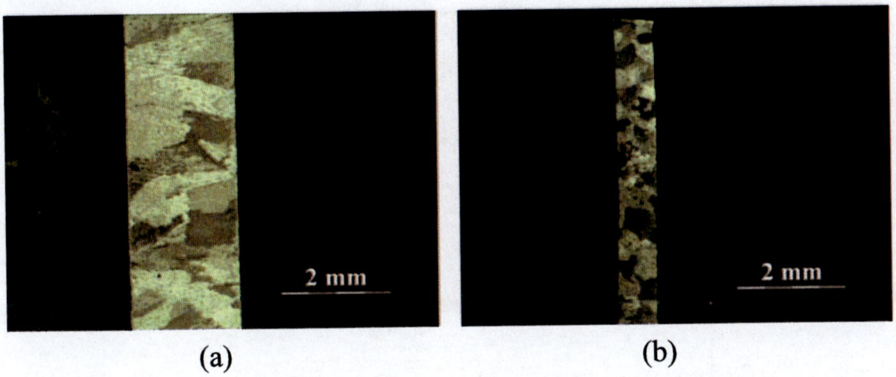

Fig. 8.18 Grain structure of different thickness sheet samples: **a** 2 mm, **b** 1 mm

The process parameters and sampling sites are consistent with Fig. 8.16. As can be seen from the figure, the grain size in the 1 mm sheet sample is relatively small, with an average of 2 –3 grains in the cross-section, while the grain size in the 2 mm one is relatively coarse and the number of grains in the cross-section becomes 1–2.

With the decrease in thickness of the casting, the modulus of the sheet and the solidification time are continuously reduced and the cooling rate of the melt becomes faster. Thus, the supercooling degree at the solid–liquid interface increases, the energy fluctuation required for nucleation decreases, and the formation rate of the nucleation cores increases. As the number of nucleation cores increases, the grain size decreases accordingly.

(2) Influence of gas flow rate on castings

Figure 8.19 shows the metallographs of 2 mm sheet samples with different gas flow rates of 3 KPa/s, 4 KPa/s, and 5 KPa/s, respectively. It can be seen that the central grains of the casting are still equiaxial with the increase of gas flow rate, but the size gradually increases.

The average grain size of each sample was measured and the average value was calculated. The trend of the average grain size with the gas flow rate is shown in

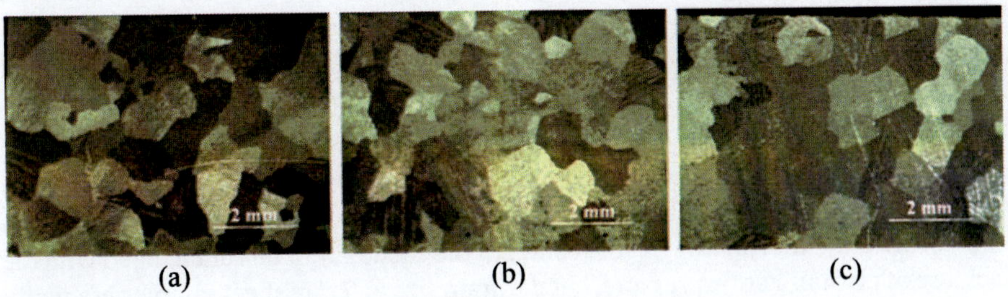

Fig. 8.19 Grain structure of different gas flow rate in 2 mm sheet samples: **a** 3 KPa/s; **b** 4 KPa/s; **c** 5 KPa/s

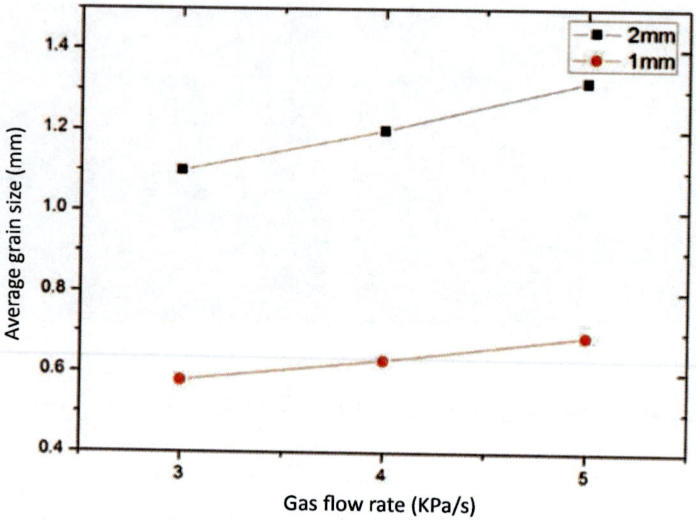

Fig. 8.20 Average grain size versus gas flow rate of sheet samples

Fig. 8.20. For a 1 mm sheet sample, the average grain diameter is 0.58 mm when the gas flow rate is 3 KPa/s; the average grain diameter is 0.63 mm when the gas flow rate is 4 kPa/s; the average grain diameter is 0.71 mm when the gas flow rate is 5 kPa/s, and the grain size changes by more than 15%. For a 2 mm sheet sample, the average grain diameter is 1.1 mm when the gas flow rate is 3 kPa/s; the average grain diameter is 1.2 mm when the gas flow rate is 4 kPa/s; the average grain diameter is 1.42 mm when the gas flow rate is 5 kPa/s, and the grain size changes by about 29%.

In the adjusted pressure casting process, the gas flow rate determines the mold filling speed of molten metal. When the gas flow rate rises, the time spent on the mold filling process is shortened. As a result, the contact time of the molten metal with the mold is shorter and the temperature drop is smaller during filling. In this way, the solidification time of molten metal is extended, leading to solidification crystals stay at the growing stage for a long time and the grain size becomes larger. For 1 mm sheet samples that experience a cooling rate significantly higher than that of the 2 mm one, the change in solidification time caused by the change of gas flow rate is small, and the degree of grain coarsening is reduced.

(3) Influence of crystallization holding pressure on castings

Figures 8.21 and 8.22 show the metallographs of 2 mm and 1 mm sheet samples, respectively, under different crystallization holding pressures. The crystallization pressures considered are 100, 300, and 600 kPa. As can be seen from the two figures, the grain morphology in the center of the casting remains unchanged with the increase of crystallization pressure, but the grain size of 2 mm sheet samples gradually decreases, while the grain size of the 1 mm one does not change significantly.

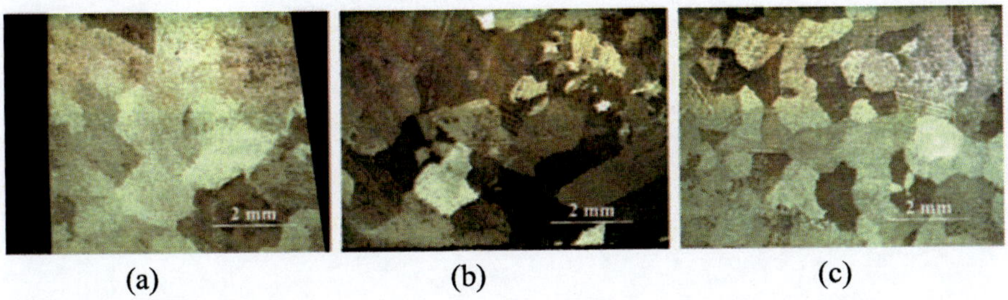

Fig. 8.21 Grain structure of different crystallization pressure in 2 mm sheet samples: **a** 100 kPa; **b** 300 kPa; **c** 600 kPa

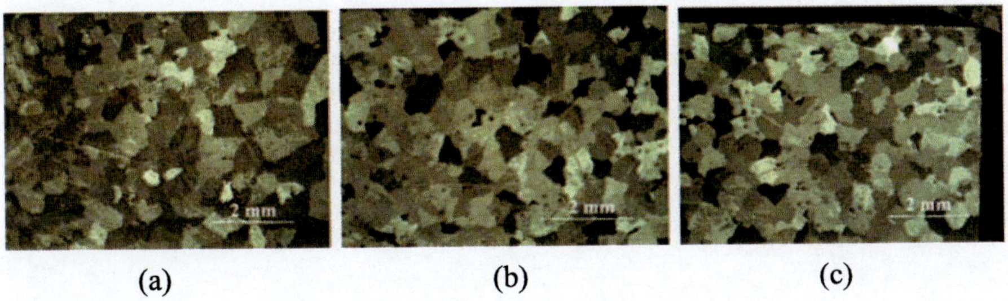

Fig. 8.22 Grain structure of different crystallization pressure in 1 mm sheet samples: **a** 100 kPa; **b** 300 kPa; **c** 600 kPa

The trend of grain size with the crystallization pressure of different samples is shown in Fig. 8.23. For 1 mm sheet samples, the curve is nearly flat and the grain size reduction is less than 15%. For 2 mm ones, when the crystallization pressure increases from 100 to 300 KPa, the grain size changes significantly with a reduction of more than 30%. And when the crystallization pressure further increases, the grain size continues to decrease up to about 30%.

During the solidification process of adjusted pressure casting, the molten metal on the casting surface shrinks after solidification and an air gap will form between the casting surface and the mold. The heat transfer capacity of the air gap directly influences the cooling rate of castings, thus affecting the solidification process. The increase in crystallization pressure leads to an increase in the pressure and density of the gas in the air gap, which results in an increment of the gas heat conduction capacity. As a consequence, the heat dissipation speed of the casting increases, leading to the change of the casting grain size.

The filtration capacity of molten metal caused by pressure can be expressed by

$$\Delta G = \frac{K\Delta P}{\mu\Delta L}F\Delta t \tag{8.14}$$

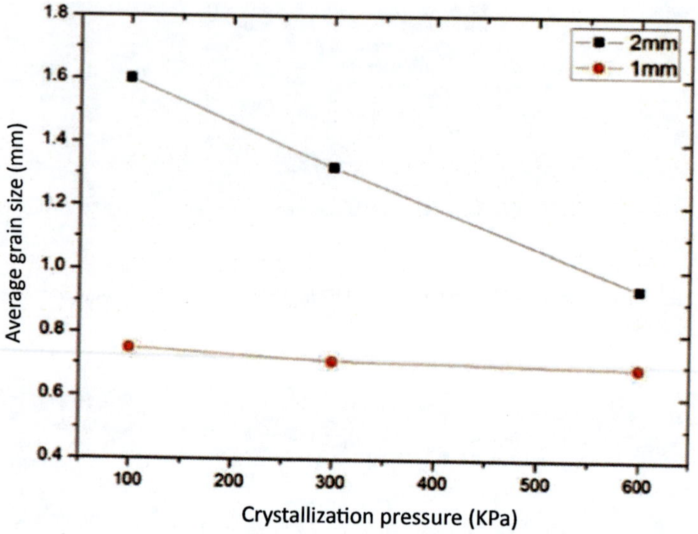

Fig. 8.23 Average grain size versus crystallization pressure of sheet samples

In the above equation, ΔG is the metal volume filtered into the shrinkage under pressure, ΔP is the pressure difference applied to the molten metal, ΔL is the depth of the filtration, Δt is the filtration time, K is the permeability coefficient, μ is the dynamic viscosity coefficient, and F is the filtration area.

The increase of crystallization pressure will improve the filtration effect of molten metal, making the incompletely solidified metal flow into the interdendritic channel. If the dendrite skeleton of the solidified part has low strength, it may be broken by the flow of interdendritic molten metal, which increases the number of nucleation sites and thus refines the grain structure.

Besides, the solidification temperature and supercooling degree of the alloy will be increased under the crystallization holding pressure and the crystal grains of castings will be refined.

However, the inert gas source of the equipment affects the crystallization holding pressure and gas flow rate; it takes a long time to reach the prescribed pressure. For a 1 mm sheet sample, the molten metal is completely solidified before the pressure reaches a high level. As a result, it is less affected by the crystallization holding pressure, and the grains are hardly refined.

8.5.1.5 Influence of Casting Process on the Secondary Dendrite Spacing of Castings

The secondary dendrite arm spacing (SDAS) refers to the distance between the secondary dendrite arms, whose size directly affects the mechanical properties of castings, as well as the composition segregation of alloys. At the same time, the distribution of the second phase and the distribution of micro shrinkage cavity and porosity are also influenced by SDAS, which is one of the important factors affecting the quality of the casting. The relationship between the SDAS and the cooling rate and the local solidification time of the casting can be expressed by the following equation:

$$\lambda_2 = at_f^{\frac{1}{2}} = bR^{-\frac{1}{2}} \tag{8.15}$$

In this equation, R is the cooling rate and t_f is local solidification time.

It can be known from Eq. (8.15) that the secondary dendrite arm spacing decreases with increasing the cooling rate or decreasing the local solidification time.

During the growth of dendrites, there is a certain degree of solute enrichment in the liquid phase beside dendrite arms. The solute concentration changes with the change of the secondary dendrite arm radius. The larger the dendrite arms, the higher the solute concentration in the surrounding liquid phase. Because secondary dendrite arms have different radii, there is a certain concentration gradient in the solute, which tends to promote diffusion from coarse dendrite arms to fine dendrite arms along the direction of the concentration gradient, whereas the solvent has a diffusion tendency in the opposite direction. This tendency will cause the dissolution of fine dendrite arms and the growth of coarse dendrite arms. With the decrease of cooling rate and the increase of local solidification time, alloys remain in the solid–liquid two-phase region for a longer time. As a result, the dendrite coarsening process proceeds more fully and the secondary dendrite arm spacing also increases.

(1) Influence of position on the secondary dendrite arm spacing of castings

Figure 8.24 shows the distribution of dendrites at different positions of the 2 mm sheet sample. The experimental process parameters for this group are chosen as a gas flow rate of 5 KPa/s and a crystallization holding pressure of 300 KPa. It can be seen from Fig. 8.24 that the dendrites are gradually refined along the flow direction of the molten metal and the arrangement direction tends to be disordered. The dendrites at the top of the sheet are densely arranged and the direction of them is not obvious. However, the dendrites at the bottom are relatively coarse and arranged in the same

(a) (b) (c)

Fig. 8.24 The dendrite morphology in different positions of 2 mm sheet samples: **a** bottom, **b** middle, **c** top direction. As a whole, the secondary dendrite arms pacing gradually decreases along the molten metal flow direction. Because of the lower temperature at the edge of the mold and the faster heat dissipation, the secondary dendrite arm spacing at the top of the sheet is finer than the center. However, the bottom of the sheet is connected to the thick casting system, and the temperature drops slowly. It is the last solidification part of the sheet and the secondary dendrite arm spacing is relatively large.

Figure 8.25 shows the trend curve of the secondary dendrite arm spacing at different positions along the flow direction of the molten metal. It can be seen that as the thickness decreases, the secondary dendrite arm spacing in the center of the casting tends to decrease. The maximum difference between the average SDAS of the 1 mm sheet sample and the 2 mm one is about 50% and the difference between the SDAS of different thicknesses decreases with the increase of the gas flow rate. While the distribution of the secondary dendrite arm spacing is similar at different heights, there is a certain change in the secondary dendrite arm spacing at different gas flow rates.

For castings, the thicker is the wall, the smaller the heat dissipation area is per unit volume and the slower the cooling rate is. Consequently, the dendrite coarsening occurs more thoroughly during solidification and a large number of fine dendrites melt as well as coarse dendrites grow, which lead to larger secondary dendrite arm spacing.

However, the increase of gas flow rate will increase the temperature of molten metal when it enters the sheet cavity, which reduces the difference in solidification time between sheets of different thickness and alleviates the problem of different SDAS caused by different cooling rates.

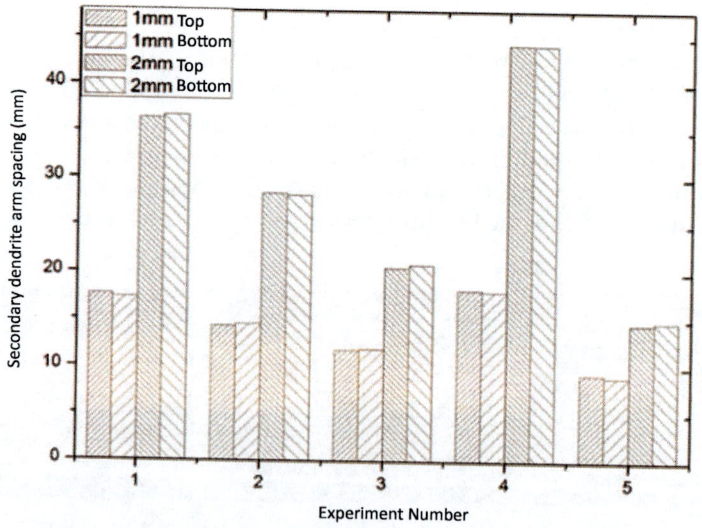

Fig. 8.25 Secondary dendrite arm spacing versus the height of 1 and 2 mm sheet samples

Because the height difference between different sheets of the same casting is small, the filling speed of the molten metal is fast and the filling interval between different heights is small. Meanwhile, the thickness of runners is large, which has a better heat preservation effect on the molten metal, and the temperature difference is small when the molten metal reaches the entrance of the sheet metal of different heights. The heat dissipation conditions at the corresponding positions of different heights are the same and the cooling speed is similar, leading to little changes of the secondary dendrite arm spacing with a height variation.

(2) Influence of gas flow rate on the secondary dendrite arm spacing of castings

Figure 8.26 shows how the secondary dendrite arm spacing changes with the gas flow rate. It can be seen that the secondary dendrite arm spacing decreases as the gas flow rate increases.

When the gas flow rate changes, the time for molten metal to reach the cavity changes and the temperature of the molten metal and the shell changes as well, both of which lead to variation in the cooling and solidification process of molten metal, affect the cooling rate and the temperature gradient and then cause the change of secondary dendrite spacing.

Due to the long cooling time of the mold before casting, the change in filling time caused by the gas flow rate is relatively short and the change of mold temperature is small, while the temperature of molten metal during filling is greatly affected by the filling time. Therefore, the increment of the molten metal temperature is higher than that of the mold shell when the gas flow rate increases. As a consequence, the

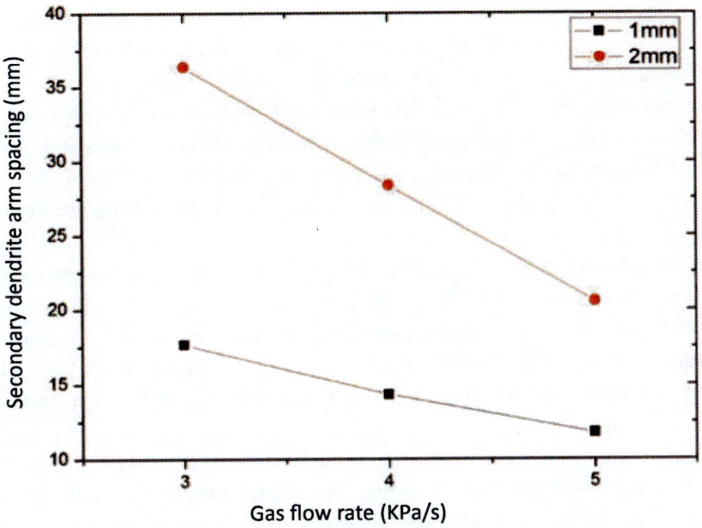

Fig. 8.26 Secondary dendrite arm spacing versus gas flow rate of sheet samples

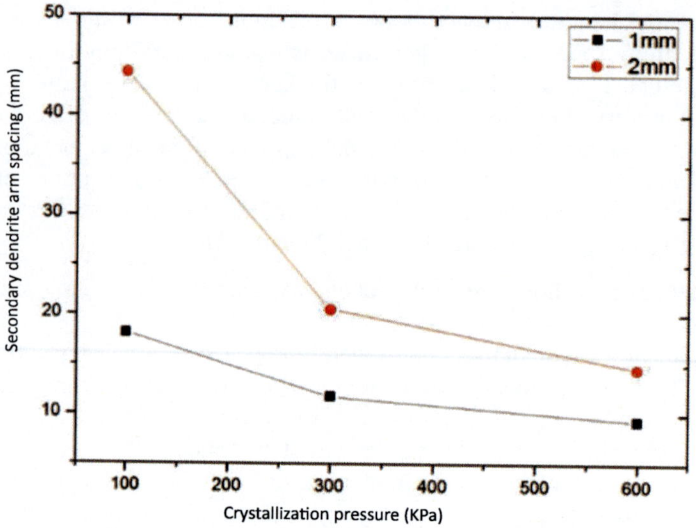

Fig. 8.27 Secondary dendrite arm spacing versus crystallization pressure of sheet samples

temperature gradient between the two increases, which is beneficial to the refinement of the secondary dendrite arm spacing of the casting.

(3) Influence of crystallization holding pressure on the secondary dendrite arm spacing of castings

Figure 8.27 shows the variation of the secondary dendrite arm spacing with the crystallization holding pressure. It can be seen from the figure that when the crystallization pressure increases, the secondary dendrite arm spacing decreases.

The increase of crystallization pressure improves the increase of pressure and density of the air gap, which results in the increment of the gas thermal conductivity and thus accelerates the cooling rate of castings. It can be seen from Eq. (8.15) that increasing the cooling rate will lead to a decrease in the secondary dendrite arm spacing of castings.

It can be known from Eq. (8.14) that the existence of crystallization pressure will force the molten metal to flow for feeding, reduce the concentration gradient of the solute in the liquid phase, and change the solute distribution in the liquid phase at the front of dendrite growth. This will affect the growth speed of dendrites and promote the secondary dendrite arm to grow, leading to the refinement of secondary dendrites.

Meanwhile, the increase of the crystallization pressure will enhance the ability of filtration and promote the flow of the molten metal along capillary channels between dendrites. This flow will generate a certain shear force on secondary dendrite arms. If the strength of secondary dendrite arms is insufficient, plastic deformation, bending, and even fracture will occur, thereby increasing the secondary dendrite arm spacing. Besides, the release of latent heat of crystallization may cause the fusing of fine dendrite arms, which increases the secondary dendrite spacing to a certain extent.

The change of the final secondary dendrite arm spacing is due to the superposition of these effects. Because the influence of crystallization pressure on the cooling rate of the casting is dominant, the secondary dendrite arm spacing of the final casting decreases with crystallization pressure increasing.

However, since the equipment is affected by an inert gas source, the speed of crystallization holding pressure is limited and it takes a long time to reach the predetermined pressure. Therefore, when the pressure increases to 600KPa, the secondary dendrite arm spacing of the 2 mm sheet sample is not obvious. For 1 mm one, the molten metal is completely solidified before the pressure reaches a high enough level. As a result, it is less affected by the crystallization pressure and the refinement degree of secondary dendrite arm spacing is low.

8.5.1.6 Influence of Casting Process on the Hardness of Castings

Mechanical properties are the ultimate embodiment of casting quality and the most important assessment standard. After analyzing the influence of the casting process on the structure of castings, the relationship between the casting process and mechanical properties of castings is further studied. Because the sheet itself is thin, it is easy to deform in the cleaning process. As a result, it is hard to machine the tensile samples and the hardness property of sheet castings is investigated. The hardness is an important standard for investigating the mechanical properties of alloys. Meanwhile, the measurement method is simple and can be easily converted to mechanical properties such as strength. Table 8.3 list the average hardness measured at different positions of the as-cast alloy.

(1) **Study on Hardness of Castings at Different Positions**

Figure 8.28 shows the hardness distribution of the castings at different locations. The process parameters are a boosting speed of 3 kPa/s and a crystallization holding pressure of 500 kPa. It can be seen from the figure that the hardness is less affected by the change in vertical height, and the hardness is continuously increasing along the direction of metal liquid flow, and the rising trend is similar under different process parameters. The hardness of a 1 mm thick sheet is higher than that of a 2 mm thick sheet. Different process parameters reflect this situation.

Due to the slight difference in the vertical height of the casting and the rapid filling speed of the metal liquid, conditions of the metal liquid, such as the temperature and element distribution when they reach thin plates of different heights, are not much different and the shell temperature at different locations is the same. Besides, there are no major differences in the heat dissipation environment and element distribution and consequently, the hardness is less affected by changes in vertical height.

However, the heat dissipation conditions of the shell along the direction of the molten metal flow are different. The average microstructure parameters such as the average grain size and the secondary dendrite spacing of the casting are reduced,

Table 8.3 Average hardness of castings at different positions

No	Thickness (mm)	Vertical position	Horizontal position (flow direction of molten metal)		
			Bottom hardness	Center hardness	Top hardness
1	1		328	348	359
	2	top	307	319	337
	2	bottom	304	320	342
2	1		335	378	393
	2	top	316	338	362
	2	bottom	318	330	367
3	1		348	392	409
	2	top	330	347	375
	2	bottom	325	344	369
4	1		370	409	432
	2	top	360	374	394
	2	bottom	363	380	410
5	1		334	357	376
	2	top	316	328	365
	2	bottom	318	324	357

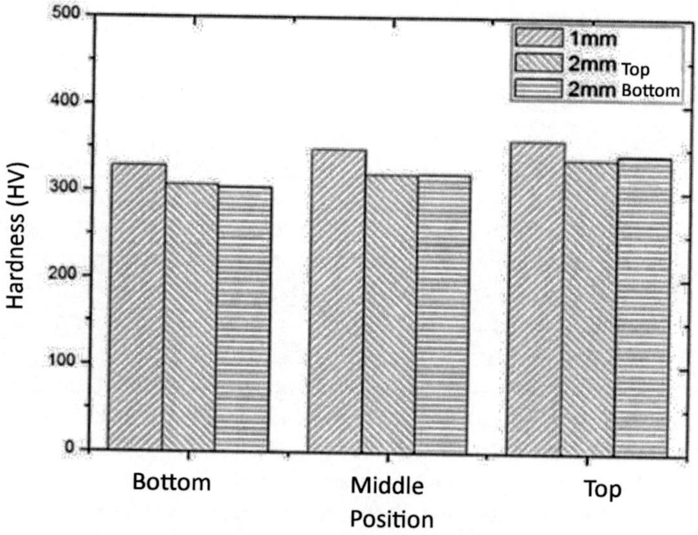

Fig. 8.28 Average hardness versus position of 1 and 2 mm sheet samples

which results in a certain change in the hardness and increases in the direction of the molten metal flow.

The thermal conditions of castings with different thicknesses are different. The 1 mm thick sheet has a relatively small average grain size and secondary dendrite spacing, resulting in an increment in its hardness. The 2 mm thick plate has a relatively long solidification time, and the element segregation and oxidation are more serious, which reduces its mechanical properties to a certain extent.

(2) **Effect of boosting speed on casting hardness**

Figure 8.29 shows the hardness distribution of castings at different boosting speeds. It can be seen from the figure that the hardness of castings increases with the boosting speed and the rising trend of castings with different thicknesses is similar. From the previous analysis, it can be known that the increase of the boosting speed will reduce the secondary dendrite spacing of the casting, whereas the decrease of the secondary dendrite spacing will improve the mechanical properties of the casting and make the casting hardness increase.

(3) **Effect of crystal holding pressure on casting hardness**

Figure 8.30 shows the hardness distribution of castings under different holding pressures. It can be seen from the figure that the casting hardness increases with the holding pressure. From the previous analysis, it can be known that the increase of the holding pressure will reduce the average grain size and the secondary dendrite spacing of the casting, thereby improving the mechanical properties of the casting.

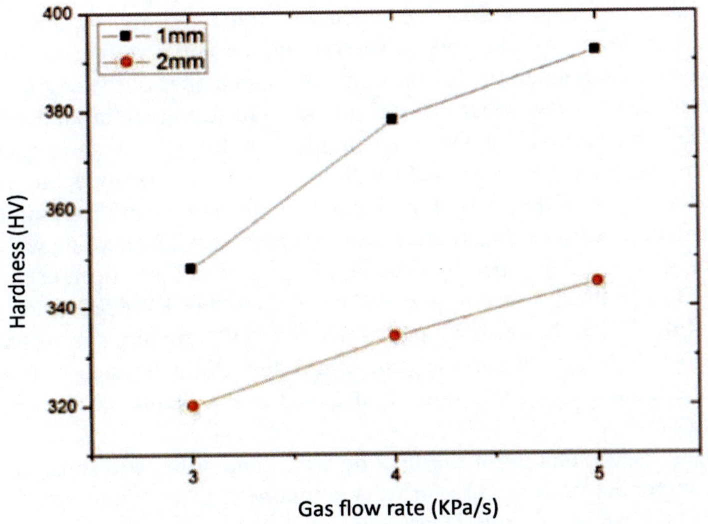

Fig. 8.29 Average hardness versus gas flow rate of K4169 of sheet samples

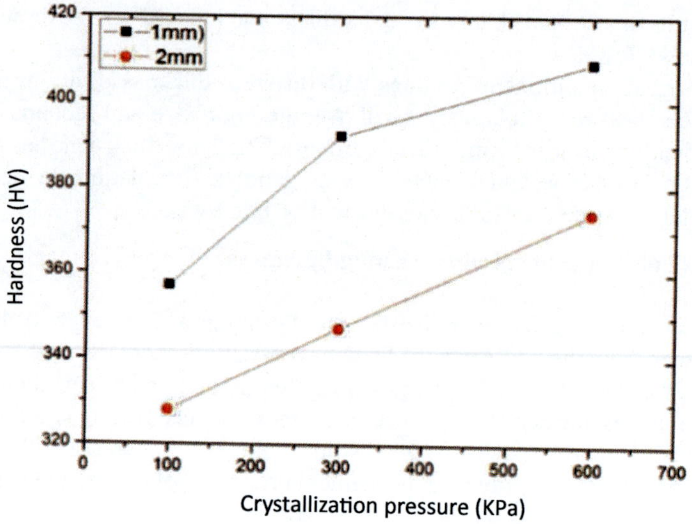

Fig. 8.30 Average hardness versus crystallization pressure of K4169 of sheet samples

8.5.2 *Simulation Study on Regulating Casting of Superalloy Floating Wall Tiles*

8.5.2.1 Floating Wall Flame Tube

With the continuous advancement of the aviation industry, the requirements for the performance of aviation engines are becoming higher and higher. One of the most important performance indicators of an aero-engine is the weight-to-weight ratio, and increasing the temperature of the combustion chamber of the engine is a major measure to increase the weight-to-weight ratio. In the breakthrough of the push ratio of the combustion chamber of the engine from 8 to 10, the temperature of the combustion chamber was increased by 200 °C, which is mainly achieved through the gas atomization of the head of the flame tube, the composite cooling of the wall surface, and the floating wall structure design technology. The floating wall structure flame tube was first developed by Pratt & Whitney, which has been used in V2500 engines, F119 military engines, and PW4000, PW6000, PW8000, and other civil engines. Rolls-Royce has also adopted a floating wall structure design on a variety of civilian engines. At present, the structural design of the floating wall flame tube has become the mainstream design of advanced engine flame tubes in Europe and the United States.

The main design feature of the floating wall flame tube is to separate the load-bearing component from the heat bearing component of the flame tube, in order to form a double-walled structure. The interior is a floating wall (heat bearing component). The floating wall generally adopts the circumferential direction and diameter.

Fig. 8.31 **a** Floating wall of the combustion chamber, **b** Pad unit

The division of the grid format is composed of a certain number of floating tiles and is fixed on the outer load-bearing components. The block structure allows appropriate gaps between the tiles, which allows the tiles to have a certain degree when being heated. The expansion space reduces thermal stress at high temperatures. The flame tube structure and tiles are shown in Fig. 8.31.

Based on this structure and due to the different cooling methods used, the floating wall tile has a variety of structural designs. The current mainstream is the impact + dense hole divergent gas film cooling structure, or completely covered gas film cooling, which is to design a large number of regularly arranged dense microporosity on the surface of the arc-shaped floating wall tile. The holes are inclined at a certain angle along the airflow direction. On the one hand, the cold air outside the floating wall can directly cool the floating wall by impact and can flow into the microholes, forming a continuous protective air mold on the inner wall of the floating wall. The floating wall of this structure design has the advantages of lighter weight and higher cooling efficiency, etc., and has a wide range of applications in the current advanced engine combustion chamber.

The shape of the floating wall tile is obtained by dividing the curved surface of the flame tube by circumferential and radial cutting. Since the curved surface of the flame tube is an approximately conical surface, the tile obtained in this way has a certain radian and is an arc-shaped strip. But the arcs on the upper and lower sides of the arc-shaped bar are not the same, and the length is slightly different. As a result, the four corners of the arc-shaped bar are not right angles. The floating wall tile itself does not bear the load. Due to the requirements of weight reduction, its wall thickness is thin, which is easy to hang on the bearing component. There are five column bolts on the thin wall surface to connect with the bearing component. The middle bolt mainly plays a positioning role. Floating wall tile impact and dense hole divergent air film cooling structure. It is an arc-shaped strip with a size of about 150 $\times$ 60 mm. The five-column bolts are distributed diagonally. The wall thickness of the tile is required to be 1.2 mm and densely packed, the number of microporosity is 200, the diameter is 1 mm, and the concave sides near the upper and lower edges have elongated protrusions with a height of 1 mm.

8.5.2.2 Simulation Analysis of the Filling Process

Figure 8.32 shows the optimized floating wall tile of the casting system model. The casting system is provided with a boss for placing the gate, the total height is about 220 mm, and the estimated filling time is 3.5 s. When filling, the maximum pressure parameter is set to 25 kPa and the characteristic of boosting speed during filling is the increase of castings.

Figure 8.33 shows the simulation results of the filling process of the casting.

(1) Enter the runner. After the molten metal enters the sprue, it keeps rising smoothly and there is no turbulence. When it reaches the supporting round rod, a small amount of molten metal enters the round rod.
(2) Reach the gate. The molten metal reaches the gate and enters the upper thick frame of the casting under the combined effect of gravity and filling pressure. The metal liquid entering the cavity from the lower round rod is only concentrated in the thick frame of the lower edge due to gravity, flows to both sides of the frame, and does not enter the thin-walled cavity of the casting, which successfully prevents the metal liquid flow from converging inside the casting.

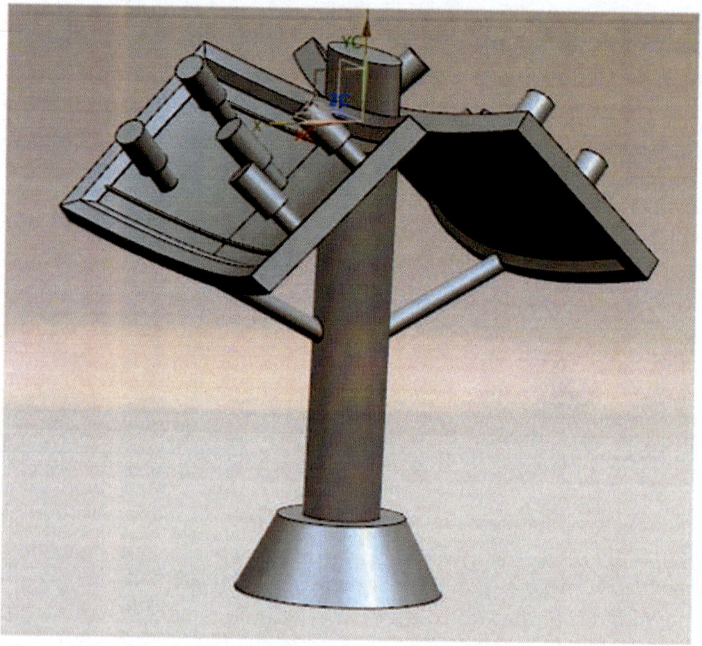

Fig. 8.32 Casting system model after optimization

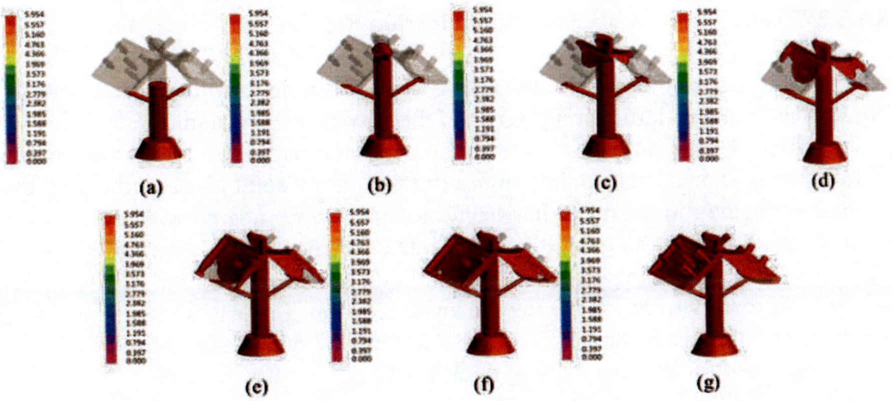

Fig. 8.33 Simulation of the casting filling process: **a** 2.0169 s **b** 2.2843 s **c** 2.4025 s **d** 2.5129 s **e** 2.6423 s **f** 2.7162 s **g** 2.9055 s

(3) Thin-wall filling. After entering the gate, the metal liquid stays for a while at the thick frame, and then enters the thin-walled area of the casting under the driving force of the mold. It flows in a fan shape from the gate and the liquid flow is relatively stable. The molten metal at the lower edge still gathers in the thick frame and flows to both sides, which does not enter the casting cavity.

(4) Side border drainage. The fan-shaped metal liquid continues to flow downward and there will be a certain hindrance when passing through the column. When the metal liquid flow in the upper frame reaches the side, it starts to flow downward, which is faster than the liquid flow in the thin-walled cavity. The speed of the flow complements it. The metal liquid in the bottom frame does not change much.

(5) The upper and lower parts of the liquid metal flow merge. At 2.6423 s, the middle part of the upper fan-shaped liquid flow and the lower liquid flow merge at the lower edge of the thin-walled area of the casting, which is located outside the required casting. The liquid flow on both sides also reached the bottom at this time. After the metal on the side edge entered the thin-walled area slightly, it merged with the fan-shaped liquid flow. Since the thin-walled area was expanded outward in the design of the new casting system, the convergent area is still outside the required casting.

(6) The final confluence point of the liquid flow in the thin-walled cavity. Located outside the casting, only the cylinder and riser are not filled.

(7) Filling is completed. The filling was completed at 2.9055 s.

8.5.2.3 Simulation Analysis of Solidification Process

After the filling is completed, the casting system begins to solidify after a while. Figure 8.34 is the solidification process of the new casting system.

At 6.0457 s after the completion of filling, some parts of the thin-walled area of the casting showed zero points in the fraction of the solid phase, indicating that in these parts the molten metal has begun to solidify. Compared with the previous casting system, the starting solidification time of the new casting system is delayed by nearly 3 s, so that the pressure during solidification can be raised higher, which prevents the casting from solidifying at a lower pressure. The places where the casting begins to solidify are randomly distributed in thin-walled areas, and these areas can be compensated for the surrounding thick parts.

After the filling is completed at 46,000 s, the thin-walled part of the casting is solidified, all of which are grayish-white. Meanwhile, the solid phase fraction of the thick frame and riser around the casting is about 50%, and the sprue and sprue gate has not started to solidify.

After the filling is completed at 281.0000 s, the casting has completely solidified except for the flow gate part. The solidification time of the flow gate part is slower compared to the other casting part, which may cause the casting part to have more shrinkage holes near the gate part.

After the filling is completed at 2996.0000 s, the casting system was completely solidified.

The simulation of the solidification process shows that the casting system can also meet the requirements of sequential solidification. The thin wall of the casting is first solidified and the freezing point is randomly distributed. The entire thin wall begins to solidify at the same time and the solidification is completed almost simultaneously. Compared with the solidification process of gravity, the main force during anti-gravity solidification is the applied pressure and the metal liquid is subjected to more

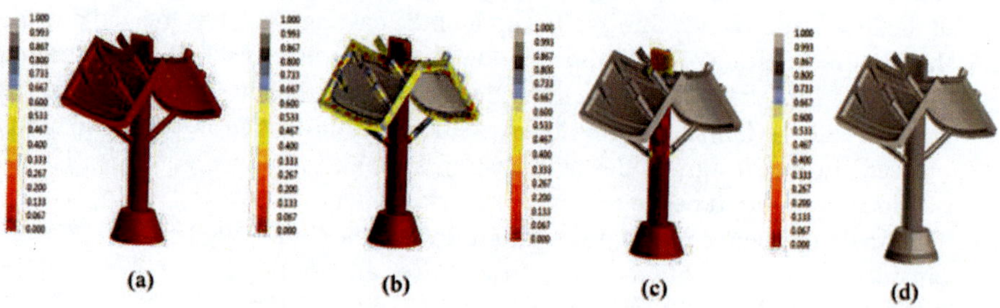

Fig. 8.34 Simulation of solidification process: **a** 6.0457 s **b** 46.0000 s **c** 281.0000 s **d** 2996.0000 s

uniform force during solidification. Therefore, the island-like areas where the thin-walled areas begin to solidify are almost randomly distributed. Thin-walled regions are also solidified almost simultaneously. Such thin-walled grain size and secondary dendrite spacing are more likely to obtain relatively uniform results. There is a clear difference in the solidification of the thin-walled areas of gravity casting.

8.5.2.4 Simulation Analysis of Defects

Figure 8.35 shows the shrinkage distribution. The shrinkage in the figure is mainly concentrated in the corner of the column riser and thick frame and the part where the metal liquid enters. In comparison, the casting itself does not have many shrinkage defects. Figure 8.36 shows the distribution of thermal nodes. The distribution of thermal nodes is consistent with the shrinkage distribution, most of which are concentrated on the pouring system. Figure 8.37 shows the distribution of microscopic shrinkage probability based on the Niyama criterion. Compared to other parts of the casting system, the Niyama criterion value of the casting part is lower, indicating that there is still a greater possibility of microscopic shrinkage.

According to related works of literature, for thin-walled castings with large differences in wall thickness, the critical value of the Niyama criterion for thin-walled areas and other areas are not the same. Currently, there is no reliable unified method for determining this critical value. In fact, due to the existence of interdendritic flow supplementation, the micro-constriction in the thin-walled area predicted by the Niyama criterion is likely to be condensed by the post-solidified thick metal liquid and transferred to the thick portion.

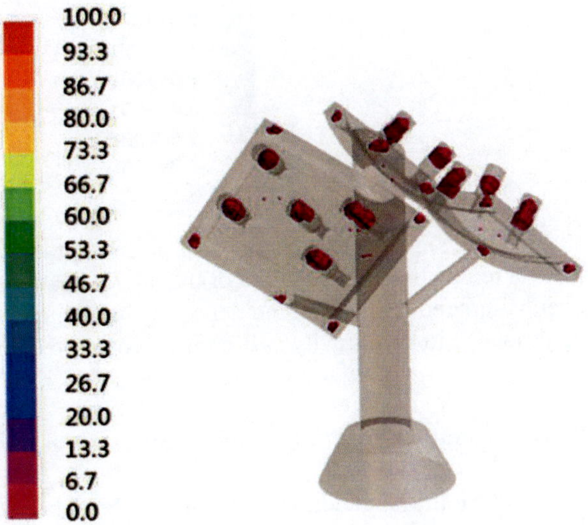

Fig. 8.35 Distribution of shrinkage

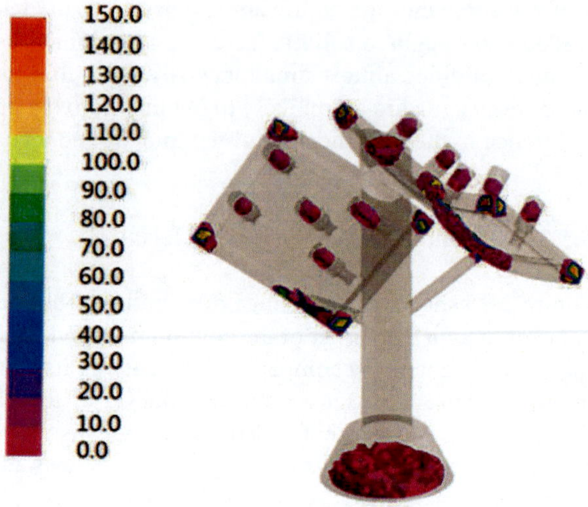

Fig. 8.36 Distribution of hot spot

Fig. 8.37 Niyama criterion distribution of micro-segregation/(°C·s)$^{1/2}$ cm^{-1}

From the shrinkage distribution of anti-gravity casting, the pouring system does make a role in the feeding effect of the micro-shrinkage of casting thin walls. Moreover, compared to the high incidence of shrinkage in gravity casting, the incidence of shrinkage in anti-gravity casting has decreased significantly, mostly below 0.2.

8.5.2.5 Simulation Analysis of Macro-Segregation

In the M951 alloy, in addition to the matrix Ni element, the most important elements are Cr, Co, and Al, whose compositions are 9 wt.%, 5 wt.% And 5.9 wt.%, among

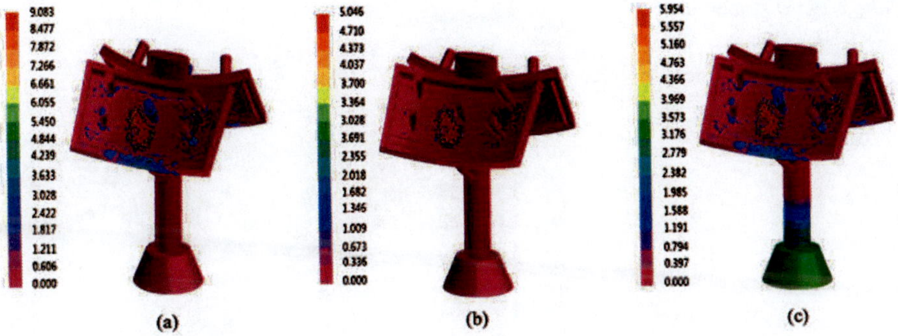

Fig. 8.38 Macro-segregation of Cr, Co and Al: **a** Cr/wt.%, **b** Co/wt.% and **c** Al/wt.%

which Cr, Co are the main elements to strengthen the high-temperature oxidation resistance of the alloy and Al is the main strengthening element of the high-temperature mechanical properties of the alloy. As a high-temperature alloy, M951 has a wide temperature solidification range and is prone to component segregation.

Figure 8.38 shows the macro-segregation of three elements in the simulation. It can be seen that the solute partition coefficients of the three elements in the alloy are all less than 1, and the content of the element contained in the alloy in the first solidification region is extremely poor, and the element solidification occurs in the solidified part. In the thin-walled area of the casting, due to the fast solidification, the solute can only diffuse to a certain extent in the thin-walled area, and there is no time to communicate with the thick frame and the runner. Therefore, during the solidification of thin-walled casting, there are fewer solute elements in some areas, while in the other area, such as the red area shown in the Fig. 8.38, there is element enrichment. This situation divides the required casting part into two parts, the compositions of which exhibit great difference, implying possibly a significant effect on the casting. Compared with the macro-segregation of gravity casting, the thin wall of anti-gravity casting is more uniform and almost solidified due to the applied force, and the elements have not diffused. There are both element-rich regions and element-depleted regions at the thin walls. Almost all these elements of gravity casting are enriched in the pouring system.

In addition, it is found that the behaviors of Cr and Al are consistent in macrosegregation, and their distribution on the casting is almost the same. The distribution of Co is similar to the former.

8.5.2.6 Simulation Analysis of Micro-Segregation

The microstructure of the casting has a direct impact on the performance of the casting. Numerical simulation analysis gives the grain size and secondary dendrite spacing distribution map of each part of the casting system. Figure 8.39 shows the grain size distribution of the casting system. As can be seen from the figure, the

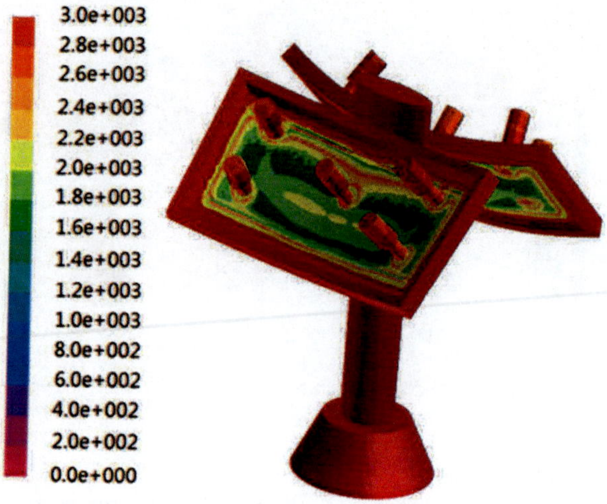

Fig. 8.39 Distribution of grain size/μm

grain size of the thin-walled part and the thick frame of the casting are completely different. There are two band-shaped regions in the middle of the thin wall. The grain size is about 1,600 μm. In the thick frame and runner section, the grain size exceeded 3,000 μm, and the grains were significantly coarse. Grain sizes from fine to coarse are consistent with the solidification sequence of the casting system. Generally, the grain size of anti-gravity casting is smaller than that of gravity casting, and it is more uniformly distributed in the thin-walled area of the casting.

Figure 8.40 shows the distribution of secondary dendrite spacing. It is shown that the secondary dendrite spacing at the thin wall of the casting is much smaller than that of the thick frame and runner. In the middle of the thin-walled region, the

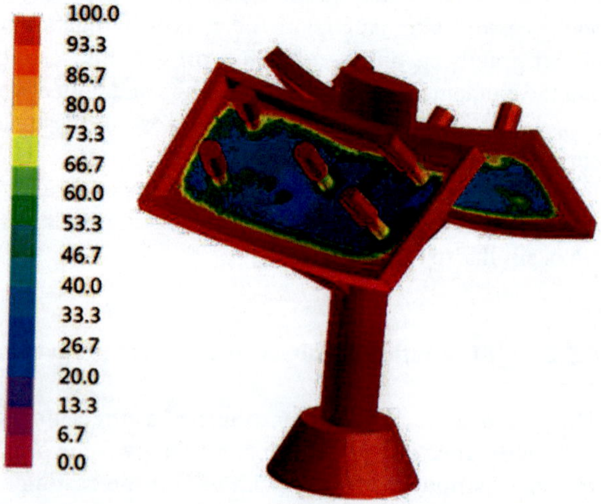

Fig. 8.40 Disruption of secondary arm spacing/μm

secondary dendrite spacing reached a minimum of 26 μm. The closer is to the edge, the larger the secondary dendrite spacing gets and the value of the secondary dendrite spacing of almost the entire thin-walled region is below 50 μm. At the edge of the thin-walled area, the value of the secondary dendrite spacing increases sharply and the secondary dendrite spacing of the thick frame and the runner are all larger than 100 μm. Compared with the gravity casting condition, the secondary dendrite spacing of the casting obtained by gravity casting is significantly reduced, slightly more than half of that. Simulation analysis shows that the use of pressure-regulated precision casting technology to form superalloy floating wall tiles has obvious advantages compared to the gravity casting.

References

1. Z.W. Chen, Q.T. Hao, W.Q. Jie, Study on counter-gravity casting of complex thin wall A357 alloy castings%A357. Casting **053**(012), 988–991 (2004)
2. B. Mondal, S. Kundu, A.K. Lohar et al., Net shape manufacturing of intricate components of A356/SiC composite through rapid prototyping integrated investment casting. Mater. Sci. Eng. A **498**(1), 37–41 (2008)
3. R.Z. Hong, Research for foundry process of complicate thin-wall aluminum alloy castings. Mater. Eng. **07**, 47–48 (2000)
4. Y.Z. Lu, Y.H. Wei, X.Q. Zeng et al., Development of the forming process for casting magnesium alloys. Casting **07**, 383–387 (2000)
5. S.M. Xiong, Research progress of magnesium alloy die casting forming technology. Aeronaut. Manuf. Technol. **02**, 32–35 (2006)
6. Z. Liu, The latest research progress of magnesium alloy casting forming. Prog Mater China **02**, 10–15 (2011)
7. Y.G. Zhou, W.D. Zeng, X.Q. Li et al., Toughening mechanism of titanium alloy under high temperature deformation. Acta Metall. Sinica **01**, 45–48 (1999)
8. X. Xue, L. Xu, Numerical simulation and prediction of solidification structure and mechanical property of a superalloy turbine blade. Mater. Sci. Eng. **499**(1), 69–73 (2009)
9. S. Wang, P. Guo, L. Yang, Centrifugal precision cast TiAl turbocharger wheel using ceramic mold. J. Mater. Process. Technol. **204**(1), 492–497 (2008)
10. J.R. Wood, P.A. Ruso, M.F. Welter et al., Thermomechnical processing and heat treatment of Ti-6Al-2Sn-2Zr-2Mo-Si for structure applications. Sci. Eng. A **243**(1), 109–118 (1998)
11. S.L. Xiao, Y.Y. Chen, H.Y. Zhu et al., Recent advances on precision casting of large thin wall complex castings of titanium alloys. Rare Metal Mater. Eng. **35**(5), 678–681 (2006)
12. T. Noda, Application of cast gamma TiAl for automobiles. Intermetallics **6**(7), 709–713 (1998)
13. W.Q. Liu, *Study on Bottom Injection Vacuum Casting of TA15 Alloy* (Harbin Institute of Technology, Harbin, 2010).
14. X.C. Ye, *Basic Research on Bottom Injection Vacuum Casting Technique for Small Thin-Walled Alloy Parts* (Harbin Institute of Technology, Harbin, 2010).
15. S. Misha, R. Ranjana, Reverse solidification path methodology for dewaxing ceramic shells in investment casting process. Mater. Manuf. Process **25**(12), 1385–1388 (2010)
16. J. Zhang, Z.H. Zhang, *Magnesium Alloy and Its Application* (Chemical Industry Press, Beijing, 2004).
17. W.H. Chen, R.Z. Chen, Development of aerospace investment casting technology. J. Aeronaut. Mater. **01**, 57–69 (1992)
18. X.F. Yu, G.Z. Zhang, X.Y. Wang et al., Non-equilibrium microstructure of hyper-eutectic Al-Si alloy solidified under super high pressure. J. Mater. Sci. **34**, 4149–4152 (1999)

19. Z. Liu, Z.G. Wang, Y. Wang et al., Study on pressure filling and solidification process of Magnesium alloy. Chin. J. Mater. Res. **13**(6), 641–644 (1999)
20. D.F. Zhang, Z.T. Fan, H.B. Wu et al., Filling ability of magnesium alloys in the lost foam casting (LFC) with vacuum and low pressure. Special-Cast Non-Ferrous Alloys **25**(2), 115–117 (2005)
21. L. Yan, M. Wang, Z.F. Shan et al., Research on filling capacity of adjusted pressure casting of magnesium alloys. Foundry Technol. **26**(10), 914–915 (2005)
22. Z.B. Zhou, J.S. Li, H.C. Kou et al., Effects of processing parameters on filling behavior of thin-walled casting in low pressure casting. Special-Cast Non-Ferrous Alloys **28**(1), 23–25 (2008)
23. X. Dong, N. Huang, S. Wu, Newly developed vacuum differential pressure casting of thin-walled complicated Al-alloy castings. China Foundry **2**(2), 102–107 (2005)
24. Z.L. Wang, Influence of gas in mold cavity on mold-filling capacity for thin-walled aluminum alloy castings. Foundry **61**(7), 768–773 (2012)
25. W.M. Jiang, Z.T. Fan, D.J. Liu, Influence of process parameters on filling ability of A356 aluminum alloy in expendable-patterns hell casting with vacuum and low pressure. Int. J. Cast Metal. Res. **25**(1), 52 (2012)
26. G.X. Liang, E.D. Wang, S.Y. He et al., Microstructure characteristics of aluminum alloy under pressure solidification. Ordnance Mater. Sci. Eng. **15**(1), 15 (1992)
27. B. Tang, Z.T. Fan, Z. Zhao et al., Effects of pressure field on structure and properties of loot foam casting ZL101 alloy. Special-Cast Non-Ferrous Alloys **29**(7), 629–641 (2009)
28. G.F. Mi, S.Y. Zeng, Effect of solidification condition on secondary dendrite arm spacing of the A357alloyundercounter-pressure casting. J. Wuhan Univ. Technol. Mater, 119–122 (2009)
29. R.W. Gregorutti, J.E. Grau, C.I. Elsner, Microstructural mechanical and electrochemical characterization of biomaterial ASTM F745castbyvacuum. Mater. Sci. Technol. **28**(6), 742–747 (2012)
30. Q. Yan, H. Yua, Z. Xua, Effect of holding pressure on the microstructure of vacuum counter-pressure casting aluminum alloy. J. Alloys Compd. **501**, 351–357 (2010)
31. S. Shendye, B. King, P. McQuay, Mechanical properties of counter gravitycastIN718. TMS **124**, 133 (2005)
32. B.W. Xiong, Q. Zaho, Effects of sample thickness on microstructures and mechanical properties of A357alloy. Int. J. Cast Metal. Res. **25**(6), 374–378 (2012)
33. Y.J. Wang, Counter gravity casting technology of aluminum alloy. Foundry Technol. **25**(5), 361–362 (2004)

Index

B. Sun et al., *Precision Forming Technology of Large Superalloy Castings for Aircraft Engines*, https://doi.org/10.1007/978-981-33-6220-8

F

G

H

I

K

L

M

N

W

X